CAMERON VOLUME ON
Unconventional Mineral Deposits

Wayne C. Shanks, III
editor

sponsored by
SOCIETY OF ECONOMIC GEOLOGISTS

published by
Society of Mining Engineers
of the
American Institute of Mining, Metallurgical, and Petroleum Engineers Inc.
New York, New York • 1983

Printed in the United States of America
by Guinn Printing Inc., Hoboken, New Jersey

Library of Congress Catalog Card Number 83-71422
ISBN 0-89520-414-2

Eugene Nathan Cameron

Eugene Nathan Cameron was born in Atlanta, Georgia, on Aug. 10, 1910, the second of three sons of Nathan M. and Jessie R. Cameron. His father's involvement in the lumber business led to many moves of the family when Gene Cameron was a youth. He moved to Roanoke, Virginia, at age 6, Baltimore, Maryland, at age 7, Jacksonville, Florida, at age 9, and Baltimore again at age 12, where he stayed long enough to complete high school, and to New York City after his senior year in high school.

Gene Cameron worked for two summers at the Chase National Bank in New York City, while still in high school but living with an uncle in New York, and this led indirectly to his career in geology. Dr. Benjamin M. Anderson, chief economist at the bank, liked Cameron's work as a page at the bank. When he learned that Cameron wanted to attend New York University to study biology but had no money to do so, he personally collected $250 in donations from the bank's officers to allow Cameron to attend college full time for one year. But by the time the money was at hand Cameron was late for registration at New York University and found that all of the biology classes were filled. He registered instead for a course in geology, and this was to prove a blessing in disguise. In the introductory geology course and subsequent courses he came under the inspiring influence of Leslie E. Spock, Jr., and George I. Findlay who recognized Cameron's abilities and encouraged him in that field. Cameron soon knew that he had found an exciting career that would challenge his abilities.

Financing his education continued to be a problem after the first year, and for the next three years Gene Cameron worked full time on the night shift at Chase National Bank and was a part-time student during the day. During his fifth year at NYU he received a teaching assistantship and was able to quit the bank and to complete his B.S. degree in geology (1932).

Gene Cameron then went to study under Paul F. Kerr at Columbia University for his M.A. (1934) and Ph.D. (1939) degrees. His Master's thesis was an optical study involving the properties of the synthetic resix Hyrax and its use as a mounting medium for rock thin sections in applications favored by its high refractive index. His Ph.D. thesis was a field and laboratory study of the geology and mineralization of the northeastern Humboldt Range in Nevada. Although the thesis was finished in 1936, Columbia University at that time required publication of the thesis prior to issuance of the Ph.D. degree. The Nevada Bureau of Mines had agreed to publish the thesis but ran out of funds, so the paper had to be rewritten in Geological Society of America (GSA) style for an eventual publication in 1939. Gene Cameron was supported at Columbia while a student by teaching assistantships 1932–1934, a research assistantship 1934–1935, and the Samuel Franklin Emmons Memorial Fellowship 1935–1936.

Among the graduate students working for a Master's degree under Professor Kerr at that time was a lovely young lady from Ridgewood, New Jersey. Gene Cameron married Adrienne Marie Macksoud in 1939. But he soon lost her services as a good field assistant when to their union were born Beatrice A. in 1941, poet and linguist, James N. in 1944, professor of marine biology, and Donald E. in 1954, economic geologist.

From 1936 to 1942 Gene Cameron served on the junior faculty at Columbia University as an assistant in mineralogy 1936–37, lecturer in geology 1937–40, and instructor in geology 1940–

42. This experience convinced him that he wanted to be a teacher, but these plans had to be shelved while he put his training into practice for the government during World War II.

Gene Cameron joined the US Geological Survey (USGS) in 1942 as an associate geologist within the Strategic Minerals Group. He was promoted to geologist in 1944 and senior geologist in 1946. He was in charge of pegmatite mineral investigations in Connecticut, 1942–1943; in New England, 1943–1944; and throughout the United States, 1944–1945. His charge was to help the development and mining of mica, beryl, and columbite-tantalite minerals, all of which were in short supply as a result of the war in progress and were on the critical minerals list. This work led to five publications with his co-workers in the period 1945 to 1949, including the important Monograph 2 of *Economic Geology* on the internal structure of granitic pegmatites. In 1946–1947 he took an administrative position with the USGS as commodity geologist in charge of Industrial Minerals Investigations. This introduced him for the first time to the broad field of nonmetallic minerals, in which he developed a special interest, and this experience also led in 1947 to an offer to resume his teaching career as an associate professor at the University of Wisconsin.

R. C. Emmons, chairman of the Department of Geology at Wisconsin 1945–1950, was faced with the problem of rebuilding the departmental faculty after the retirement of C. K. Leith, A. N. Winchell, and W. H. Twenhofel, and the resignations of N. D. Newell and H. E. McKinstry. Emmons was seeking an economic geologist to replace Leith and McKinstry who would be knowledgeable in nonmetallic ores as well as the more visible and popular metallic ores. Gene Cameron was qualified, and he was eager to resume a teaching and research career. He joined the Department of Geology at Wisconsin in the fall of 1947, was promoted to professor in 1950, served as departmental chairman 1955–1960, became Charles R. Van Hise Distinguished Professor of Geology in 1970 and Emeritus Professor in January of 1981.

Gene Cameron quickly developed a strong program in economic geology at Wisconsin. He took the attitude that any worthwhile problem within the entire field of economic geology was suitable as a student thesis. Students were allowed to choose their own topics, if they wished, and to publish their results independently. Cameron's publications (74 and still counting) were a result of his own research, and it is interesting to note that he is either the sole or first author on all but 13 of his papers. For many years he could be found working in his office in Science Hall day and night, and his productivity led quite naturally to the students naming him "Supergene" in the departmental skits. As his reputation grew, many students came to Wisconsin to study with him. The philosophy that Gene Cameron imparted to his students was that both field and laboratory work were indispensable in research on mineral deposits, and that an understanding of the genesis of different types of rocks and of the ore bodies within them was also essential in the exploration for ores. Gene Cameron supervised 38 Ph.D. theses and 57 Master's theses for a total of 90 individuals during his career. An impressive number of university professors, industry presidents, vice-presidents, and chief geologists are included among this group of graduates, 28 of whom returned for his retirement dinner.

Much of Gene Cameron's familiarity with ore deposits around the world came as a result of 12 fruitful years (1951–1962) spent as a consultant for Union Carbide Corp. He investigated carbonatites in Uganda, Brazil, and Canada as potential sources for columbium; pegmatites in Namibia, Swaziland, Madagascar, Mozambique, and Brazil; chromite deposits in Cuba, Zimbabwe, South Africa, the Appalachian, Sierra Nevada, and Klamath Mountains, Alaska, and Manitoba; titanium-rich beach sands in Florida and Utah; rutile deposits in Arkansas; and cobalt, gold, and tungsten deposits in Morocco. The chromite deposits within the Bushveld complex of South Africa intrigued him in particular, and he has spent 32 years (and still counting) in the field and in the laboratory studying different aspects of the stratigraphy of the lower Bushveld, the origin of the chromite seams, and the mineralogy and petrogenesis of the entire complex.

Cameron set up an ore microscopy course and laboratory when he first came to Wisconsin. He found the theory of reflected light microscopy to be incompletely developed, and spent a number of years doing research on its theory and application. This led in 1961 to his useful book *Ore Microscopy*. He employed reflected light techniques a great deal in his study of Bushveld chromites and later (1968–1971) as a principal investigator of lunar samples after having served 1962–67 as a consultant to NASA. He helped organize the International Commission on Ore Microscopy in 1963 and served as its USA delegate, 1963–1968.

Gene Cameron has been a fellow of four geological and mineralogical societies and has been especially active in the Society of Economic Geologists (SEG). He served as secretary of SEG, 1961–1964, vice president 1969–1970, president-elect 1972–1973, president 1973–1974, and past president 1974–1975, as well as secretary of the SEG Foundation 1968–1973, a member of its Board of Trustees 1972–1974 and 1976–1978, and a member of the Board of Directors of the Economic Geology Publishing Co. 1976–1978. He was the Thayer Lindsley Lecturer for SEG in 1977 and the Henry Krumb Lecturer for the Society of Mining Engineers of AIME 1980–1981. He was also served as a member of the US National Committee on Geology 1973–1977 and was chairman of the US National Committee on the International Geological Correlation Programme, 1973–1976.

Cameron did not shirk his university duties as a result of all of his outside activities. He taught eight courses in economic geology on an alternate year basis, served on numerous departmental committees, and as departmental chairman 1955–1960. On the university level he served on 14 different committees. Among those that were the most demanding of his time was a three-year term, 1966–1968, on the prestigious University Committee that served as a liaison between the faculty and the administration and to whom the most troublesome university problems were referred for solution. The last two years of this term, for the last one of which he was committee chairman, the university was rocked by student protests and riots in connection with the Vietnam War. The committee literally worked seven days a week for that period in meeting with state legislature, faculty, university administration, University Regents, and various student organizations. The University Committee attempted to define the issues, develop the factual background, and recommend appropriate faculty action. Despite the trauma of the times, the University held its ground, and the protests ended a year later when the main body of students realized that the violence had escalated too far.

Gene Cameron became especially interested at Wisconsin in following up the work of C. K. Leith on strategic mineral resources and governmental policies dealing with them. As a member of the Materials Advisory Board of the National Research Council, he served as chairman of the Panel on Beryllium Raw Materials 1955–1956, deputy chairman of the Panel on Chromium Raw Materials 1957–1958, member of an NAS-NRC,NSF,NMAB ad hoc committee on Materials Policy 1972, and chairman of its Panel on Commodities Abundances and Problems Affecting Future Primary Supplies, member of the NSF-NRC Committee on Mineral Resources and the Environment 1972–1975, and chairman of its Panel III on Reserves and Resources of Uranium in the United States. In all of his committee work Gene Cameron was recognized as a clear and incisive thinker. He could organize the facts, cut through the morass of conflicting or immaterial arguments, and come to logical conclusions. And he could be very persuasive in presenting his reasoning and his conclusions to his committee colleagues. It was his participation in the committees dealing with mineral resources and his interest in mineral resources of the future that led to his suggestion for the present Symposium on Unconventional Mineral Deposits, which appropriately has been named in his honor.

S. W. Bailey
Madison, WI

SELECTED BIBLIOGRAPHY OF EUGENE N. CAMERON

Cameron, E. N., 1939, "Geology and Mineralization of the Northeastern Humboldt Range, Nevada," *Bulletin 50*, Geological Society of America, pp. 563–634.

Cameron, E. N., Larrabee, D. M., Page, J. J., and McNair, A. H., 1945, "Structural and Economic Characteristics of New England Mica Deposits," *Economic Geology*, Vol. 40, pp. 369–393.

Cameron, E. N., Jahns, R. H., McNair, A. H., and Page, L. R., 1949, "Internal Structure of Granitic Pegmatites," *Economic Geology Monograph 2*, 115 pp.

Cameron, E. N., and Green, L. H., 1950, "Polarization Figures and Rotation Properties in Reflected Light and Their Application to the Identification of Ore Minerals," *Economic Geology*, Vol. 45, pp. 719–754.

Bailey, S. W., and Cameron, E. N., 1951, "Temperatures of Mineral Formation in Bottom-Run Lead-Zinc Deposits of the Upper Mississippi Valley Deposits, as Indicated by Fluid Inclusions," *Economic Geology*, Vol. 46, pp. 626–651.

Cameron, E. N., 1951, ''Geology of the Mount Prospect Intrusive Complex,'' Bulletin 76, Connecticut Geology & Natural History Survey, 44 pp.

Cameron, E. N., 1951, ''Feldspar Deposits of the Bryson City District, North Carolina,'' Bulletin 62, North Carolina Dept. of Conservation and Development, 100 pp.

Cameron, E. N., 1955, ''Concepts of the Internal Structure of Granitic Pegmatites and Their Applications to Certain Pegmatites in Southwest Africa,'' *Transactions*, Geological Society of South Africa, Vol. 58, pp. 45–70.

Cameron, E. N., and Emerson, M. E., 1959, ''The Origin of Certain Chromite Deposits of the Eastern Part of the Bushveld Complex,'' *Economic Geology*, Vol. 54, pp. 1152–1213.

Cameron, E. N., and Weis, P. L., 1960, ''Strategic Graphite, A Survey,'' Bulletin 1082-E, US Geological Survey, pp. 201–321.

Sorem, R. K., and Cameron, E. N., 1960, ''Manganese Oxides and Associated Minerals of the Nsuta Manganese Deposits, Ghana, West Africa,'' *Economic Geology*, Vol. 55, pp. 278–310.

Cameron, E. N., 1961, *Ore Microscopy*, John Wiley and Sons, New York, 292 pp.

Cameron, E. N., Davis, J. H., Guilbert, J. M., Larson, L. T., Mancuso, J. J., and Sorem, R. K., 1961, ''Rotation Properties of Certain Anisotropic Ore Minerals,'' *Economic Geology*, Vol. 56, pp. 569–583.

Cameron, E. N., 1963, ''Structure and Rock Sequences of the Critical Zone of the Eastern Bushveld Complex,'' Special Paper 1, Mineralogical Society of America, pp. 93–107.

Cameron, E. N., 1964, ''Chromite Deposits of the Eastern Bushveld Complex,'' *Geology of Some Ore Deposits of Southern Africa*, Vol. 2, pp. 131–168.

Cameron, E. N., and Desborough, G. A., 1969, ''Occurrences and Characteristics of Chromite Deposits in the Eastern Bushveld Complex,'' *Economic Geology Monograph 4*, pp. 23–40.

Cameron, E. N., 1969, ''Compositions of Coexisting Phases in the Eastern Part of the Bushveld Complex and Their Petrogenetic Significance,'' *Symposium on the Bushveld Igneous Complex and Other Layered Mafic Intrusions*, Special Publication 1, Geological Society of South Africa, pp. 46–58.

Cameron, E. N., 1970, ''Opaque Minerals in Certain Rocks from Apollo 11,'' *Proceedings*, Apollo 11, Lunar Science Conference, Vol. 1, pp. 221–245.

Cameron, E. N., 1971, ''Opaque Minerals in Certain Lunar Rocks from Apollo 12,'' *Proceedings*, 2nd Lunar Science Conference, Vol. 1, Massachusetts Institute of Technology Press, Cambridge, pp. 193–206.

Cameron, E. N., 1971, ''Problems of the Eastern Bushveld Complex,'' *Fortschritte der Mineralogie*, Vol. 48, pp. 86–108.

Cameron, E. N., 1973, ''The Contribution of the United States to National and World Mineral Supply,'' *The Mineral Position of the United States, 1975–2000*, E. N. Cameron, ed., University of Wisconsin Press, Madison.

Cameron, E. N., 1975, ''Postcumulus and Subsolidus Equilibration of Chromite and Coexisting Silicates in the Eastern Bushveld Complex,'' *Geochimica et Cosmochimica Acta*, Vol. 39, pp. 1021–1033.

Cameron, E. N., ''A Mineral Policy for the United States, 1977,'' *Mining and Mineral Resources*, College of Engineering, University of California-Berkely, Report 78-3, pp. 69–93.

Cameron, E. N., 1977, ''Chromite in the Central Sector of the Eastern Bushveld Complex, South Africa,'' *American Mineralogist*, Vol. 62, pp. 1082–1096.

Cameron, E. N., 1978, ''The Lower Zone of the Bushveld Complex in the Olifants River Trough,'' *Journal of Petrology*, Vol. 19, pp. 437–462.

Cameron, E. N., 1979, ''The Nature of Reserves,'' *Materials and Society*, Vol. 3, pp. 19–26.

Cameron, E. N., 1980, ''Evolution of the Lower Critical Zone, Central Sector, Eastern Bushveld Complex, and Its Chromite Deposits,'' *Economic Geology*, Vol. 75, pp. 845–871.

Cameron, E. N., 1982, ''Nonfuel Mineral Problems of the United States,'' *Resources and Conservation*, Vol. 9, pp. 1–16.

Cameron, E. N., 1982, ''The Upper Critical Zone of the Eastern Bushveld Complex-Precursor of the Merensky Reef,'' *Economic Geology*, Vol. 77, pp. 1307–1327.

Foreword

Ten years ago, Gene Cameron organized an SEG-sponsored symposium entitled "The Mineral Position of the United States, 1975–2000." In the introduction to the publication of the papers of this 1972 symposium, Elburt Osborn, then president of SEG and former director of the US Bureau of Mines, cited the National Mining and Minerals Policy Act of 1970: "Congress declares that it is the continuing policy of the Federal government in the national interest to foster and encourage private enterprise in the orderly and economic development of domestic mineral resources, reserves and reclamation of metals and minerals to help assure satisfaction of industrial, security and environmental needs . . ." A decade later, Congress began investigating minerals crises and produced the "National Materials and Minerals Policy, Research and Development Act of 1980," which declares that, "the Executive Office shall coordinate the responsible departments and agencies to identify material needs and assist in the pursuit of measures that would assure the availability of materials critical to commerce, the economy, and national security."

Meanwhile, the minerals export-import imbalance in the US has continued to worsen and, as Gene Cameron pointed out in his 1980 Henry Krumb lecture, the discouraging figures do not include metals imported as finished products.

President Reagan on Apr. 5, 1982, issued the first presidential report on minerals in 30 years. This report stated that the importance of mineral information has been consistently identified by every major mineral policy study commissioned by the federal government in the last 30 years and that such studies have urged better analytical capacity to improve the integration of mineral concerns into the policy process. The report continues, "Mineral information and analysis can provide guidance on policies affecting the strategic stockpile, land use, tax and tariff, trade, investment, research and development, environmental protection, and other aspects of domestic and foreign policies."

The Cameron Symposium on Unconventional Mineral Resources was held in Dallas, Texas, on Feb. 15 and 16, 1982, during the annual meeting of the American Institute of Mining, Metallurgical, and Petroleum Engineers and Constitutent Societies and the Society of Economic Geologists. The symposium was sponsored by the Society of Economic Geologists.

The symposium originated from a suggestion by Gene Cameron for a symposium to investigate the possibility of subsidized efforts to recover unconventional and strategic minerals. The SEG Research Committee following a discussion with Gene Cameron and SEG President Paul Bailly broadened the concept to global mineral resources and narrowed the concept to the geological aspects of unconventional mineral resources, with the exception of a paper by William Van Rensburg to provide the global economic background for the symposium. The concept was approved by the SEG Council and the Program Policy Committee.

The Cameron Symposium on Unconventional Mineral Resources included 20 talks (see the Table of Contents and Symposium Addendum, p. ix, x), which addressed 18 continental and oceanic environments and one extra-terrestial environment. Occurrences of at least 21 metallic and several nonmetallic elements were described. The symposium provided a summation of analytical data on metals and environments as diverse as tin in porphyritic magmatic rocks or cobalt in deep sea nodules to base metals in the "smokers" of the East Pacific Rise and platinum group elements in captured asteroids. Many of the metal-bearing environments represent strategic resources. In time, all may be of economic importance.

Of the 20 topics in the Cameron Symposium, 16 have been expanded to full-length chapters and are included herein. In addition, Gene Cameron has added an Introduction. All chapters have received at least two peer reviews, and revisions have been made accordingly.

The Volume has been arranged into four general sections. The first section, "Unconventional Mineral Deposits: Recognition, Exploration, and Economics," includes the keynote chapter by Paul Barton and an overview of mineral economics by William van Rensburg.

The second section, "Unconventional Deposits in Low Grade Environments," is the largest and includes papers on deposits which are unconventional principally because they are low grade. Most of these occurrences have been known for many years, and this section includes up-to-date assessment of the resource and exploration potential of the many diverse environments. Within the section, papers are arranged as follows: three chapters (Duke and Naldrett, Gross, Lipin) on deposits which have conventional mineralogies compared to currently active deposits; three chapters (Everhart, Hagni, Hatch) which explore future byproducts and coproducts of existing economic deposits; and two chapters (Desborough and Poole, Murray) which deal with deposits of unconventional mineralogy.

The third section of the volume includes three chapters on deposits which are unconventional because of physically inaccessible environments. Finally, the fourth section, "Unconventional Exploration Concepts," contains three chapters which focus on exploration rationales for new unconventional deposits which may carry fairly high grades. Thus the distinction from the chapters of the second section.

We hope that the Cameron Volume on Unconventional Mineral Resources will serve as an up-to-date catalog of presently known unconventional mineral deposits, will focus additional exploration activity on unconventional occurrences, and will stimulate additional geologic research on unconventional exploration approaches. In addition, we hope this compedium of future mineral resources will stimulate research in mining and recovery techniques for such deposits. Finally, we believe the Cameron Volume will aid and influence not only forward-thinking geologists and mining and metallurgical engineers, but also political and economic planners concerned with future land use and availability of mineral resources.

We gratefully acknowledge the assistance and support of AIME and the SEG Council. Paul Melancon, SEG Program Chairman, and Ruth Orologio, Meetings Manager SME-AIME, contributed substantially to the successful incorporation of the symposium into the framework of the annual AIME meeting. Marianne Snedeker, Manager of Publications for SME-AIME, has been most helpful and effective in the production of the volume. Professor S. W. Bailey, official and unofficial historian for the University of Wisconsin Geology and Geophysics program, provided the sensitive and insightful biography of Gene Cameron. *In particular, we thank Gene Cameron for inspiring us to the type of high quality geologic endeavor and responsiveness to the general public need which can, uniquely, result in the production of a volume of this importance.* Finally, we wish to thank the authors and reviewers, who obviously have made the most substantive and lasting contributions to the successful completion of the Volume.

Wayne C. Shanks, III
Editor

James J. Eidel
Symposium Chairman

March 25, 1983

Table of Contents

Symposium Addendeum

(papers presented but not included in the Volume)

Introduction

From time to time, the Society of Economic Geologists has sponsored symposia dealing with various aspects of mineral resources and mineral resource policy. The first symposium, in 1972, examined "The Mineral Position of the United States, 1975–2000" (Cameron, 1973). The second, in 1974, had as its theme "Politics, Minerals, and Survival" (Marsden, 1975). Now we have a third on "Unconventional Mineral Deposits." It deals with mineral deposits that are not presently available to man but might become important sources of minerals in the future.

The framework within which unconventional mineral resources must be discovered and appraised is discussed in papers by Barton and van Rensburg. Barton reviews the means and events through which unconventional resources in the past have become conventional. One important category of events is the discovery of new types of deposits. Barton points out that thus far most such discoveries have been accidental, but that the concept of the geochemical cycle could provide a purposeful basis for future exploration. If it is used intelligently, accidental discovery could be largely replaced by discoveries that result from selective exploration of the systems in which the formation of mineral deposits has taken place. Van Rensburg has reviewed for us the various factors, technologic, economic, political, and social, that encourage or discourage the transfer of unconventional resources into the conventional class, and also the obstacles that must be overcome. Only when the balance of these factors is favorable can new mineral resources become available.

At various points in the geochemical cycle there come into operation systems in which mineral deposits are formed. Two of the systems, the porphyry system and the epithermal system, are discussed in detail by Sillitoe and by Berger and Eimon. Knowledge of both systems is now well advanced, but the authors warn us that there is still a great deal of work to be done.

Sillitoe's paper calls attention to the great variety and widespread distribution of porphyry deposits and their potential as sources of metals other than copper and molybdenum. It is evident from his paper that much is to be gained by further investigation of porphyry systems. The potential for tin and tungsten is already impressive, and some porphyry deposits of those metals have passed into the conventional class. World reserves of the two metals are not overly large. Major additions through development of new sources would be most welcome in the future.

One of the more exciting developments of recent years is the revival of interest in the epithermal gold-silver deposits that furnished many of the great bonanzas of the past. In the United States, we have seen a long period during which these deposits contributed very little to supplies of precious metals. Now they have returned to the limelight as targets of a new approach, mass mining of volumes of lower-grade rock associated with the rich vein deposits that were worked in earlier days. Berger and Eimon summarize the impressive progress that has been made in defining the systems in which epithermal gold-silver deposits have formed. It will be interesting to watch the application of their models to future exploration of the deposits. A copy of the paper should be sent to the economist who, in connection with the debate over the MX missile site in Utah and Nevada, averred that the mineral resources of the area have already been discovered and exhausted.

The approach that is being followed in studies of the two sorts of hydrothermal systems is especially interesting to me. When I was in graduate school, studies of epigenetic deposits laid great emphasis on structural control of mineralization. Looking back, we see these studies as forerunners of present studies of whole hydrothermal systems.

Several papers deal with mineral resources of the ocean basins. We are given a fascinating glimpse of a new world that is just opening to our view. The resources thus far known in the deep ocean basins are now entirely in the class of unconventional resources. Yet they are pieces of the geochemical cycle, and already, as indicated by Shanks' discussion of the Red Sea deposits, by Dean's paper on the geochemistry of manganese nodules, and by the paper by Field and others on the metalliferous deposits of the East Pacific Rise, the definition of deep-sea systems is under way. How many other systems are still unrecognized? It seems unlikely that the great diversity of the mineral-forming systems now recognized on the continents can be matched in deep-sea basins. On the other hand, it seems equally unlikely that at this early stage of exploration of the ocean basins all the mineral-forming systems that exist there have been found.

The technologic, economic, and political obstacles to recovery of minerals from the oceans of the world are formidable indeed. Yet no one who has followed the development of technology for recovery of manganese nodules from depths of 4000 to 5000 m or more beneath the ocean can be very pessimistic for the future. Within a short span of years the manganese nodules have passed from an outrageously unconventional resource to one that is on the threshold of the conventional sphere. The remaining obstacles are political, embodied in the protracted negotiations over what shall be the Law of the Sea. It was inevitable that this controversy would arise, given the pressure of a growing population on the world's food and mineral resources. Yet in a way, the timing is unfortunate, simply because we are now in such an early stage of determining what and how important ocean mineral resources are. The nations of the world are in a poker game in which no one really knows the value of the chips that are being shoved to the center of the table. However, any assessment of the current status of exploration of the seas, against the background of the long history of exploration, discovery, and development of mineral resources on the lands, should warn us that the stakes in negotiation of a Law of the Sea are probably far greater than we know.

At the present time, more than three-quarters of the total tonnage of minerals produced and consumed annually in the world is won from deposits in sedimentary rocks. A series of papers reminds us that the mineral potential of sedimentary basins on the continents is far from fully explored, much less developed. I find it impossible to read these papers without feeling that the future will see the transfer of some of these unconventional resources into the conventional class. The assay values given by Desborough and Poole for some of the metals in black shales are not far from ore grades. The authors point out that although the Meade Peak member of the Phosphoria formation has received much study, present knowledge of concentrations, distribution, and occurrence of metals in it is very incomplete. The same appears to be true for other black shale formations. Surely it is reasonable to think that the richest portions of these extensive formations may still be undiscovered.

Metals in coals, discussed by Hatch, are another intriguing resource, one that becomes potentially more significant as the annual consumption of coal increases. Just at the public utility generating plants of the United States, nearly 72.6 Mt (80,000,000 tons) of fly ash are annually created. Already useful nonmetallic materials and magnetite are being produced from ash, and research of methods of recovering metals in the ash is being vigorously pursued. Part of the impetus for this research arises from the need to develop economically feasible and environmentally acceptable means of disposing of the ash, and from the prospect of worse problems if the Environmental Protection Agency (EPA) classifies the ash as toxic waste. Similar considerations are stimulating research into recovery of metals from flue dusts of smelters and refineries. If this is extended to tailings and other mill and smelter products of Mississippi Valley-type ores, more of the metals reported by Hagni from those ores may be recovered. Additions to conventional resources will result. Uranium in phosphorites, discussed by Everhart, already straddles the line between unconventional and conventional deposits.

Five papers concern unconventional resources that are of special interest to the United States, because all four metals involved are high on the strategic and critical minerals list. Duke and Naldrett describe the low-grade nickel deposits in Archaean terranes. It appears that only in some distant future, when nickel in laterites, in sulfide deposits of other types, and in manganese nodules have been depleted, will these deposits have a chance to become conventional.

Low-grade resources of manganese are examined by Gross. He stresses the association of many of these with volcanic activity, over a time span ranging from Archaean to Tertiary. The mode of separation of iron and manganese in sedimentary basins is still a problem. From the economic standpoint, for the future availability of manganese for industry, the most discouraging conclusion is that except under very special conditions the separation of iron and manganese is very incomplete. High-grade sedimentary manganese deposits appear to be unusual occurrences in nature. Most of the low-grade resources are iron-manganese deposits, with silica as a further contaminant. Major advances in technology of extraction, in metallurgy of steel, or in prices for manganese would seem required to resolve the problem of availability of manganese for the longer-range future.

The paper by Cannon and Force addresses itself to one of the great frustrations of North American geology, namely, the failure to find a single major deposit of high-grade manganese ore north of the Mexican border. There is no lack in this area of Precambrian iron formations, either of the Algoma type or the Lake Superior type. Sedimentary formations of Tertiary and Cretaceous age are of vast extent and great sedimentological diversity. In every one of the other continents, except icebound Antarctica, important manganese formations have been found. It is difficult to believe that none were ever formed in Canada or the United States. The authors of the paper point to the outcrop belt of mid-continent Cretaceous rocks and the Gulf and Atlantic Coastal Plains as the most promising regions. It would be scientifically satisfying to have a major discovery result from definition of the sedimentary systems of these terranes, but meanwhile the United States would gladly settle for a first-class accidental discovery.

A major concern of the United States and the other industrialized countries of the Free World is the extraordinary maldistribution of the world's chromite reserves and near-economic resources. Thus far, as pointed out by Lipin, chromite has been mined almost exclusively from high-grade chromite deposits in alpine peridotites and in layered mafic intrusions, but there has been much interest, in the United States and elsewhere, in finding lower-grade, large tonnage deposits that might be worked. One could wish that Lipin's findings were more encouraging. Lipin concludes that the most promising low-grade resources are in the iron-nickel laterite deposits, small comfort to the United States.

Murray has given us a comprehensive report on unconventional resources of aluminum available within the United States. The problems of bringing them into the conventional class have been exacerbated by the rising cost of energy, and they have not yet been resolved by the intensive research that has been in progress for some years. It seems that there is still no substitute for weathering in a steamy tropical climate.

Some of the deposits described in the symposium, such as the epithermal gold-silver deposits, seem to require only the extension and perhaps refinement of present-day techniques of extraction. Encouragement from higher and more stable prices for minerals would certainly help. Others, however, such as the black shale deposits, seem to call for significant technological innovations; the decline in levels of support for research, pointed out by van Rensburg, is thus a matter of serious concern. All the deposits require political and social climates favorable to discovery and development of mineral resources. For most of the deposits discussed, and here we note van Rensburg's remarks again, rising costs of energy may prove a serious obstacle. Clearly the mining industry has a major stake in efforts to develop new energy technologies.

Finally, we are all greatly indebted to the organizing committee and to the contributors for an excellent and thought-provoking symposium. They have given us a look into the world's mineral future. For my part I am honored indeed to have my name attached to this symposium, and I thank Jim Eidel, the members of his committee, and the Council of the Society of Economic Geologists very, very much.

Eugene N. Cameron
University of Wisconsin
Madison, Wisconsin

August 23, 1982

REFERENCES

Cameron, E.N., ed., 1973, *The Mineral Position of the United States 1975–2000,* University of Wisconsin Press, Madison, 159 pp.

Marsden, R.W., ed. 1975, *Politics, Minerals, and Survival,* University of Wisconsin Press, Madison, 86 pp.

1

Unconventional Mineral Deposits: Recognition, Exploration, and Economics

1

Unconventional Mineral Deposits: A Challenge to Geochemistry

Paul B. Barton, Jr.

US Geological Survey, Reston, VA

ABSTRACT

Unconventional mineral deposits are those that differ significantly from productive deposits in mineralogy, grade, or geologic setting. Thus, the initial representatives of each deposit type are, by definition, "unconventional" until they are exploited. The "unconventional" status changes to "conventional" because of economic shifts (e.g., bulk silver ores), because of technological innovations (e.g., froth flotation), or simply through the recognition of, and production from, novel mineral concentrations. Heretofore the discovery of new deposit types has been almost entirely accidental, but the long-established concept of the geochemical cycle provides a nearly untouched opportunity and challenge for the application of geochemistry in predicting novel, potentially economic, mineral concentrations.

INTRODUCTION

Over the past half century, this nation has moved steadily from self-sufficiency toward dependency for most mineral needs. There are many reasons for this, some of them involving the simple geologic fact that not all of the world's best ore deposits are within the United States. High production costs, limited access to new ground for exploration, and the rapid growth of the environmental movement have compounded the problem. Also, we must not ignore the spectre that we just may be trying to live beyond our mineral means, and it is totally unrealistic to assume that the rest of the world will dedicate its resources just to sustain us.

The exposed, pristine deposits that the prospector could recognize by sight or panning are nearly an extinct breed in this country today, and they are an endangered species worldwide as the earth's endowment of easy discoveries is nearly depleted. We are in the stage where scientific insight has become as important as blind good luck. Future mineral resources will become harder and harder to find as the earth is successively combed for this or that type of recognizable deposit. But, to paraphrase Sheldon and Raup (1981), we must be encouraged by the fact that each new deposit type, each new geologic setting, each new interpretation that permits deeper testing of ground, and each advance in exploration or extraction technology opens new, and reopens old, frontiers for exploration.

DEFINITION OF "UNCONVENTIONAL"

This symposium volume is devoted to forward looks into "unconventional" deposits. Let me note at the outset that I have no neat, clean, exceptionless definition of "unconventional."

Everyone knows what it means, but prior to this volume, few have tried to define it. One effort is that of Tooker (1982) who has defined "unconventional *future* ores" as "naturally occurring or manmade accumulations of metallic and (or) nonmetallic materials that are not technologically and (or) economically recoverable at present, but are expected to be recoverable ultimately." Although this definition is entirely reasonable for Tooker's own discussion, it does not include novel, but undiscovered, high-grade deposits, and thus it is incomplete for my present purpose.

Each type of ore deposit was once "unconventional;" that is, the first deposit, by definition, was distinct from anything else that had come along before. However prospectors from ancient times found, and when possible mined, whatever valuable materials they could without much concern for whether or not they were "unconventional." For example, Sudbury was discovered as a result of a lucky choice for a railroad route, and it produced millions of tons of fine ore—everyone concerned being completely oblivious to the possible "unconventional" origin of the nickel eruptive as an astrobleme. Thus "unconventionalness" in a genetic sense has nothing at all to do with economic viability. Most deposit *types* in production today were found by simply discovering mineral values that proved economic—whether or not the deposit was "peculiar" came later. In fact, only when the mining profession began to classify ore types could the concept of "unconventional" exist. Of course, once a deposit type is recognized, its characteristics become identified and the probability of additional similar discoveries increases. For example, many porphyry coppers were recognized once Bingham was shown to be economically viable and set the pattern for exploration elsewhere.

For the purpose of this discussion, let us define "unconventional" deposits as those which are significantly different from productive deposits in mineralogy, grade, or geologic setting. Thus, unconventional deposits would go unrecognized by persons focusing only on known deposit types. That does not necessarily mean that such persons would overlook ore-grade material if it were presented to them, but at least they would overlook the probability of ore in that particular geologic setting, or of that particular grade or mineralogical composition. For example, how many of us 20 years ago—or even today—would walk over the Utah bertrandite beryllium deposit without recognizing the economic mineralization, whereas we would quickly notice beryl in a pegmatite?

Another example: an eminent mining geologist a few years ago turned down a minor lead prospect in the Mojave without noticing that the wallrock for the lead vein was what today is the baritic bastnasite ore of the Mountain Pass rare-earth deposit. We all too often see only that which we are *looking*.

Thus, papers in this symposium should provide encouragement, incentive, and direction to *look for new* things; and this paper, too, will offer some suggestions later. But first let us discuss some of the factors that interact to influence mineral reserves. Van Rensburg's paper (Chapter 2, this Volume) deals with the situation in much more depth.

RESOURCES, RESERVES, AND UNCONVENTIONAL MINERAL DEPOSITS

A simplified version of the famous "McKelvey box" (Fig. 1) provides an apt classification of resources in terms of *Certainty of Existence and Profitability* (McKelvey, 1972). There are many modifications of McKelvey's diagram, but they all share three components: *Reserves, Undiscovered Economic Deposits,* and *Non Economic Deposits,* the total of which constitutes *Resources.**

Reserves constitute materials that are proven to exist and that are economic to extract under the conditions prevailing at the time of determination. In the usual terminology, they are "ore" (known materials that can be mined at a profit). However, in today's world the profit

*The US Bureau of Mines and the US Geological Survey (USGS, 1980) have subdivided resources in considerably more detail than does Fig. 1, but except for the "Reserve base" (which includes reserves, plus marginally economic deposits, plus that part of known, noneconomic material that may become economic in the future) used in compiling Fig. 2, this discussion will use the less precise, but simpler classification shown in Fig. 1. For our purpose it is unimportant whether the size of the *Reserve* compartment changes to include (or exclude) additional deposits or whether the position of the deposits moves into (or out of) the *Reserve* compartment; the cartoon (Fig. 1) is drawn in terms of the former choice.

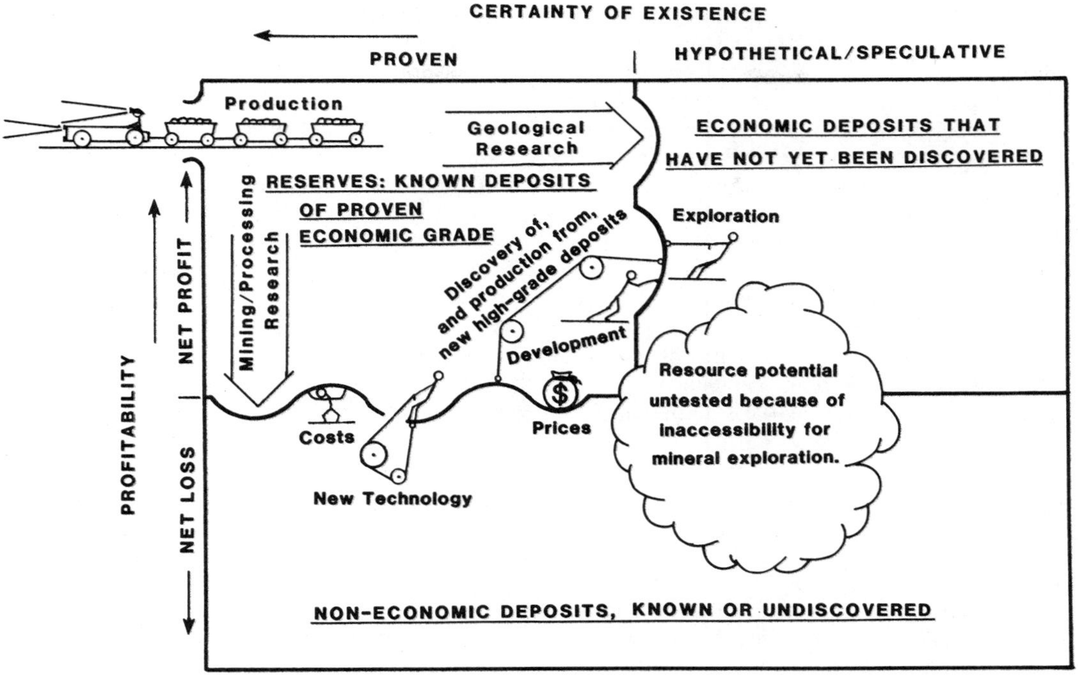

FIG. 1. Simplified classification of resources, and a summary of the forces that tend to enlarge or diminish the quantity of reserves. Modified and embellished from McKelvey (1972) and US Geological Survey Circular 831 (1980).

criterion must be modified; for a variety of reasons many individual mines are now operated at a net loss. An example of such is a mine operated at a loss in the hope of maintaining staff, facilities, or markets during an economic recession that is hoped to be temporary. Another example is the operation of nationalized properties despite losses in order to provide fuller employment, or to supply foreign exchange, or to secure a domestic supply of a critical commodity. For the purpose of this discussion the *Net Profit–Net Loss* boundary separates what could be chosen to be produced from that which could be chosen *not* to be produced; the final profit/loss balance sheet is tallied at levels ranging from an individual mine operation to an entire national economy.

Now let us consider how the contributions to this symposium relate to the McKelvey box. The development of technology (e.g., heap leaching) and increases in prices (e.g., precious metals) have transferred some known noneconomic deposits into the "ore" category, and such factors have encouraged the search for a recovery of additional products from existing deposits. Examples of the technological/price development are the large-tonnage, low-grade, epithermal silver-gold deposits described by Berger and Eimon, the recovery of uranium from phosphorite described by Everhart, the sphalerite in cleats in coal by Hatch, the wide variety of minor metals in Mississippi Valley-type ores by Hagni, and Sillitoe's description of scarce metals in "porphyry" settings. Several of the other papers deal with environments that have remained relatively untested because of their physical inaccessibility deep below the sea. These include: Dean on nickel-copper-cobalt-bearing deep-sea manganese nodules; Shanks on the hot metalliferous oozes of the Red Sea deeps; and Field, Rye, Dymond, Whelan, and Senechal on deep-sea metalliferous sediments. All of the articles (including those so far unmentioned: Desborough and Poole on a variety of metals in lithified, organic-rich shales; Gross on low-grade manganese deposits related to Algoma-type iron formation; Duke and Naldrett on low-grade nickel resources; Lipin on low-grade chromium; Cannon and Force on a novel model for sedimentary manganese deposits; and Murray on nonbauxite alumina) seek to identify and characterize deposit types located in the *Hypothetical/Speculative* part of the McKelvey box so that subsequent work may bring them to the left into the *Proven* class by quantifying their properties. Further movement toward converting any, some, or all of these *Resources* to *Reserves* will then depend on demand (= price) and technological advances. Examples of demand/technology actions are the Climax molybdenum and Mountain Pass rare-earth deposits in which it was first demonstrated that a great abundance of something existed and then markets and technology were developed.

Although the vertical boundary between *Proven and Hypothetical/Speculative* is a bit fuzzy because of uncertainties inherent in exploration and development, the interface can move only to the right—deposits do not often become lost. Of course, *Production* steadily drains the amount of *Reserves*. In contrast to the vertical boundary, the horizontal interface between ore and waste moves up and down in response to many factors. For example, successful exploration and development may lead to production from a new high-grade deposit thereby putting an older, less economic producer out of business. The current metals industry is no stranger to this concept.

The rest of this report will deal with the probability that there may be many more kinds of unconventional mineral deposits, and we shall develop a geochemical approach to identify and locate them.

THE MINERAL ENDOWMENT OF THE CRUST

Let us consider the magnitude of the mineral endowment of the earth's crust compared with the resources thus far identified. Fig. 2 compares the amounts of metals in the upper kilometer of the earth's continental crust with the total metal known in reserves *plus* identified subeconomic materials of possible eventual utilization. This diagram is based on one by McKelvey (1960); it has many interesting attributes, but we will discuss only three:

(1) Only a very small proportion of the metal in the upper kilometer of continental crust is concentrated into known deposits. Even for the most strongly concentrated metals, lead, bismuth, and tellurium, there is more than 10,000 times as much metal in the upper kilometer of ordinary rock as there is in all known deposits. For tantalum, only 1 part in 100 million of the crustal endowment is concentrated into known deposits.

(2) Either ore-forming processes are extremely inefficient or rare, or both. At any rate, known deposits only scratch the surface relative to total crustal endowment; thus the metal available for additional ores is large, thereby offering opportunities for unconventional deposits.

(3) Even the largest mineral concentrations, such as gold of the Witwatersrand; the copper of southern Arizona, Chile, or Zambia; the zinc of the Selwyn Basin; the lead of Missouri, the molybdenum of Colorado; the tin of Malayasia, or the phosphate of Idaho could be "lost" in the uncertainties of background metal concentrations present in the crust in the general vicinity of the deposits. Stated another way, if you stirred up the entire crust beneath each mineral-rich province, the bulk metal content of a metal-rich province might not be significantly different from that of a metal-poor province. Thus, metallogenic provinces may be regions where the concentration *Process* functioned well rather than regions of intrinsically high metal content, a point also made by Lehmann (1982) and indirectly by Holland (1972) and Burnham (1981).

Fig. 3 indicates how the crustal endowment is believed to be distributed. This figure shows a smoothed, hypothetical, grade-tonnage histogram for what Skinner (1976) calls the "abundant" elements. Statistically, it may well constitute a normal distribution. Present ores are at the high-grade end, and as they become exhausted, the ever-increasing amounts of slightly lower grade materials can meet resource demands. There is no significant geologic limit to such resources—only economic, environmental, and energy concerns.

In contrast, Fig. 4 shows the "scarce" elements which Skinner (1976) points out may well show multimodal grade-tonnage curves so that an ever-increasing supply is not necessarily guaranteed simply by lowering grade.

We must also consider the mineralogical limit. Above the limit, much of the metal is present in discrete "high-grade" minerals. The mineralogical limit is highly sensitive to the geochemical character of the metals and the physical-

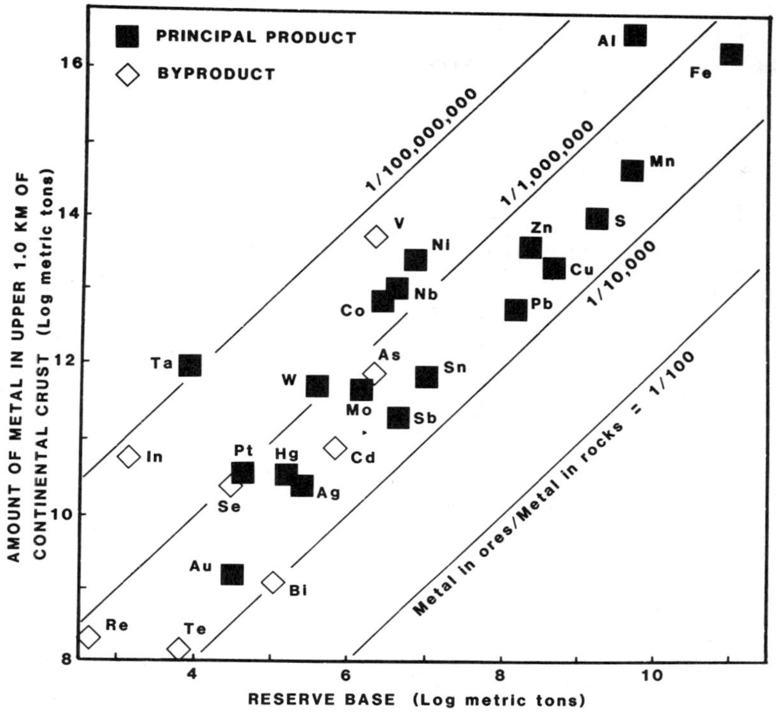

FIG. 2. Comparison of the amounts of metals in known economic (or near economic) deposits with the amounts of metals in the upper kilometer of Continental Crust. Modified from McKelvey (1960). The reserve-base data are from the US Bureau of Mines (1980). The concentrations in the crust are from various sources including Erickson (1973) and Parker (1967).

chemical environment of the host. Thus, one rock may have most of its lead in galena whereas another rock of equal lead content may have most of it in feldspar. If Fig. 4 were to represent lead, galena would be present to the right of the mineralogical limit. In theory at least, we could produce a high-purity galena concentrate from any lower-grade materials which are above the mineralogical limit. On the other hand, below the mineralogical limit, lead is present only in trace amounts in rock-forming minerals such as feldspar. We could make no high-grade concentrate, but would have to smelt the entire rock.

The symmetrical, bell-shaped "normal" distribution represents a measure of the probability that any randomly chosen sample lies within a

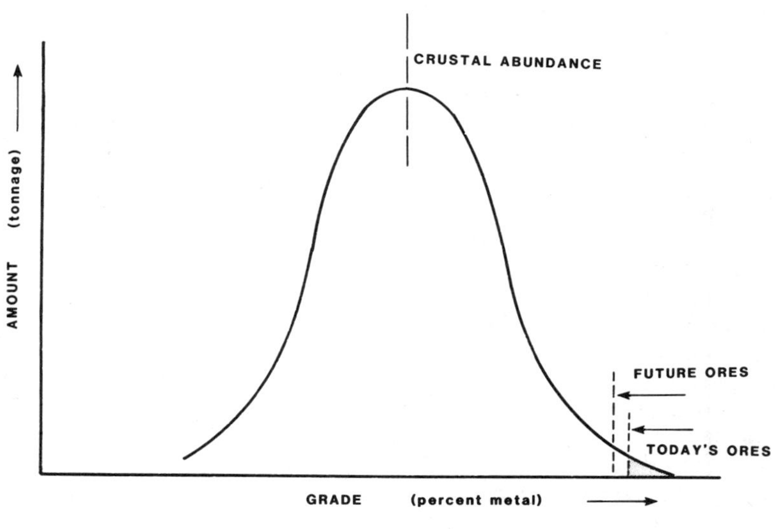

FIG. 3. Idealized grade-tonnage curve for abundant elements. Modified from Skinner (1976). Typical examples would be iron or aluminum.

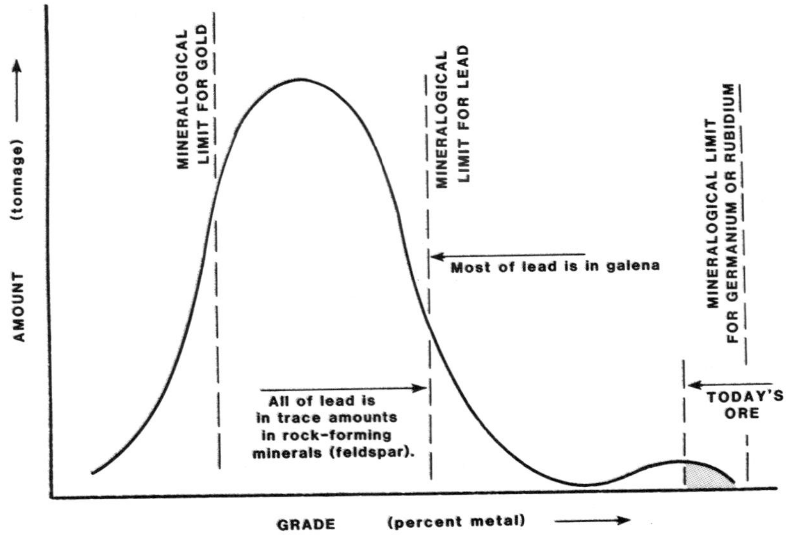

FIG. 4. Grade-tonnage curve for scarce elements. Modified from Skinner, 1976. Typical examples in addition to those on the figure would be lead, copper, tungsten, or chromium.

certain range. For the "abundant" elements the normal distribution is thought to represent a plot of frequency vs. concentration on a *linear* scale. For the "scarce" elements, a normal distribution is thought to require that the logarithm of the concentration be used (Ahrens, 1965). An adequate explanation for why element concentrations should follow any sort of normal distribution is absent; indeed, we even lack the proof of a rigorously normal or, for that matter, even monomodal, distribution for any element. Even for uranium which has by far the best data set, the situation is not unambiguous (Holland and Petersen, 1981; Deffeyes and MacGregor, 1980). Nevertheless, the *concept* of a smooth, bell-shaped distribution is well established.

To further complicate the picture, the distinction between those elements following either the monomodal "abundant" or the multimodal "scarce" element curve may not simply be one of abundance but instead may be linked to the positions of the mineralogical limits. Thus, although rubidium and chromium have comparable crustal abundances, rubidium probably has the monomodal pattern and chromium the multimodal, because rubidium hides in potassium minerals, never forming its own species, whereas chromium often forms the Cr-rich species, chromite. This relation is supported by B. R. Lipin's observation (this volume) that chromium abundances in ultramafic rocks appear to be bimodal. Uranium seems to fit the monomodal pattern (Holland and Peterson, 1981) probably because uranium is seldom present as uraninite, but instead is found as "high-level" traces in apatite, organic materials, zircon, allanite, monazite, and so on (Deffeyes and MacGregor, 1980).

UNCONVENTIONAL DEPOSITS AND EXPLORATION

Chance discovery has always been important; here is a modern example that might fit the "unconventional" critieria. The Pena Blanca uranium deposits in Chihuahua contain uranium in a new mineral, margaritasite—$Cs_2(UO_2)_2V_2O_8 \cdot nH_2O$—in altered volcanic rocks. Very sketchy data suggest that the cesium content is a few tenths percent in the uranium ore; the cesium has even greater value than does the uranium. Today, cesium comes from pegmatitic pollucite and it is used in small quantities in electronics and medical applications, but it also has potential use in magnetohydrodynamic power generation. Is this another chance discovery of an unconventional deposit just waiting for market development?

What about predicting unconventional processes of mineral deposition? Fig. 5 shows an example based on a hydrologic model (Steven and Eaton, 1975; Barton et al. 1977; Bethke and Rye, 1979) developed for the Creede silver-base metal ores in volcanic rocks. A hot pluton in-

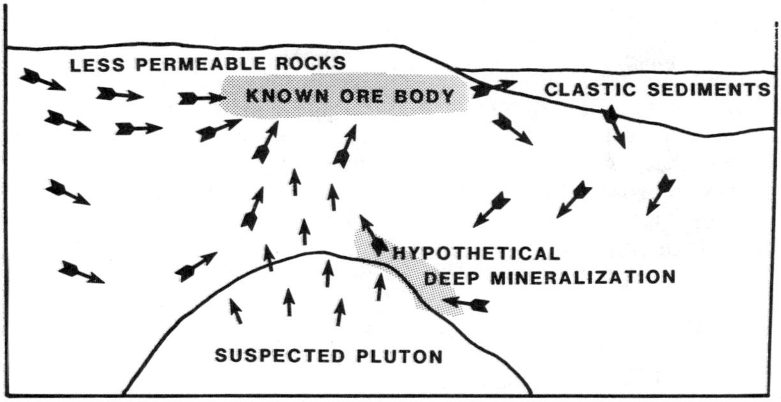

FIG. 5. Hydrologic model for the Creede Mining District, Colorado. Generalized from Bethke and Rye (1979), Steven and Eaton (1975), and Barton et al. (1977). Except for the clastic sediments to the upper right, all of the rocks are flat-lying calc-alkaline volcanic or subvolcanic rocks. The full arrows indicate schematically the paths of flow of meteoric or hybrid solutions; the arrow heads indicate possible magmatic fluid from the hot pluton.

terrupts a regional hydrologic flow and initiates its own convecting hydrothermal system which deposits ore at the top where the fluid is coolest and the minerals are least soluble. So far, all is conventional. However, consider the descending part of the cell. We know that some minerals have thermally retrograde solubilities and may be expected to be concentrated in hotter rather than cooler regions—calcite, anhydrite, and molybdenite are examples. Furthermore, the intricate chemical variations that fluids may experience as they change temperature and react with different rock types may provide solubility minima at places other than the coolest part of the system for minerals whose solubilities are conventionally viewed as prograde. Thus there is an unconventional rationale for deeper exploration.

Another unconventional process model can be described using this same (Fig. 5) cross section, but let us imagine the wallrocks to be sedimentary instead of volcanic rocks. We know that at temperatures just warm enough to prevent biological activity, above perhaps 100°C, solutions containing metals plus sulfate plus organic matter may exist metastably for long periods of time. If such metastable solutions were to be drawn into a convecting system around a pluton, they might be heated to the point of reaction between the sulfate and the organic constituents (probably about 300°C) thereby precipitating sulfides in a geologic setting very similar to that of a contact metamorphic aureole (see discussions in Barton, 1967, 1981).

Let us consider yet another example of unconventional exploration rationale, in the Spor Mountain region in western Utah. In addition to the bedded bertrandite, this district contains several vertical pipes of uraniferous, earthy fluorspar cutting and replacing gently dipping Paleozoic carbonate rocks. From a geochemical point of view, these pipes appear to be the products of a very low pressure, fluoride-bearing steam which was unable to transport the base metals, the barite, the silica, the carbonate, etc., which so often accompany fluorspar in vein deposits. The earthy character of the fluorspar attests to a fumarolic environment with a lack of *liquid* hot water to promote crystal growth. Perhaps these deposits formed in vapor-dominated geothermal systems. One would expect that such pipes would pass downward into pebble dikes, intrusion breccias, and finally into rhyolitic cupolas similar in composition to the parent magma bodies of the high-silica rhyolite that makes up Topaz Mountain to the east. We might ask whether the process that permitted the boiling off of the fluoride-bearing steam might not have left behind (in the vicinity of the point where the steam could have been throttled, or where a deep, dense, aqueous brine body may have boiled) other materials which were not readily transported in the vapor. One might anticipate such materials as molybdenum, tungsten, or tin. If such a deposit exists, and if it could be found, and if it proved economic, would it constitute an "unconventional" deposit? Probably so, but one could also argue that boiling as a depositional mechanism is well ingrained into geologic lore, and that the only unconventional aspect is the rationale for exploration.

THE GEOCHEMICAL CYCLE: A GUIDE TO MINERAL OCCURRENCE MODELS

The foregoing examples have a common feature—each is based in one way or another on genetic mineral occurrence models. A given element may participate in dozens of mineral occurrence models, each applicable to specific geologic processes. Moreover, there are processes that destroy, as well as processes that create, mineral concentrations. The systematically arranged, integrated product of geochemical processes and reservoirs is the *classic geochemical cycle,* an example of which is given for manganese (Fig. 6). Here, reservoirs are shown as labeled boxes. The shaded areas denote potential ores, and they form where the mobility of manganese relative to potential dilutants becomes low. Careful consideration of such cycles can identify otherwise unforeseen potential deposits. The critical parameter for manganese is its dramatically contrasting behavior as redox and pH conditions change.

As an example, let us enlarge upon one aspect of this complex cycle—the subsurface water to chemical deposit link. To do this, we use the O_2-pH diagram shown in Fig. 7b. Note that manganese has a very extensive region of high solubility under reducing, acid conditions—complemented by an insoluble, depositional region at high pH and high redox potential. This diagram has the boundary between transportation and deposition drawn at 6 ppm, but it is

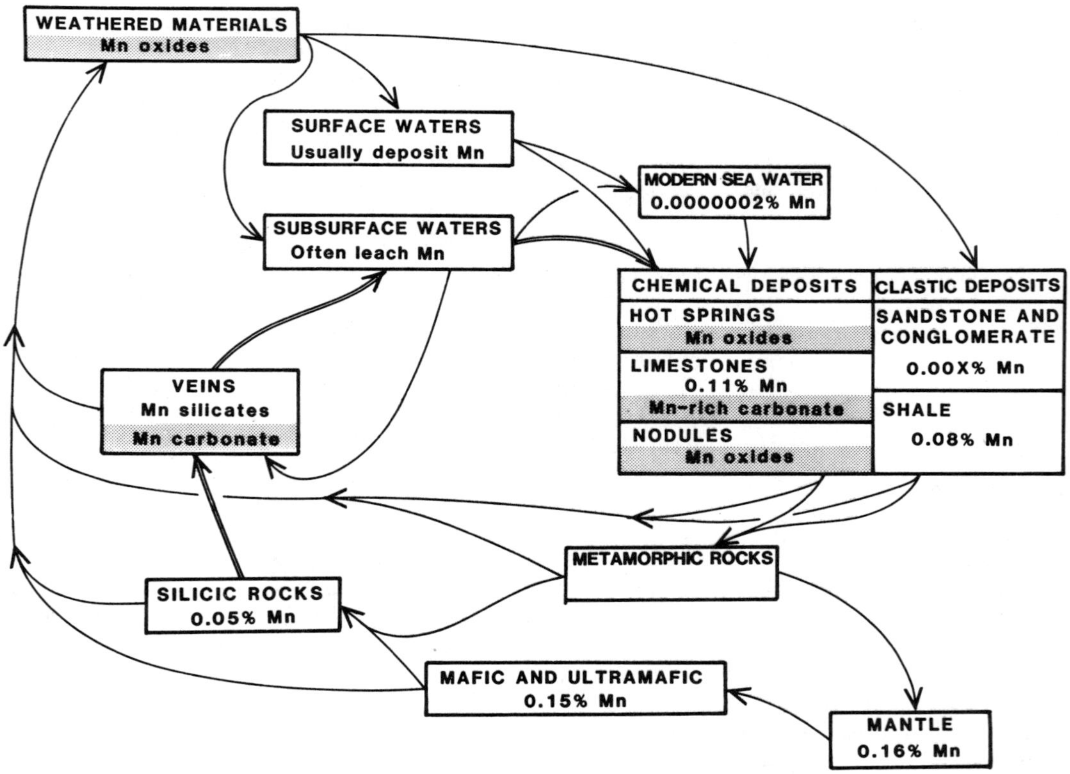

FIG. 6. Generalized geochemical cycle for manganese. Boxes denote reservoirs; shaded portions of boxes represent potential environments of manganese ore concentration. Arrows denote fluxes of materials with the double arrows marking especially important manganese transport.

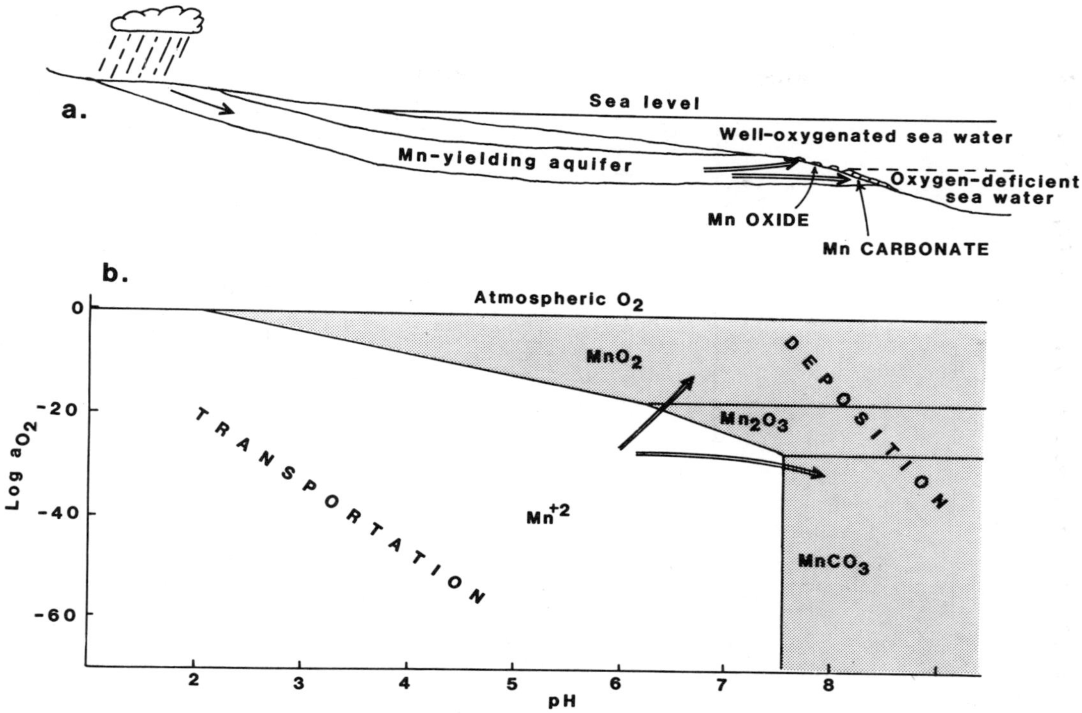

FIG. 7. Submarine spring manganese model. Important manganese flux is shown by double arrows. Fig. 7a is a schematic geologic cross section. Fig. 7b is a diagram of log a_{O_2} vs · pH distinguishing a region of magnanese transport (concentration of manganese greater than 6 ppm) from a region of deposition (shaded). Mineral stabilities are from Robie et al. (1978). Calculated for 25°C and P_{CO_2} = 0.0003 atm. The field of Mn_3O_4 is just eliminated at this P_{CO_2}.

the qualitative,[†] not the quantitative, aspect that is significant.

Now, let us look at a submarine spring (Fig. 7a) in terms of the solubility diagram. Water entering the aquifer will leach available manganese provided that the aquifer chemistry lies in the *Transportation* region. As the fluid empties into the sea, its pH will rise and its oxygen potential may lead it into the manganese oxide or manganese carbonate field. In either event, manganese will precipitate. There must be sufficient artesian head to displace the denser sea water—otherwise, the system would stagnate. Thus a high recharge elevation, or a low stand at sea level, would be favorable; the offshore, Pleistocene sediments along the Atlantic Coast might offer such a target.

Cannon and Force (this volume) give an entirely different model for sedimentary manganese. My purpose here is to emphasize the potential for using geochemical cycles in exploration. I applaud the fact that they have chosen a geochemical model (no matter that it differs from mine; it is nevertheless built on the chemistry shown in Fig. 7b) to arrive at a very tantalizing interpretation for manganese sedimentation.

Let us now consider the geochemistry of uranium as represented in Fig. 8b. In this model cool, oxidized, carbonate-bearing ground water reacts, probably by feldspathization or zeolitization, with tuffaceous lake sediments thereby releasing uranium to solution. The intrusion of a hypabyssal pluton (Fig. 8a) provides a localized heat engine that drives a convecting hydrothermal system and also provides a reducing environment as, for example, by supplying H_2S-bearing fluids. When the uranium-bearing so-

[†]It is interesting that although we almost universally give praise to *quantitative* treatments of problems, the real quantum leaps in understanding are often *qualitative*.

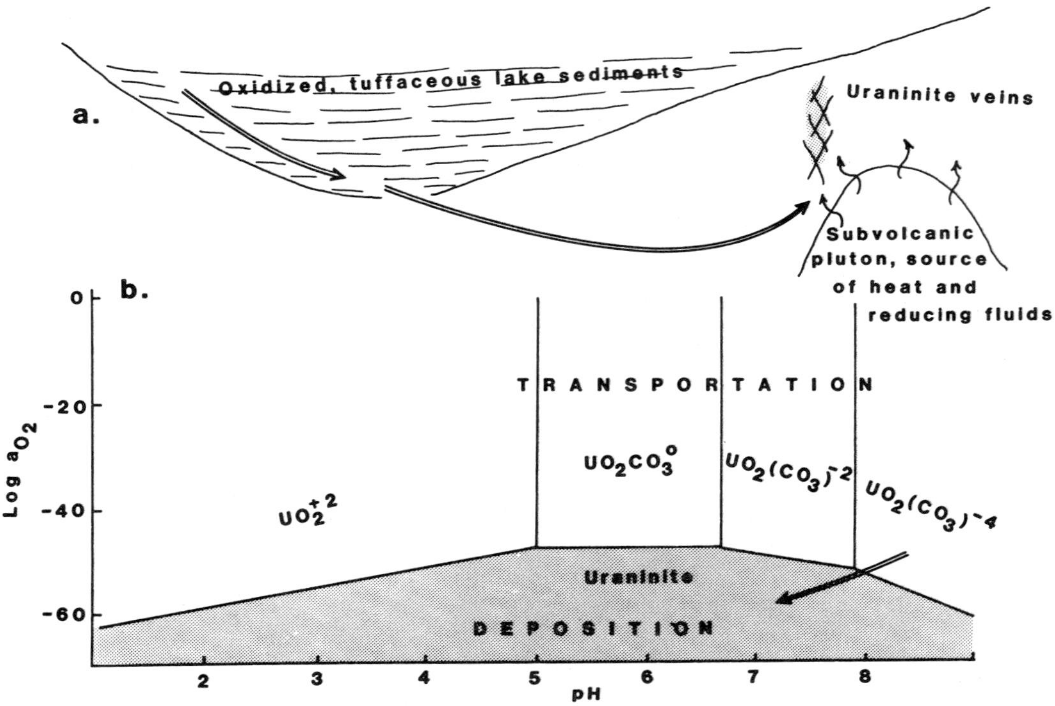

FIG. 8. Model for hydrothermal uraninite veins associated with tuffaceous lake sediments. Double arrows indicate hydrologic and chemical fluid paths. Fig. 8a is a schematic geologic cross section. Fig. 8b is a diagram of log a_{O_2} vs · pH distinguishing a region of uranium transport (concentration of uranium greater than 0.2 ppm) from a region of deposition (shaded). Calculated for 25°C and P_{CO_2} = 0.01 atm.

lutions encounter the reducing environment they precipitate uraninite.

A final example (Fig. 9) of a hypothetical unconventional model concerns variants of the typical Basin and Range "no-see'em" gold deposit. The oxygen-pH diagram shows a window of maximum solubility. In terms of sulfur chemistry, the window is superposed over the triple junction of the $SO_4^{=}$, HS^-, and H_2S fields (the solid triangle in Fig. 9b). Any departure from that junction, be it oxidation, reduction, basefication, or acidification, results in lower gold solubility.

A Carlin-type deposit may form from reduction by organic matter from the host rock as in a reaction such as that represented by path 1. Under other circumstances, leakage of the fluid toward the surface may result in oxidation such as illustrated by path 2 with the formation of mineralization below a hot spring. An unconventional target is produced by a reaction represented by path 3 wherein ferrous iron in a mafic rock serves as reductant, giving ore in an altered basalt. A reaction similar to type 3 is basic to Coveney's suggestion (1981) to explain the favorability of serpentinized wallrock in gold veins of the Allegheny district, California. An additional unconventional target might be generated (path 4) if gold bisulfide solutions encounter an oxidant in the form of redbeds. Lower sulfur concentrations in the ore fluid will sharply decrease the amount of gold in solution, but even very low gold concentrations can produce ore if the large volumes of solution pass through a chemical "sieve" that selectively drops out the gold.

Further consideration of this overall model suggests that perhaps SO_4^{-2}- and H_2S-rich hot brines may exist with sufficient gold contents to be processed directly for gold without waiting for nature to precipitate the metal in solid form. Finally, one should not forget that, at temperatures of perhaps 200°C and lower, sulfur and carbon redox reactions can be very sluggish and

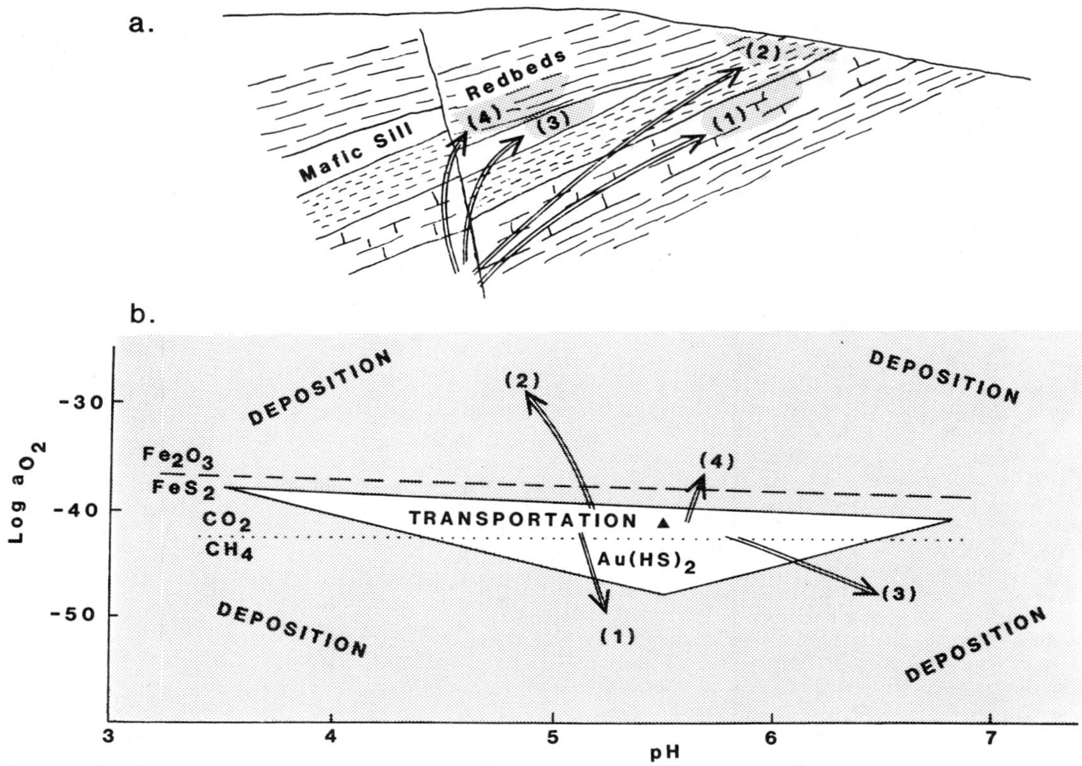

FIG. 9. Model for "no-see-'m gold deposits. Fig. 9a is a schematic geologic cross section. Fig. 9b is a diagram of log a_{O_2} vs · pH distinguishing a region of gold transport (concentrations of gold greater than 0.1 ppm) from a region of gold deposition (shaded). The calculations are taken from S. B. Romberger (personal communication, 1982) and Robie et al. (1978) and are for 200°C and a sulfur concentration of 100 ppm. Under these conditions the maximum gold concentration is 3 ppm and is located at the solid triangle amidst the "transportation" field. The numbered double arrows indicate hydrological and chemical fluid paths for four different deposit types. Path 1 typifies a Carlin-type environment and is represented by a reaction such as $4Au(HS)_2^- + C + 4H^+ + 2H_2O = 4Au + 8H_2S + CO_2$; the length of the arrow shown in Fig. 9b is exaggerated and should only extend from the high solubility region near the solid triangle down to the $CO_2 - CH_4$ boundary (dotted) because that boundary will be the general position for carbonaceous matter. Path 2 represents the oxidation of the sulfide ore fluid in a shallow, hot-spring environment with consequent deposition of gold according to a reaction such as $4Au(HS)_2^- + 15O_2 + 2H_2O = 4Au + 8SO_4^{-2} + 12H^+$. Path 3 is for an unconventional deposit localized in an altered mafic rock by reactions such as $4Au(HS)_2^- + Fe_2SiO_4 = 4Au + 2FeS_2 + SiO_2 + 2H_2O + 4HS^-$. Path 4 is for the deposition of gold in pyritized red beds through reactions such as $2Au(HS)_2^- + Fe_2O_3 + 2H^+ = 2Au + 2FeS_2 + 3H_2O$.

that metastable behavior might well overprint the simple equilibrium patterns we have discussed.

This Symposium Volume represents a stride forward wherein we begin to predict from whence the next generation of nonfuels resources will come, rather than simply relying on good luck to carry us. I hope that I have been able to show that there is a real opportunity for the application of predictive geochemistry to mineral deposits. So far as I am aware, the ideas I have noted are "unconventional," but I also realize that geologists are very imaginative and that the chances are excellent that someone else has already foreseen, and maybe even found, deposits representing my "hypothetical" examples.

In closing, I wish to thank Bruce Lipin, Bill Kelly, Ed Tooker, Pat Shanks, and Gene Cameron for helpful reviews; Ed Tooker for sharing his at-the-time unpublished thoughts on unconventional future ores; and Jim Eidel and the other organizers of this Symposium for the opportunity to air some heretofore unventilated thoughts about mineral deposits. Let us note that the fine collection of papers that follow in this volume honoring an outstanding geologist and teacher, Gene Cameron, demonstrate that the constructive imagination of man is our most valuable possession and it is the most critical resource for our continued national well being.

REFERENCES

Ahrens, L.H., 1965, *Distribution of the Elements in Our Planet,* McGraw-Hill, New York, 110 pages.

Barton, P.B., Jr., 1967, "Possible Role of Organic Matter in the Precipitation of the Mississippi Valley Ores," Economic Geology Monograph 3, *Genesis of Stratiform Lead-Zinc-Barite-Fluorite Deposits,* J.S. Brown, ed., Economic Geology Publishing Co., pp. 371–378.

Barton, P.B., Jr., 1981, "Physical-Chemical Conditions of Ore Deposition," *Physics and Chemistry of the Earth,* Vol. 13–14, pp. 509–526.

Barton, P.B., Jr., Bethke, P.M., and Roedder, E., 1977, "Environment of Ore Deposition in the Creede Mining District, San Juan Mountains, Colorado: III. Progress toward Interpretation of the Chemistry of the Ore-Forming Fluid for the OH Vein," *Economic Geology,* Vol. 72, pp. 1–24.

Bethke, P.M. and Rye, R.O., 1979, "Environment of Ore Deposition in the Creede Mining District, San Juan Mountains, Colorado: IV. Source of Fluids from Oxygen, Hydrogen and Carbon Isotope Studies," *Economic Geology,* Vol. 74, pp. 1832–1851.

Burnham, C.W., 1981, "Covergence and Mineralization—Is There a Relation?," *Memoir 154,* Geological Society of America, pp. 761–768.

Coveney, R.M., Jr., 1981, "Gold Quartz Veins and Auriferous Granite at the Oriental Mine, Alleghany District, California," *Economic Geology,* Vol. 76, pp. 2176–2199.

Deffeyes, K.S. and MacGregor, I.D., 1980, "World Uranium Resources," *Scientific American,* Vol. 242, pp. 66–76.

Erickson, R.L., 1973, "Crustal Abundance of Elements, and Mineral Reserves and Resources," Professional Paper 820, *United States Mineral Resources,* US Geological Survey, pp. 21–26.

Holland, H.D., 1972, "Granites, Solutions, and Base Metal Deposits," *Economic Geology,* Vol. 67, pp. 281–301.

Holland, H.D. and Petersen, U., 1981, "Element Dispersion, Element Concentration, and Ore Deposits," *Mineral Resources: Genetic Understanding for Practical Applications,* US National Research Council, pp. 39–46.

Lehmann, B., 1982, "Metallogeny of Tin: Magmatic Differentiation Versus Geochemical Heritage," *Economic Geology,* Vol. 77, pp. 50–59.

McKelvey, V.E., 1972, "Mineral Resource Estimates and Public Policy," *American Scientist,* Vol. 60, pp. 32–40.

McKelvey, V.E., 1960, Relation of Reserves of the Elements to Their Crustal Abundance," *American Journal of Science,* Vol. 268-A, pp. 234–241.

Parker, R.L., 1967, *Data of Geochemistry,* "Composition of the Earth's Crust," Chap. D, Professional Paper 440-D, US Geological Survey, 17 pp.

Robie, R.A., Hemingway, B.S., and Fisher, J.R., 1978, "Thermodynamic Properties of Minerals and Related Substances at 298.15 K and 1 Bar (10^5 Pascals) Pressure and Higher Temperatures," Bulletin 1452, US Geological Survey, 456 pp.

Sheldon, R.P., and Raup, O.B., 1981, "Research in Sedimentation as an Ore-Forming Process," *Mineral Resources: Genetic Understanding for Practical Applications,* National Academic Press, 92–101.

Skinner, B.J., 1979, "Earth Resources," Proceedings, National Academy of Sciences USA, Vol. 76, pp. 4212–4217.

Skinner, B.J., 1976, "A Second Iron Age Ahead?," *American Scientist,* Vol. 64, pp. 258–269.

Steven, T.A., and Eaton, G.P., 1975, "Environment of Ore Deposition in the Creede Mining District, San Juan Mountains, Colorado: I. Geologic, Hydrologic, and Geophysical Setting," *Economic Geology,* Vol. 70, pp. 1023–1037.

Tooker, E.W., 1982, "Unconventional Future Ores," *Encyclopedia of Materials Sciences and Engineering,* M. Brown ed., Pergamon Press, New York.

US Bureau of Mines, 1980, *Mineral Commodity Summaries,* 190 pp.

US Geological Survey, 1980, Principles of a Resource/Reserve Classification for Minerals, Circular 831, 5 pp.

2

Global Economics Relating to the Exploration of Unconventional Mineral Resources

W. C. J. van Rensburg

The University of Texas at Austin, Austin, TX

ABSTRACT

Exploration of unconventional mineral resources is stimulated by uncertainties about the cost and security of conventional resource supplies. Costs of the latter are likely to increase as a result of inflation, higher energy costs, declining productivity, a lack of sufficient investment in politically volatile areas, or through cartel-like action. Exploration of unconventional resources is likely to be delayed by these same factors, insufficient research and development facilities, and a decline in capital formation. World economic growth is likely to be slow, so there will be more emphasis on material conservation. Therefore, the incentive to explore for unconventional resources may be reduced.

INTRODUCTION

The Western World is experiencing a period of great economic uncertainty. After more than four decades of more or less liberal economic policies in the United States, there is now growing concern about soaring federal deficits, chronic inflation, and near record unemployment rates. Economic growth appears to have stagnated, and attempts to restore the value of the dollar and control inflation have led to unacceptably high interest rates. These interest rates have so far had a negative effect on the Reagan Administration's efforts to revive economic growth by cutting income taxes and reducing federal spending. They have also had an adverse effect on investments in nonfuel minerals exploration and development. By comparison with the 1950s and 1960s, we now face a situation where inflation, unemployment, and interest rates are very high, and economic growth rates considerably lower. Of equal concern are recent declines in labor productivity. Most western European countries, and even Japan, are facing similar problems.

The extent to which the oil crises of 1973–74 and 1979–80 caused these adverse economic conditions is in dispute. However, these developments with their concomitant much higher energy prices have made the processing and transport of minerals more expensive. Energy is now a very significant factor in the cost of production of a wide range of minerals. This, together with increasingly stringent environmental regulations have had a strongly adverse effect on the processing of minerals in the major consuming countries. Large land withdrawals in the US federal domain have reduced the areas available for exploration. This forced some US companies to turn to foreign countries for exploration and for mineral supplies. Hence, the nature of this country's mineral import dependence has changed. Where previously we imported ores and concentrates, and protected our own processing industries by imposing stiff import tariffs on imported processed minerals, economic and social realities now increasingly force us to import minerals in processed forms. This has negative implications for the consuming countries as far as balance-of-payments, employment, and security of supplies are concerned. It also limits the number of countries from which a particular mineral can be obtained. This trend has been enhanced by high labor costs in the industrial countries and by pressures by governments in developing countries for further processing of these minerals before export.

The decline of mining and metallurgical research facilities and of the level of government support for such research means that this is a situation which will be very difficult to reverse.

Low economic growth rates, the higher production costs for many minerals, and efforts to conserve energy by, for instance, producing smaller and more fuel-efficient cars, have had the effect of reducing the demand for many important minerals, and of changing the mix of minerals used. For example, in the automobile industry the move is away from the use of mild steel and towards increased use of high-strength low alloy steels, aluminum, plastics, and new materials such as carbon fibers. There are suspicions that the demand for some minerals, such as lead and zinc, may have peaked, and increased efforts at recycling of metals place an even bigger question mark on the future demand for primary metals.

The mining industry has generally done a poor public relations job. Protection of the environment, limiting economic growth, and hostility towards large, multi-national corporations are some of the realities with which the industry must now live. These developments have made it much more difficult, costly, and time-consuming to develop new mineral ventures in the industrialized countries.

The call for a New International Economic Order, expropriations, nationalization, forced changes of agreements, increased taxation, and greater host government interference, as well as a lack of physical and social infrastructure are problems which have to be faced when mineral investments are considered in developing countries.

Lead times for the development of new mineral ventures have been growing in almost all countries, and the mineral industry is faced with a smaller number of investment opportunities with far greater front-end loads than in the past. This means greater uncertainty with greatly reduced opportunities for risk-spreading. High levels of inflation combined with relatively low rates of return on investment have adversely affected the debt:equity ratios of many mining companies. The traditional mining companies were faced with the danger of being absorbed by the financially more powerful oil companies, although this trend is now reversing as the oil companies face the cyclical nature and capital intensiveness of new mineral ventures.

This is the environment that executives of mining companies have to live with when they decide on their exploration strategies. However, they also appreciate that the most important assets of any mining company are its reserves of economically extractable minerals. In order to survive, grow, and make a profit they must continue to explore. The question is: what do they explore for, at what level of expenditure, and where? As opportunities to explore for conventional sources of minerals in the politically more stable parts of the world decline, as they eventually must, companies will be faced with the choice of searching for conventional sources of the same minerals in politically volatile countries, or to develop unconventional sources of these minerals in politically more stable areas. This becomes a trade-off between sociopolitical risks in the former case with technical and economic risks in the latter. The choice may well depend on the risk preference of the particular company.

For most minerals the danger is not so much that we will physically run out of supplies on a worldwide basis as that we may not invest enough in exploration, the development of both conventional and unconventional sources of minerals, and in infrastructure so as to ensure sufficient productive capacity for the future. There is also a growing risk of politically-induced shortages.

Even in a free-enterprise system, the role of governments in providing the necessary geological base maps, and mining and metallurgical research facilities as an aid to exploration cannot be overemphasized. Unfortunately, in the absence of a strong public commitment to mineral development, which we lack, these services may be inadequate.

The overall economic environment for mineral exploration in the Free World does not appear to be very attractive, and the major incentive for exploring for and developing unconventional mineral resources in politically stable areas or on the deep seabed may well be disenchantment with the political risks of exploring in the Third World.

UNCONVENTIONAL MINERAL RESOURCES

Unconventional mineral resources represent materials of possible economic importance which occur in geological or physical environments that have not been successfully exploited for the particular mineral. A wide range of such resources is covered in this volume. Physical environments include the outer continental shelf, the deep seabed, outer space, and Antarctica; and new geological environments include nonbauxitic sources of alumina, marine black shales, subsurface brines, and deep-sea manganese nodules. We also include the recovery of byproducts from phosphorites, porphyry copper, and coal in this category, if such byproducts have not previously been economically extractable.

Unconventional mineral resources are therefore characterized by the fact that they are either low in grade, complex metallurgically, difficult to mine, or that they occur in a hostile unexploited environment. For any one, or a combination of these reasons, there is some question about the present economic viability of exploiting them. In terms of the McKelvey classification of mineral resources and reserves, we would have to place them in the paramarginal or submarginal categories. For some, so little exploration has been done that we should probably regard them as hypothetical or even speculative resources.

We do not exploit mineral resources. Resources have to be transformed by an orderly process into reserves. This requires capital, knowhow, and time. Incentives are required to induce such efforts to transform unconventional mineral resources into reserves.

Most unconventional mineral resources require technological innovations in mining or metallurgy before they can be successfully transformed into reserves. The process of technological innovation embraces a sequence of activities by which technical knowledge is translated into physical reality and becomes used on a scale having substantial societal impact. It includes initiation of a technical idea, acquisition of necessary knowledge, its transformation into usable hardware or procedure, its introduction to society, and its diffusion or adoption to the point where its impact is significant. The parts of this process do not take place in a vacuum, but in a complex environment of nontechnical factors.

There are various stages in the development of a technological innovation. For unconventional mineral resources the following stages may be considered:

Stage 1

Technological innovation starts with a technical suggestion, by discovery, or by the recognition of a need or opportunity. For instance, it is recognized that the United States has limited resources of bauxite, but large resources of high-alumina clays.

Stage 2

Is the proposal of a theory or proposal of design concept. For instance, an alkaline-leach process is proposed for the extraction of alumina from clays.

Stage 3

Is the verification step in which the validity of the proposed theory or design concept is confirmed. The alkaline-leach concept is confirmed. The alkaline-leach process is designed in more detail.

Stage 4

Is the laboratory demonstration of application. The alkaline-leach process is tested on a bench scale.

Stage 5

Is the larger scale approach that may eventually become a small prototype of the innovation. The alkaline-leach process is tested in a pilot plant.

There are many failures between stages 4 and 5.

Stage 6

Is the commercial introduction or first operational use step.

Stage 7

Is the widespread adoption or commercial acceptance step.

Stage 8

Is the proliferation stage.

There is frequently a 9th stage, namely an obsolescence step.

The full process of technological innovation usually takes upward of 10 years, with 20 to 25 years being more likely. However, failure to recognize these various steps may lead us to a false sense of security. For example, alunite and nepheline syenites did not turn out to be viable alternative sources of alumina.

A company contemplating a technological innovation leading to the commercialization of an unconventional mineral resource therefore has to face all the risks and uncertainties associated with an investment which may only yield a positive cash flow 10 to 25 years beyond the initial investment. Forecasting capital and operating costs, prices, and markets so far into the future is a hazardous undertaking. Faced with the prospect of so many years of negative cash flows, the project would have to show promise of eventual high profitability.

With such long lead times any company would clearly only make the necessary commitment in a favorable, stable political environment. In a two-party system this would probably require at least some bipartisan elements in the mineral policies of the various political parties, and the expectation that a high level of political morality would be maintained. Uncertainties about security of tenure over unconventional mineral deposits (such as those prevailing in the deep sea), frequent changes in environmental and other regulations, a history of government interference in the mineral industries, and numerous changes in mining taxation would be strong disincentives to such investments.

INCENTIVES FOR EXPLORATION AND COMMERCIALIZATION OF UNCONVENTIONAL MINERAL RESOURCES

Some unconventional mineral deposits may be discovered accidentally during exploration for conventional deposits. Others, such as ocean manganese nodules, oil shales, and various nonbauxitic sources of alumina were discovered long ago during routine scientific investigations. Continued purely scientific geological research is a prerequisite for future discoveries of such deposits, and the importance of state and national geological surveys, and of university research, cannot be overemphasized in this regard.

The further delineation, evaluation, and development of unconventional mineral deposits requires some promise that they will become economically exploitable, or incentives that would virtually guarantee their economic viability. Such incentives may take many different forms. They may include loan guarantees, low-interest loans, price guarantees, purchase agreements, direct or indirect subsidies, and import tariffs or restrictions. Their aim is to improve the economic viability of the particular unconventional mineral venture, and to reduce the risks and uncertainties associated with its development. The longer term economic viability of unconventional mineral resources almost invariably requires the development of new markets, or technological innovations related to their mining, processing, or utilization. For instance, the development of the argon oxygen decarburization (AOD) process made possible the use of low-grade, high-carbon ferrochrome, produced from low-grade chromite ores in South Africa, in the production of stainless steel. Previously these ores were considered to be of chemical grade only.

Very often there are political or security incentives that lead to the development of unconventional mineral resources. Lacking adequate domestic resources of bauxite, the Soviet Union, for instance, used nepheline syenite as a source of alumina. Threatened by an oil embargo, and lacking any domestic oil resources, South Africa developed its oil-from-coal industry. Similar incentives have caused a revival of interest in oil shales in the United States.

In most instances, an unconventional mineral resource has to have some advantage over its conventional counterpart to make it economically attractive. It may be located closer to the market, a factor which could become particularly important if freight rates increase. This is what has made Texas lignite attractive relative to Western coals. It may be recoverable as a byproduct of another mineral, which could pay for most mining and beneficiation costs. This is what made the low-grade uranium deposits of the Witwatersrand system in South Africa exploitable. It may be larger and more consistent in grade than its conventional counterpart, thereby allowing economies of scale. It may already have been mined and comminuted for some other purpose, thereby allowing the elimination of two costly process steps from the economic evaluation. The use of coal ash as a possible source of alumina is an example of this.

In the broader context, countries which maintain centers of excellence in geological, mining, and metallurgical research and development have a higher probability of making the necessary technological breakthroughs that would enable the commercialization of unconventional mineral resources than those that do not. A reasonable proportion of research programs should be devoted to projects having a low probability of success, but a high financial return if success is achieved. There are many examples from the US Bureau of Mines, the South African National Institute for Metallurgy, and the Australian Mineral Development Laboratories which illustrate this point. The public and private sectors both have a role to play in this regard. The lead times in the development of many unconventional mineral deposits are too long to make them attractive to private companies. However, governments having the responsibility for long-term adequacy and security of mineral supplies should be able to justify such research. In this regard, there is a critical need for the revitalization of the US Bureau of Mines.

ECONOMIC OUTLOOK FOR EXPLORATION OF UNCONVENTIONAL MINERAL RESOURCES

Exploration of unconventional mineral resources is stimulated by uncertainties about the cost and security of supplies of conventional sources of energy and nonfuel minerals. Costs of the latter are likely to increase in real terms as a result of inflation, higher energy costs, threats of war, declining productivity, the depletion of high-grade deposits, a lack of sufficient investment opportunities in politically volatile areas, or through cartel-like actions. These factors would appear to make the exploration of unconventional mineral resources more attractive. Unfortunately many of these factors could also have an adverse effect on the development of unconventional deposits.

In the short to medium term, most of the major mineral-consuming countries face the prospect of depressed economic growth rates, high rates of inflation and unemployment, high interest rates, continued high oil-import bills, and balance-of-payments problems. All these factors are likely to have an adverse effect on the demand for minerals. At the same time, the costs of new mineral ventures have soared, and nonfuel mineral prices are depressed. This does not constitute a favorable economic climate for the exploration and development of conventional mineral resources, let alone unconventional ones.

The expectation of continued high energy prices makes the prospects for exploration of unconventional sources of energy, such as tar sands, oil shales, and coal-based synthetics somewhat more favorable. However, the high cost of energy has made the exploitation of low-grade unconventional metal deposits less attractive.

Political and security interests, based on growing fears about the aims of Soviet imperialism and increasing tension between the superpowers, have led to a new arms buildup, particularly in the United States, but perhaps to be emulated by our allies. This calls for increased and secure supplies of a wide range of strategic minerals. In such an environment purely economic considerations may fall by the wayside, and demands for greater domestic self-sufficiency of mineral supplies and the diversification of foreign sources of supply may mandate greater exploration for and development of a range of unconventional mineral resources in politically stable and defensible areas.

Leading members of the Reagan Administration believe that the United States and its allies are already engaged in a "resource war" with the Soviet Union, and that this accounts for the high level of unrest in the two major resources battlefields—the oilfields of the Middle East and the minerals treasure troves of central and southern Africa. Further disruptions in these areas could turn public opinion in favor of the rapid development of unconventional mineral deposits. However, in recent months concerns about the massive projected budget deficits of the Reagan Administration have reduced support for increased defense spending.

One therefore has to conclude that the economic incentives for the exploration of unconventional mineral resources are relatively poor, and that strong political incentives do not yet exist. In general, therefore, the outlook for exploration of unconventional mineral resources does not appear particularly bright. However, the prospects for increased interest in unconventional sources of energy appear to be considerably more favorable.

2

Unconventional Deposits in Low Grade Environments

3

Lower Grade Nickel Sulfide Resources

J.M. Duke

Economic Mineralogist, Geological Survey of Canada, Ottawa, Ont. Canada

A.J. Naldrett

Professor, Department of Geology, University of Toronto, Toronto, Ont. Canada

ABSTRACT

Significant resources grading less than about 1% nickel occur in magmatic sulfide deposits of two types. In the Mt. Keith type, extensive zones of disseminated nickel sulfides occur in the central portions of certain synvolcanic dunite-peridotite sills of komatiitic affinity in Archean greenstone terranes. In the Duluth type, disseminated copper-nickel sulfides occur near the basal margins of intrusions of flood basaltic affinity associated with intracontinental rifting. In both types, the parent magmas became saturated with sulfide following the onset of fractional crystallization and the separation of the immiscible sulfide melt was accompanied by the precipitation of a relatively large proportion of silicates. The abundant silicate crystals impeded the coalescense and accumulation of the sulfide droplets with the resulting formation of large but low grade deposits.

INTRODUCTION

Magmatic Ni-Cu sulfide deposits have been classified into four main groups according to the petrologic affinity and tectonic setting of the igneous bodies with which they are associated (Naldrett and Macdonald, 1980). These groups are (1) noritic rocks associated with an astrobleme, (2) intrusive equivalents of flood basalts associated with intracontinental rifting, (3) rocks emplaced early in the evolution of Precambrian greenstone terranes including (a) synvolcanic tholeiitic intrusions and (b) komatiitic lavas and synvolcanic intrusions, and (4) Phanerozoic synorogenic intrusions. The grade distribution of identified nickel sulfide resources subdivided according to petrotectonic setting is illustrated in Fig. 1. We have omitted the very low grade but immense nickel resources recoverable as a byproduct of platinum production in the Bushveld complex. It is apparent from this compilation that the flood basalt and komatiite associations dominate the lower grade (i.e., less than about 1% Ni) nickel sulfide resources, and in this chapter we discuss some genetic aspects of lower grade deposits of these two classes. Although most deposits of the types we describe are currently subeconomic, they nevertheless represent significant resources that might be developed under improved economic or technological conditions in the future.

Most conventional Ni-Cu sulfide ores may be described as massive to heavily disseminated

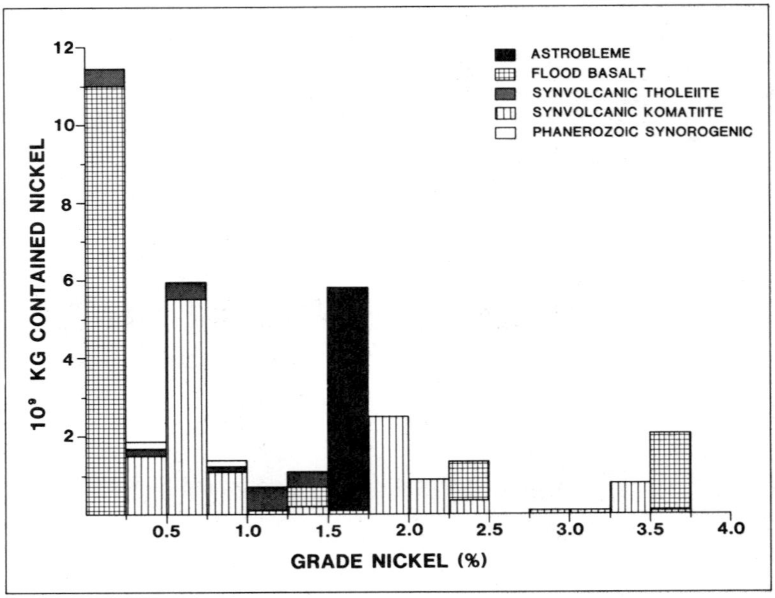

FIG. 1. Frequency distribution of grades for identified nickel resources contained in magmatic sulfide deposits. Grades are averages for single deposits or camps. Resource data are mainly from Ross and Travis (1981), and Naldrett and Duke (1980).

magmatic segregation deposits that occur at or near the base of mafic to ultramafic intrusions or lava flows. In almost every case it is clear that the sulfide phase formed initially as an immiscible liquid and this implies that the magma was at least locally saturated with sulfide for a finite period of time. The molten sulfide acted as a collector of the ore metals and the tenor of the sulfide was determined by the concentration of the metals in the parent magma, the degree to which the metals partitioned into the sulfide and the relative proportions of sulfide, silicate liquid and crystalline phases that equilibrated (Rajamani and Naldrett, 1976; Duke and Naldrett, 1978; Campbell and Naldrett, 1979). Of course, sulfide saturation in itself did not lead to the formation of a deposit of economically exploitable grade. It was necessary that some mechanism operated to accumulate the immiscible sulfide, and the ore grade reflects both the efficiency of this mechanism and the tenor of the sulfide phase.

A simplified phase diagram (Fig. 2) provides a useful framework for consideration of sulfide saturation in mafic and ultramafic magmas. The solubility of sulfide in such magmas is variable and depends upon magma composition, temperature, pressure, and the fugacities of oxygen and sulfur (Haughton et al., 1974; Shima and Naldrett, 1975; Buchanan and Nolan, 1979). Although available experimental data do not permit an accurate estimation of the solubility in a given magma, solubilities in most of the magmas in question are probably in the range of 0.1 to 2.0% sulfide, with higher values characteristic of hotter, more ultramafic magmas. The solubility of sulfide is expected to decrease as a magma cools and crystallizes, and a magma that is initially undersaturated in sulfide may eventually attain saturation as a result of the solubility decrease and the increase of the sulfur concentration in the residual liquid. For example, a magma might follow a path such as A → B → C in Fig. 2. That is, a sulfide-undersaturated magma plotting at A within the primary phase field of olivine would fractionate towards B on the olivine-sulfide boundary. Upon reaching B, an immiscible sulfide liquid would appear and the separation of olivine and sulfide together would cause the magma composition to move along the boundary curve towards C. The relative proportion of sulfide that forms is a direct function of the solubility and also the rate of decrease of solubility as crystallization proceeds. In the specific example given in Fig. 2, olivine and sulfide separate in a ratio of about 50:1 while the magma composition moves along the path B → C. Such a high proportion of olivine would militate against the coalescense and accumulation of the molten sulfide droplets, and a thesis that we will develop in this paper is that many lower grade Ni-Cu sulfide deposits

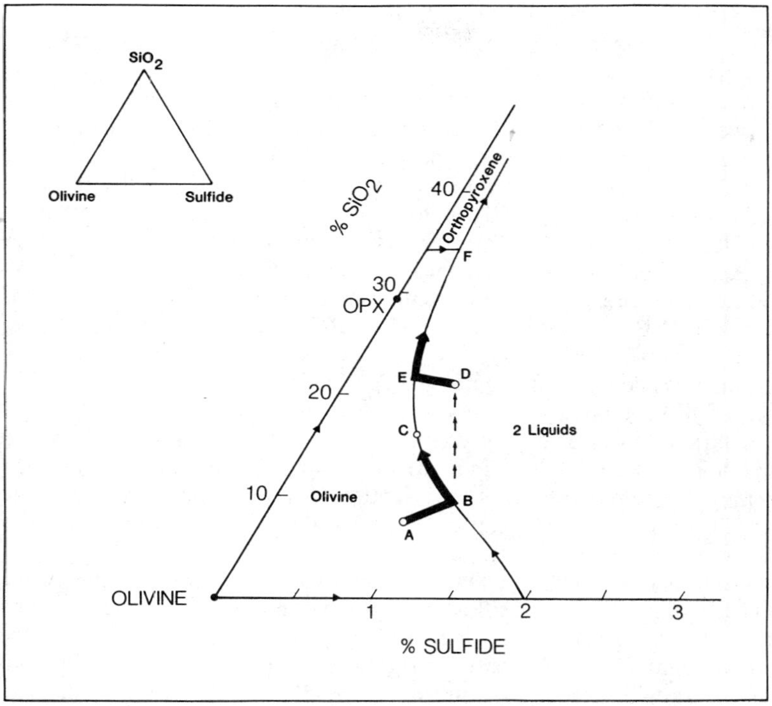

FIG. 2. Schematic phase relations for the system olivine-sulfide-silica, after Irvine (1975). The heavy arrows represent fractionation paths whereas the light arrows indicate a possible magma mixing or contamination path. Note that the horizontal and vertical scales are different.

form by the in situ accumulation of sulfide and crystalline silicates in their "cotectic" proportions.

We envisage two circumstances that would likely favor the formation of high grade sulfide accumulations. On the one hand, an immiscible sulfide melt could form along the sulfide saturation boundary prior to or during emplacement of a magma, and then segregate from the silicate crystals by some process of differentiation during flow of the magma. Alternatively, more massive bodies of sulfide could accumulate in situ if the magma composition were in the primary phase field of sulfide. The addition of sulfur to the magma by, for example, the assimilation of sulfide- or sulfate-rich sedimentary rocks is one mechanism that could move the magma into the two liquid field. Indeed, evidence from sulfur isotopes indicates that such assimilation has played an important part in the generation of some magmatic deposits. Irvine (1975, 1977) proposed a model whereby sulfide-undersaturated magmas, or magmas on the saturation boundary, could enter the sulfide field by mixing with a less primitive magma or contamination with felsic material. Accordingly, in Fig. 2, a magma with composition B might blend with a less primitive magma of composition F to produce a mixture of composition D within the two liquid field. Sulfide liquid would then separate alone, until the silicate liquid composition returned to the olivine-sulfide boundary curve at E.

LOWER GRADE NICKEL DEPOSITS OF KOMATIITIC AFFINITY

Komatiites (sensu strictu) are volcanic rocks that have crystallized from ultramagnesian magmas, that is, from melts containing more than about 20% MgO. In addition to the classic, spinifex-textured peridotite lava flows, the komatiite suite includes a variety of intrusive and extrusive rocks ranging in composition from dunite to basalt which are differentiates of such magmas. True komatiites and related rocks are restricted to Archean terranes. However, rocks generated from liquids that contained at most from 16 to 18% MgO, but which were otherwise very similar to komatiites, occur in Proterozoic and even Phanerozoic sequences. Three principal types of nickel sulfide ores occur in association with komatiitic rocks (sensu lato). The most numerous are small but high grade deposits that occur as basal accumulations in peridotitic lava flows or in the feeders to such flows. Important examples of this type occur in the Kambalda camp of Western Australia, in the Abitibi greenstone belt in Canada, and in Zimbabwe. Much larger high grade deposits occur as basal accumulations in ultramafic sills of the komatiite suite. The Archean-aged Agnew deposit in Western Australia and the Proterozoic ores of the Thompson and Ungava belts in Canada are important examples. The third type, of principal interest in the present context, includes deposits consisting of extensive zones of disseminated magmatic sulfides within the central portions of synvolcanic dunite-peridotite bodies. These deposits represent shallow level intrusive members of the komatiite suite.

The prototype of this group is the Mt. Keith deposit, one of a number of occurrences of this type in a series of dunitic lenses along a lineament extending south about 200 km from Wiluna in Western Australia. The individual lenses are as much as 20 km long and 1 km wide, and are characterized by virtually monomineralic dunite cores that grade outward to more pyroxene-rich margins (Binns et al., 1977; Naldrett and Turner, 1977). Disseminated nickel sulfides in crudely conformable zones make up 2 to 3% of the rock within the dunite core and the textures leave little doubt of the primary magmatic origin of the sulfides. The Mt. Keith deposit is reported to have 290 million tonnes grading 0.6% Ni (Burt and Sheppy, 1975). The Six Mile deposit, which is about 20 km south of Mt. Keith, contains 54 million tonnes averaging 0.6% Ni. The Epoch deposit in Zimbabwe (5 million tonnes grading 0.64% Ni) and the Dumont deposit in Canada (13.6 million tonnes grading 0.65% Ni) are included in this class, although they are much smaller than their Australian counterparts and differ in that their host intrusions embody significant mafic differentiates.

The Dumont Deposit

The intrusion that hosts the Dumont deposit is relatively undeformed and unmetamorphosed and, therefore, provides a useful field laboratory for the study of this type of mineralization. It is situated within the Abitibi greenstone belt in the Superior Province of the Canadian Shield, 62 km northeast of the city of Rouyn, Québec. The following discussion summarizes some relevant conclusions from a study by Duke (in preparation).

The Dumont intrusion is a differentiated sill about 7 km long, comprising a lower Ultramafic zone about 450 m thick and an upper Mafic zone about 250 m thick (Fig. 3). The sill dips about 60° to the northeast and also faces northeast. The Ultramafic zone has been divided into three subzones; the Lower Peridotite, which is an olivine + chromite cumulate; the Dunite, which is an olivine ± sulfide cumulate; and the Upper Peridotite, which is also an olivine + chromite cumulate. The Mafic zone comprises three subzones which are the Clinopyroxenite, a clinopyroxene cumulate; the Gabbro, a clinopyroxene + plagioclase cumulate; and the Quartz Gabbro, which includes noncumulate gabbros as well as plagioclase + clinopyroxene cumulates.

On the basis of the bulk composition of the sill as well as the compositions of the most primitive cumulus phases, it is estimated that the parent magma was a peridotitic komatiite containing about 27.5% MgO. This magma possibly carried 10% olivine phenocrysts in suspension, but given the low viscosities and yield strengths of ultramafic liquids, these phenocrysts probably settled out soon after the sill was emplaced. The Lower Peridotite subzone is believed to be an accumlation of these intratelluric phenocrysts plus a small proportion of trapped intercumulus liquid. As the suspension settled, the supernatant magma probably began to convect. The Dunite, Upper Peridotite, Clinopyroxenite, and Gabbro subzones are believed to have accumulated by a bottom crystallization process. One such mechanism is that proposed by Jackson (1961) whereby the adiabatic temperature gradient across the magma chamber is less steep than the liquidus temperature gradient and crystallization therefore takes place in a thin zone adjacent to the floor of the chamber rather than at the roof where maximum heat loss occurs. Alternatively, bottom crystallization might have resulted from a compositionally controlled liquidus temperature gradient as described by Irvine (1980). This would have required a compositional gradient in the magma which could have been the result of an influx of primitive magma, melting of roof rocks, or fractional crystallization at the roof. Noncumulate, ''liquid equivalent'' rocks in the Quartz

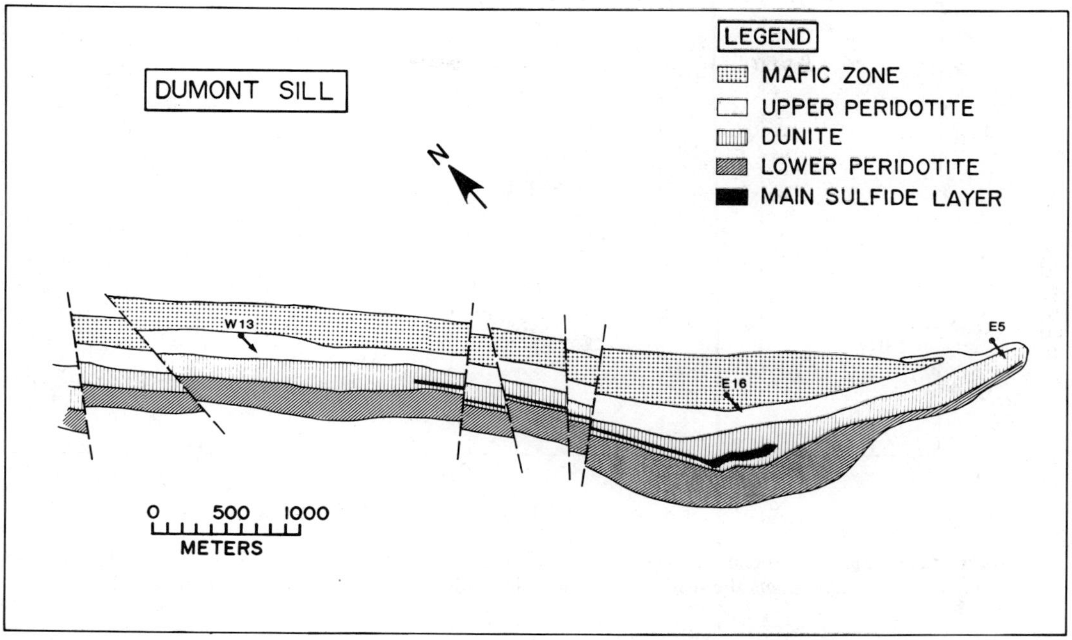

FIG. 3. Geological map of the Dumont sill. The subdivisions of the Mafic zone are not shown. The small arrows indicate the collar locations and directions of diamond drill holes referred to in the text.

Gabbro subzone have MgO contents as low as 5.5% MgO, indicating extreme differentiation of the parent peridotitic magma by fractional crystallization.

Magmatic sulfides are restricted to the Lower Peridotite and Dunite subzones. In the former unit, the magmatic sulfides are generally very fine grained, very sparsely disseminated, and likely separated from the intercumulus liquid at a relatively advanced stage in the accumulation of the subzone. Disseminated magmatic sulfides are more abundant in the central part of the Dunite. Three generally conformable, sulfide-enriched layers have been outlined and their stratigraphic positions correspond to the peaks in the assay profile in Fig. 4. The middle layer has the highest grade and is the most laterally extensive, persisting over a strike length of almost 3000 m. Its true thickness ranges from 20 to 40 m, and the overall average grade is 0.55% Ni. A zone at the eastern end of the middle layer contains about 13.6 million tonnes averaging 0.65% Ni.

The mineralized portion of the Dunite subzone is an olivine + sulfide adcumulate containing only traces of intercumulus chromite and clinopyroxene. Modal analyses of the Dunite from the intersection depicted in Fig. 4 show that the proportion of sulfide ranges from 0.7 to 3.2 wt %, averaging 1.6%. The observed proportions of olivine and sulfide are thus approximately equivalent to those in a peridotitic komatiite magma fractionating along the sulfide saturation boundary (e.g., along the path B → C in Fig. 2). The nickel content of olivine is also illustrated in Fig. 4. The marked depletion of nickel in olivine upward across the sulfide-enriched layers agrees well with the trends predicted for the co-fractionation of olivine and sulfide from komatiitic magma by the numerical model of Duke and Naldrett (1978). The degree of nickel depletion and its apparent cyclicity suggests that sulfide segregation occurred within the zone of crystallization adjacent to the upward-migrating floor of the magma chamber and that the zone was periodically replenished with fresh magma as a result of convective overturn.

The fundamental question regarding the origin of the mineralization is why sulfide saturation occurred at the particular level in the magma chamber that it did. An important clue comes from the cryptic variation of the forsterite content of olivine across the Ultramafic zone illustrated in Fig. 5. Drill holes E5 and W13 are

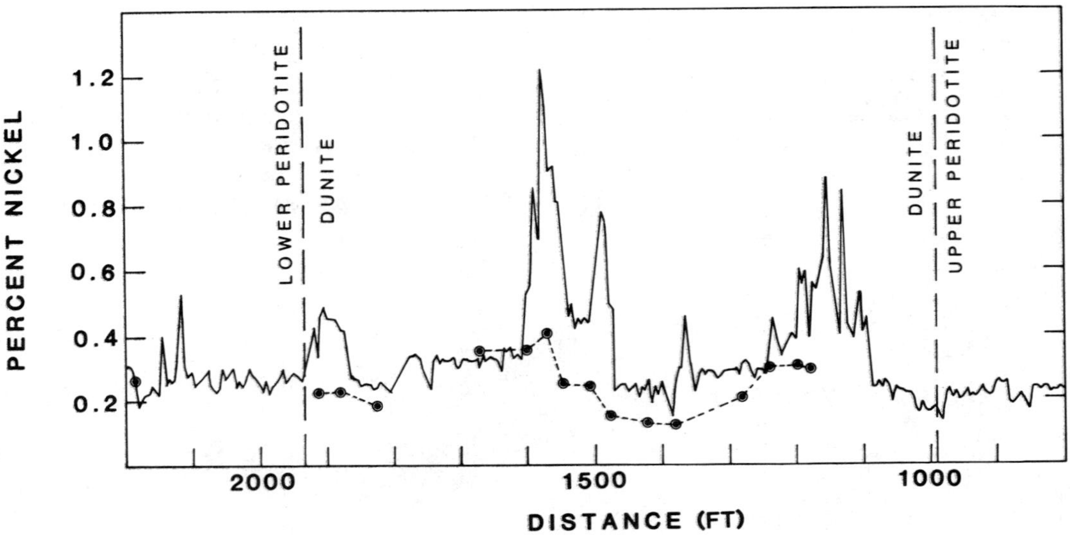

FIG. 4. Concentrations of nickel in whole-rock (solid line) and olivine (circles) for drill hole E16. The horizontal axis gives the drilled depth and also indicates increasing stratigraphic depth.

FIG. 5. Composition of olivine in the Ultramafic zone as observed in the three drill holes located in Fig. 3. The horizontal axis indicates increasing stratigraphic depth relative to an arbitrary datum. The arrows indicate the abrupt increase in the forsterite content of olivine.

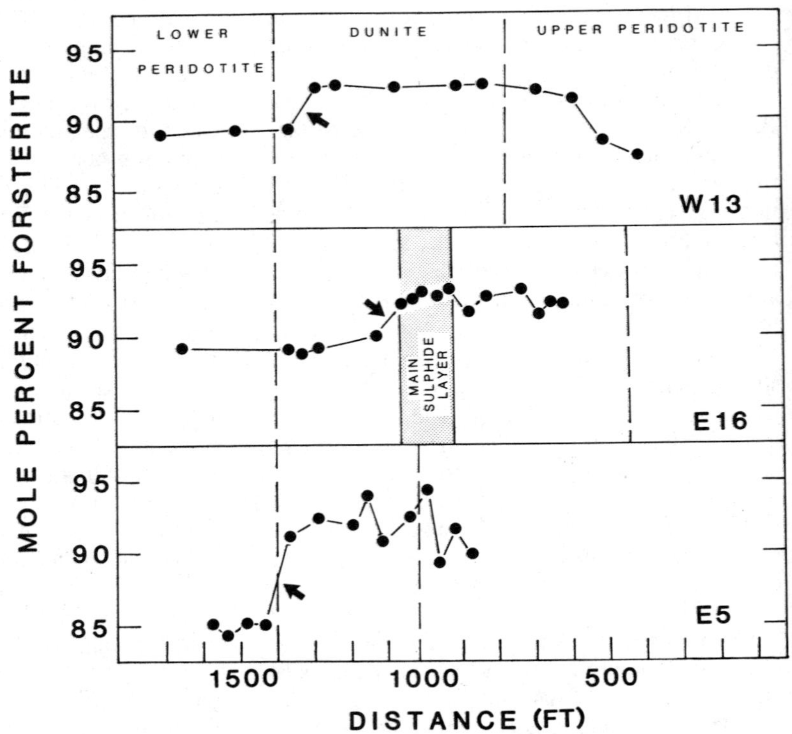

located at the eastern and western ends of the intrusion, respectively, where no magmatic sulfide occurs in the Dunite subzone (Fig. 3). Drill hole E16 penetrates the mineralized zone in the central part of the sill. In each profile, there is a gradual increase in the forsterite content of olivine upward across the Lower Peridotite subzone and then an abrupt increase in the lower part of the Dunite. The gradual increase coincides with a decrease in the proportion of intercumulus liquid trapped in the rock as calculated from incompatible element concentrations, and the trend reflects reaction between crystals and trapped intercumulus liquid. It is significant that the position of the abrupt increase corresponds to the lower contact of the Dunite subzone in profile E5, is displaced slightly upward from this contact in W13, but is displaced a significant distance upward in drill hole E16. A likely explanation for the displacement is the process of "infiltration metasomatism" proposed by Irvine (1980) whereby compaction of the cumulus pile under its own weight causes the intercumulus liquid to percolate upward, reacting with cumulus crystals along its path and, as we believe happened in the Dumont sill, to be expelled at the top of the pile where it would mix with more primitive magma in the main part of the chamber. As noted previously, Irvine (1977) predicted that mixing of primitive magma with differentiated liquid could lead to sulfide saturation. For example, referring again to Fig. 2, primitive liquid of composition A might blend with upward migrating pore liquid of composition F and thereby move towards composition C on the olivine-sulfide boundary. Upon reaching C, fractionation of 98.5% olivine and 1.5% sulfide would cause the magma composition to move along the boundary towards E. At the eastern end of the intrusion, the Lower Peridotite subzone is thin and therefore there was little compaction or filter pressing of the intercumulus liquid. The orthocumulus texture and high concentrations of incompatible elements in the Lower Peridotite in this area support this conclusion. The Lower Peridotite is thicker at the western end of the sill and grades from an orthocumulus texture at the base upward into a mesocumulate. The small displacement of the discontinuity in cryptic variation in profile W13 indicates that there was some filter pressing but it was apparently not sufficient to induce sulfide saturation in the overlying Dunite subzone. The greatest development of magmatic sulfides is in that part of the Dunite where the underlying Lower Peridotite is thickest, where the displacement of the cryptic variation trend is the greatest, and where filter pressing has therefore been the most important.

In assessing the potential exploitability of low grade deposits of this type, it is important to determine what proportion of the nickel resides in the sulfide fraction and, therefore, is recoverable by conventional ore dressing methods. We will not address this problem in detail, but will simply offer some relevant observations. All of the ultramafic rocks in the intrusion have been at least partly serpentinized, and it has been shown that the alteration process at Dumont was essentially isochemical except for the addition of up to about 15% volatiles. Prior to alteration, mineralized dunite comprised 95% or more olivine, about 3% or less sulfide, and 2% or less chromite plus clinopyroxene. The latter two minerals contain negligible quantities of nickel, hence the area between the olivine and whole-rock nickel profiles in Fig. 4 reflects the proportion of nickel resident in the sulfide fraction in the unaltered rock. This indicates that maximum recoveries of nickel from unaltered material from the middle sulfide layer would perhaps be about 50 to 75% of the amount present. We have found that the concentrations of nickel in serpentine, magnetite, and other secondary phases in mineralized samples is very much lower than that in the precursor olivine, whereas the concentrations in the magmatic sulfide grains are significantly higher than the calculated primary nickel content of the sulfide. For example, the average calculated increase of the nickel concentration in the sulfide fraction is 53% for the samples plotted in Fig. 4. Evidently much of the nickel released during serpentinization of olivine was taken up by the magmatic sulfides and consequently significantly better nickel recoveries should be achieved from altered as compared with unaltered rock.

LOWER GRADE NICKEL DEPOSITS IN INTRUSIONS ASSOCIATED WITH CONTINENTAL FLOOD BASALTS

The most important ores of this association are those of the Noril'sk-Talnakh camp in northwestern Siberia. These are extremely rich in grade and are being mined on a large scale, so they most certainly do not fall into the category of deposits forming the subject of this chapter. The most important lower grade deposits of this association are those of the Duluth complex in Minnesota, where resources of 4000 million tonnes grading 0.66% Cu and 0.2% Ni have been estimated by Listerud and Meineke (1977). Other examples include the Great Lakes Nickel Ltd. deposit in northwestern Ontario and the Waterfall Gorge occurrence in southern Africa. The former is hosted by the Crystal Lake gabbro intrusion, which may be related to the Duluth complex, and has reported resources of about 41 million tonnes grading 0.34% Cu and 0.18% Ni. The latter deposit occurs in the Insizwa complex, a Jurassic intrusion of the Karroo system, and contains about 0.54 million tonnes grading 0.5% Cu plus Ni (Scholz, 1936).

Mineralization of the Duluth Complex

The geological setting and origin of the deposits of the Duluth complex are reviewed here to illustrate the lower grade mineralization of the continental flood basalt association. The Duluth complex is exposed as an arcuate body extending about 250 km northeast from the city of Duluth, MN, to the Canadian border (Fig. 6). It is closely associated with Keweenawan basalts of the Lake Superior region and was intruded 1.12 Ga ago close to the contact of these basalts with older Precambrian rocks which are now exposed to the northwest. The older rocks to the north are a series of Archean greenstones intruded by Archean felsic plutons, such as the Giant's Range granite. The Archean rocks are unconformably overlain on the south by Biwabic iron formation, which is overlain, in turn, by black argillite, greywacke, siltstone, graphitic slate, and sulfide-facies iron formation, rocks that together form the Virginia formation.

The Keweenawan basalts are continental flood basalts, and there are significant differences between those occurring to the west of the complex and those to the east. Those to the west have a

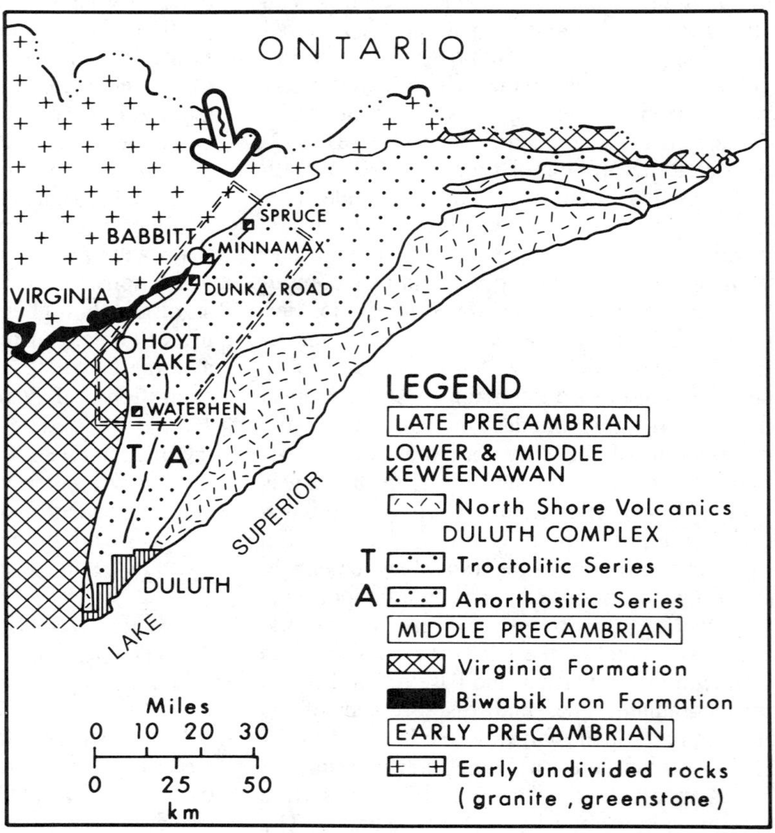

FIG. 6. Map of the western end of Lake Superior showing the location of the Duluth complex and its relationship to the country rocks. The area delineated by the dashed line is that covered in Fig. 7.

reversed remanent magnetic polarity and consist mainly of subaerial pyroxene-phyric lavas, whereas those to the east are plagioclase-phyric, and have interdigitated felsic flows and sedimentary rocks and normal remanent polarity. Citing the data of Green (1972), Weiblen and Morey (1980) described the western and eastern groups as low-Al and high-Al types, respectively.

The Duluth complex includes two major subdivisions: the anorthositic series and the troctolitic series (Weiblen and Morey, 1980). The troctolitic series intrudes the anorthositic series, particularly along the western side of the complex. The anorthositic series is a poorly exposed sequence of dominantly plagioclase cumulates, and it is not clear whether it was emplaced as a single or multiple intrusions. The troctolitic series is known to consist of numerous intrusions of which two have been studied in detail. The Bald Eagle intrusion (Weiblen, 1965) is funnel-shaped and has well developed layering which appears to be largely due to flow in a crystallizing magma. The South Kawishiwi body is spoon-shaped and is also well layered (Phinney, 1959). Weiblen and Morey (1980) have shown that the rocks of the anorthositic series could have been derived from the magma that gave rise to the low-Al tholeiitic flows west of the complex, whereas those of the troctolitic series are compatible with derivation from the high-Al magma that was parental to the eastern flows.

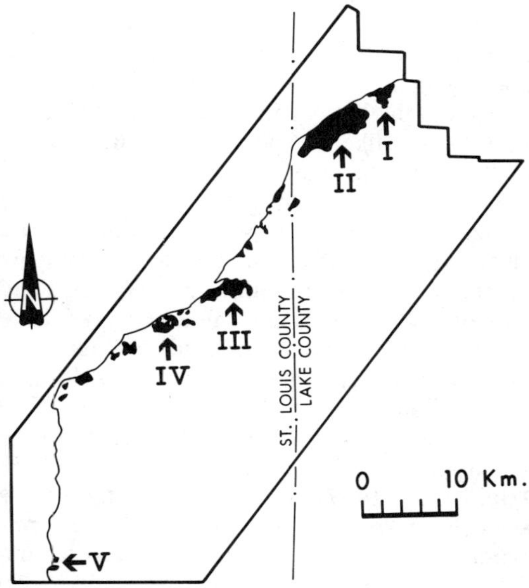

FIG. 7. Map of parts of St. Louis and Lake counties, Minnesota, showing the main areas of mineralization outlined to date and described in the text.

The Cu-Ni resources of the Duluth complex occur in five main areas along the western contact (Fig. 7). *Area I* includes the Spruce Pit deposit of INCO Ltd., which is reported to contain 248 million tonnes grading 0.46% Cu and 0.17% Ni, and the area as a whole is estimated to have resources of about 635 million tonnes on the basis of a 0.5% Cu cutoff grade (Listerud and Meineke, 1977). The mineralization, described by Wager et al. (1969), consists of disseminated sulfides in inclusion-bearing troctolite (Fig. 8). Among the inclusions are numerous and, in some cases, very large masses of Biwabic iron and Virginia formations. These are somewhat problematical inasmuch as the Giant's Range granite forms the footwall of the complex in this area and the nearest exposures of Proterozoic rocks are at least 15 km away. The majority of the mineralization occurs within 300 m of the basal contact, although one drill hole intersected 60 m of mineralization about 600 m above the contact.

Area II was investigated by INCO Ltd., The Hanna Mining Co., and Duval Corp. It is reported to have resources of nearly 2100 million tonnes of material grading at least 0.5% Cu (Listerud and Meineke, 1977). The mineralization, which forms a thinner zone than in area I, is generally not far above the basal contact and, according to Listerud and Meineke, transects the contact in places to penetrate several meters into the granite footwall.

Area III includes the Minnamax property of the Amax Corp. where a shaft has been sunk and exploration has proceeded underground. The resources quoted for the potential underground operation were 299 to 340 million tonnes grading 0.8% Cu and 0.2% Ni, whereas the resources estimated for area III using a 0.5% Cu cutoff were about 725 million tonnes (Listerud and Meineke, 1977). The mineralization occurs at or near the base of the complex, in the vicinity of a major roll in the contact. Sections presented by Listerud and Meineke indicate that some of the mineralization is separated from the footwall by several tens of meters of unmineralized troctolite.

Area IV is the Dunka Road property of the US Steel Corp. The mineralization occurs at or within 60 m of the basal contact of the complex.

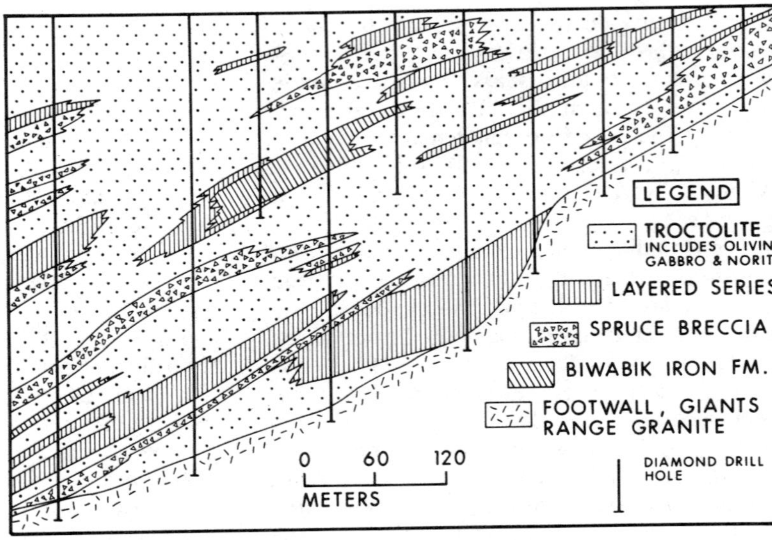

FIG. 8. Cross section through the Spruce Pit area, after Wager et al. (1969).

The footwall is Virginia formation which, in turn, overlies Biwabic iron formation.

Area V is the Water Hen occurrence described by Mainwaring and Naldrett (1977). This mineralization differs from that along much of the contact in that it is concentrated in plagioclase-free rocks, much of it is in a 60-m thick zone of ilmenite peridotite (i.e., up to about 10% ilmenite). The Cu/Ni ratio is about 2 as compared with an average of 3.3 in much of the troctolite-hosted mineralization.

Mainwaring and Naldrett (1977) drew attention to the abundance of graphite in the Water Hen mineralization, to the presence of graphite, green hercynitic spinel, cordierite, and inclusions of hornfelsed slate in the associated troctolite and peridotite, and to the ^{34}S-rich nature of the sulfur in the deposit. They suggested that sulfur was derived by interaction of the parent magma with graphite- and sulfide-rich slate of the Virginia formation for which they had also documented high $\delta^{34}S$ values.

Ripley and Rao (1980) noted that, although their sulfur isotope data suggested that considerable sulfur had been introducted into the Dunka Road deposit, 90% of the $\delta^{18}O$ values for whole rocks and plagioclase were in the range 5.6 to 6.7‰ in contrast to the range 9 to 11‰ typical of nearby metasedimentary rocks. Values greater than 9 were only found in the igneous rocks immediately adjacent to inclusions of metasedimentary rock. Rao and Ripley (1980) noted that the bulk and mineral chemistry of the rocks also preclude the possibility that wholesale assimilation has affected the composition of igneous rocks that were more than 30 m from the basal contact or 5 m from inclusions. On the other hand, the data of Ripley and Rao (1980) for $\delta^{34}S$ and δD in sulfides and biotite from the mineralized zone indicate a significant incorporation of volatiles.

A process of distillation of sulfur and water as a result of the intrusion interacting with the country rocks, coupled with diffusion of the volatiles into the hot dry intrusion, would seem to be the most likely explanation for the genesis of the mineralization.

Many of the troctolites and, in the case of the Water Hen occurrence, peridotites that host the sulfide mineralization in the Duluth complex crystallized from evolved magma. Weiblen and Morey (1980, Fig. 5) illustrated an overall range of olivine composition in 38 to 68% forsterite with the majority being less than 60%. Olivines in the Water Hen intrusion range from 50 to 65% forsterite (Mainwaring and Naldrett, 1977), and those at Minnamax fall between 45 and 57% (Hardyman, 1969). The magmas that gave rise to such iron-rich olivines cannot have been primary or even very primitive liquids (i.e., unfractionated or little fractionated mantle melts) but instead must have undergone extensive fractional crystallization prior to emplacement.

There is considerable evidence to suggest that

these fractionated magmas were intruded carrying considerable amounts of olivine and, in some cases, plagioclase in suspension. Mainwaring and Naldrett (1977) concluded that the first pulse of magma into the Water Hen complex gave rise to a 100-m layer of ilmenite peridotite with a uniform olivine composition having 63% forsterite. The top of this unit is marked by a progressive increase in the proportions of plagioclase and augite, and by a marked decrease in the forsterite content of olivine. Overlying cyclic units show considerable cryptic variation of olivine composition with the forsterite content decreasing upward. Cryptic variation of the compositions of minerals crystallizing from the cooling magma appears to have occurred in the Water Hen complex in all but the first pulse, suggesting that the uniform composition of the olivine in the basal unit was the consequence of the introduction of a large amount of olivine and ilmenite in suspension. With reference to the troctolitic series, Weiblen and Morey (1980) stressed that mineral compositions vary irregularly with stratigraphic position in many areas of the complex, that much of the layering is highly discontinuous, and that stratigraphic units cannot be traced for long distances. For these reasons they concluded that emplacement of much of the troctolitic series occurred as a partially crystalline mush.

The Virginia formation was at a very high level in the crust at the time of emplacement of the Duluth magmas, consequently the magmas did not come into contact with their sulfur source until a very late stage in their evolution and crystallization. It is likely that this late-stage introduction of sulfur is the reason for the low-grade disseminated nature of the mineralization. The settling and concentration of sulfides was probably impeded by the high proportion of silicate crystals already present when the magma became saturated with sulfide.

CONCLUSION

We have attempted to provide some insights into the origin of the very substantial lower grade nickel resources that occur in two important types of magmatic sulfide deposits.

Deposits of the Mt. Keith type are hosted by synvolcanic dunite-peridotite sills of komatiitic affinity within Archean greenstone terranes. The consistent petrographic and grade characteristics of these deposits support our hypothesis that they formed by the accumulation of olivine and sulfide in the proportions required by a fractionation path that followed the olivine-sulfide saturation boundary in a peridotitic komatiite magma system. In the case of the Dumont deposit, sulfide saturation was induced by mixing of primitive magma with more fractionated liquid expelled from underlying cumulates by filter pressing. Insufficient data are available to allow us to conclude that a similar process was responsible for sulfide saturation in other deposits of this type.

Deposits of the Duluth type occur at or just above the basal margins of intrusions of flood basaltic affinity that are associated with intracontinental rifting. The initial sulfur contents of the magmas were probably very low and sulfide saturation was achieved locally when sulfur was transferred from sulfide-rich country rocks into the magma when it was at a relatively advanced stage of fractionation of the magma.

A common aspect of deposits of both the Duluth and Mt. Keith types is that their low grade is a result of the segregation of sulfide in the presence of a relatively large proportion of crystalline silicates. The latter apparently impeded the coalescense of the immiscible sulfide liquid droplets and prevented the accumulation of relatively high grade pools.

Despite their low grade, deposits of the types that we have described contain immense resources of nickel and, in the case of the Duluth type, copper, and platinum metals as well. Only a few deposits of the Mt. Keith type have been explored sufficiently to allow their resources to be given with any confidence. However, the Mt. Keith deposit alone, with indicated and inferred resources of 290 million tonnes grading 0.6% Ni (Burt and Sheppy, 1975), contains 1.7 million tonnes of nickel. Listerud and Meineke (1977) calculated the resources for part of the Duluth complex (Fig. 6) to be about 4000 million tonnes with an average grade of 0.66% Cu and 0.2% Ni using a 0.5% Cu cutoff grade, or about 6350 million tonnes grading 0.55% Cu and 0.17% Ni using a cutoff grade of 0.4% Cu. The latter estimate amounts to nearly 35 million tonnes of contained copper and nearly 11 million tonnes of contained nickel. The possible significance of these resources is apparent when considered in the context of the current total annual world production of nickel which is approximatley 0.7 million tonnes.

ACKNOWLEDGMENTS

The constructive comments of R.I. Thorpe, O.R. Eckstrand, T.N. Irvine, and E.M. Ripley are gratefully acknowledged.

REFERENCES

Binns, R.A., Groves, D.I., and Gunthorpe, R.J., 1977, "Nickel Sulphides in Archaean Ultramafic Rocks of Western Australia," *Correlation of the Precambrian,* Vol. 2, A.V. Siderenko, ed., Nauka, Moscow, pp. 349–380.

Buchanan, D.L., and Nolan, J., 1979, "Solubility of Sulfur and Sulfide Immiscibility in Synthetic Tholeiitic Melts and Their Relevance to Bushveld Complex Rocks," *Canadian Mineralogist,* Vol. 17, pp. 483–494.

Burt, D.R.L., and Sheppy, N.R., 1975, "Mount Keith Nickel Sulphide Deposit," *Economic Geology of Australia and Papua New Guinea,* C.L. Knight, ed., Vol. 1, Australasian Institute of Mining and Metallurgy, Parkville, pp. 159–168.

Campbell, I.H., and Naldrett, A.J., 1979, "The Influence of Silicate:Sulfide Ratios on the Geochemistry of Magmatic Sulfide," *Economic Geology,* Vol. 74, pp. 1503–1506.

Duke, J.M., in preparation, "Petrology and Economic Geology of the Dumont Sill, Northwestern Quebec," Geological Survey of Canada, Ottawa.

Duke, J.M., and Naldrett, A.J., 1978, "A Numerical Model of the Fractionation of Olivine and Molten Sulfide from Komatiite Magma," *Earth and Planetary Science Letters,* Vol. 39, pp. 255–266.

Green, J.C., 1972, "General Geology of Northeastern Minnesota, North Shore Volcanic Group," *Geology of Minnesota: a Centennial Volume,* P.K. Sims and G.B. Morey, eds., Minnesota Geological Survey, St. Paul, pp. 292–332.

Hardyman, R.F., 1969, "The Petrography of a Section of the Basal Duluth Complex, St. Louis County, Northeastern Minnesota," M.S. thesis, University of Minnesota, 132 pp.

Haughton, D.R., Roeder, P.L., and Skinner, B.J., 1974, "Solubility of Sulfur in Mafic Magmas," *Economic Geology,* Vol. 69, pp. 451–467.

Irvine, T.N., 1975, "Crystallization Sequences in the Muskox Intrusion and Other Layered Intrusions—II. Origin of Chromitite Layers and Similar Deposits of Other Magmatic Ores," *Geochimica et Cosmochimica Acta,* Vol. 39, pp. 991–1020.

Irvine, T.N., 1977, "Origin of Chromitite Layers in the Muskox Intrusion and Other Stratiform Intrusions: A New Interpretation," *Geology,* Vol. 5, pp. 273–277.

Irvine, T.N., 1980, "Magmatic Infiltration Metasomatism, Double Diffusive Fractional Crystallization, and Adcumulus Growth in the Muskox Intrusion and Other Layered Intrusions," *Physics of Magmatic Processes,* R.B. Hargraves, ed., Princeton University Press, Princeton, pp. 325–384.

Jackson, E.D., 1961, "Primary Textures and Mineral Associations in the Ultramafic Zone of the Stillwater Complex, Montana," Professional Paper 358, US Geological Survey, Washington, 106 pp.

Listerud, W.H., and Meineke, D.G., 1977, "Mineral Resources of a Portion of the Duluth Complex and Adjacent Rocks in St. Louis and Lake Counties, Northeastern Minnesota," Report 93, Minnesota Department of Natural Resources, Hibbing, 74 pp.

Mainwaring, P.R., and Naldrett, A.J., 1977, "Country-Rock Assimilation and the Genesis of Cu-Ni Sulfides in the Water Hen Intrusion, Duluth Complex, Minnesota," *Economic Geology,* Vol. 72, pp. 1269–1284.

Naldrett, A.J., and Duke, J.M., 1980, "Platinum Metals in Magmatic Sulfide Ores," *Science,* Vol. 208, pp. 1417–1424.

Naldrett, A.J., and Macdonald, A.J., 1980, "Tectonic Settings of Ni-Cu Sulphide Ores: Their Importance in Genesis and Exploration," *The Continental Crust and its Mineral Deposits,* D.W. Strangway, ed., Special Paper 20, Geological Association of Canada, Waterloo, pp. 633–658.

Naldrett, A.J., and Turner, A.R., 1977, "The Geology and Petrogenesis of a Greenstone Belt and Related Nickel Sulfide Mineralization at Yakabindie, Western Australia," *Precambrian Research,* Vol. 5, pp. 43–103.

Phinney, W.C., 1969, "The Duluth Complex in the Gabbro Lake Quadrangle, Minnesota," Minnesota Geological Survey Report of Investigation 9, St. Paul, 20 pp.

Rajamani, V., and Naldrett, A.J., 1978, "Partitioning of Fe, Co, Ni, and Cu Between Sulfide Liquid and Basaltic Melts and the Composition of Ni-Cu Sulfide Deposits," *Economic Geology,* Vol. 73, pp. 82–93.

Rao, V.B., and Ripley, E.M., 1980, "Bulk Rock and Mineral Chemistry as an Aid in Evaluating the Role of Country Rock Assimilation in Sulfide Formation, Dunka Road Cu-Ni Deposit, Duluth Complex, Minnesota," *Abstracts with Programs,* Geological Society of America, Vol. 12, p. 506.

Ripley, E.M., and Rao, V.B., 1980, "Oxygen Isotopic Studies of the Dunka Road Cu-Ni Deposit, Duluth Complex, Minnesota," *Abstracts with Programs,* Geological Society of America, Vol. 12, p. 510.

Ross, J.R., and Travis, G.A., 1981, "The Nickel Sulfide Deposits of Western Australia in Global Perspective," *Economic Geology,* Vol. 76, pp. 1291–1329.

Scholz, D.L., 1936, "The Magmatic Nickeliferous Ore Deposits of East Griqualand and Pondoland," *Transactions,* Geological Society of South Africa, Vol. 39, pp. 81–210.

Shima, H., and Naldrett, A.J., 1975, "Solubility of Sulfur in an Ultramafic Melt and the Relevance of the System Fe-S-O," *Economic Geology,* Vol. 70, pp. 960–967.

Wager, R., et al., 1969, "A Comparison of the Cu-Ni Deposits of the Sudbury and Duluth Basins," *Proceedings,* 30th Annual Mining Symposium, 42nd Annual Meeting, Minnesota Section, AIME, pp. 95–96.

Weiblen, P.W., 1965, "A Funnel-Shaped Gabbro—Troctolite Intrusion in the Duluth Complex, Lake County, Minnesota," Ph.D. Thesis, University of Minnesota.

Weiblen, P.W., and Morey, G.B., 1980, "A Summary of the Stratigraphy, Petrology and Structure of the Duluth Complex," *American Journal of Science,* Vol. 280-A, pp. 88–133.

4

Low Grade Manganese Deposits—A Facies Approach

Gordon A. Gross

Geological Survey of Canada, Ottawa, Ont., Canada

ABSTRACT

Banded siliceous manganese-iron bearing sediments recognized as distinctive facies of Algoma-type iron formation constitute low-grade resources of manganese in many countries. Manganiferous facies associated with oxide and carbonate facies of iron formation have iron to manganese ratios ranging from 0.2 to 2 and commonly about one. Manganiferous facies up to 30 m thick occurring separately or with other iron-formation facies are associated with cherts, carbonates, shales, turbidites, tuffs, and lavas. Primary cherty manganiferous facies and their gondite metamorphic equivalents represent significant manganese resources and are protores for higher-grade deposits derived by secondary enrichment processes.

LOW GRADE MANGANESE DEPOSITS—A FACIES APPROACH

Most of the low-grade manganese resources distributed throughout the world are sedimentary facies of siliceous banded iron formation and formed by the same genetic processes. Manganese facies of iron formation have been mined in the past and constitute an abundant but low-grade potential source of manganese for future development.

A relationship between manganese and iron deposits is generally recognized because the chemical properties of manganese and iron are similar or analogous for many of their compounds. Many iron and manganese deposits are associated geographically and commonly occur in the same groups of rocks, but relationships are poorly defined and many genetic models previously considered have not provided a satisfactory explanation for the source of the manganese and iron and differences in the depositional environments of their occurrences. The genesis of a large number of sedimentary occurrences of siliceous manganese-bearing iron formations documented in the literature has been related to effusive hydrothermal or volcanogenic processes and can be readily defined in genetic models originally derived for the varied group of iron-formation sediments (Gross, 1965, 1980). Many of the low-grade manganese occurrences are thin-banded cherty rocks with varied proportions of manganese and iron as oxide or carbonate minerals that are intimately associated with fine-grained clastic sediments and tuff. They are essentially manganiferous facies of iron formation, and along with the copper-, zinc-, and gold-bearing sulfide facies of iron formation were formed by effusive hydrothermal processes.

Iron formation has been defined as a chemical sedimentary rock, typically thin-banded and/or finely laminated, containing at least 15% iron of sedimentary origin and commonly but not necessarily containing layers of chert (James,

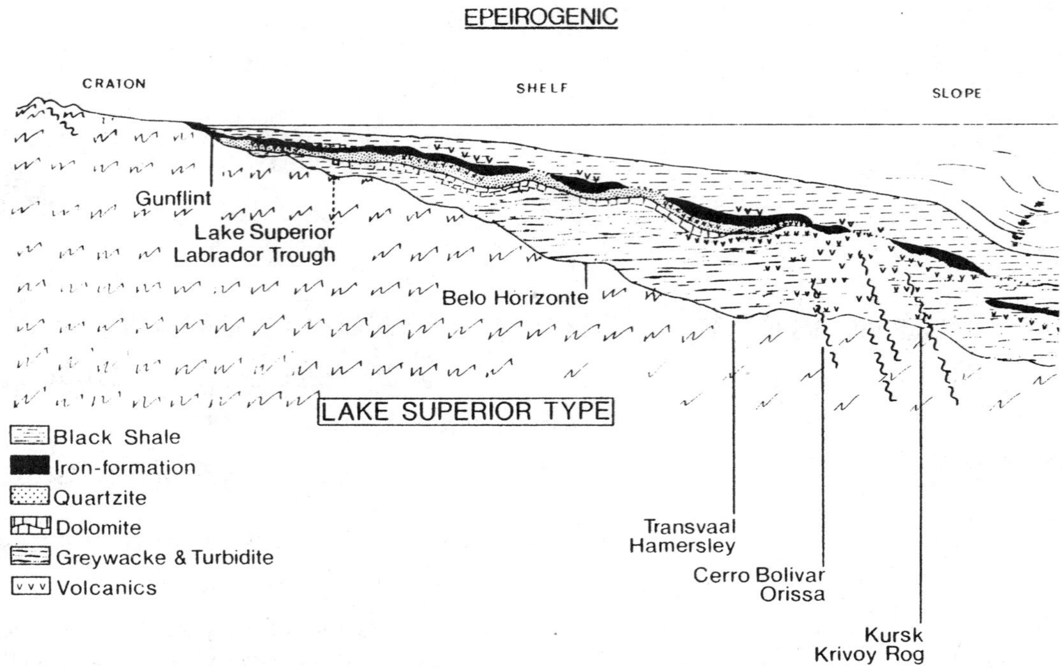

FIG. 1. Tectonic environments for deposition of iron formation.

1954, p. 239) and the predominant primary facies consist of oxide, silicate, carbonate, or sulfide mineral assemblages. Two major types of iron formation were recognized by Gross (1965) on the basis of differences in the associated rocks and in their depositional environments. The Algoma type is associated with intermixed assemblages of volcanic greywacke, turbidite, argillite, and shale rocks and formed in deeper marine environments relatively close to effusive hydrothermal and volcanic centres. The Lake Superior-type iron formations are associated with shallow-water continental shelf assemblages of quartzite, dolomite, argillite, black shale, and tuff, and are best developed in the iron ranges of the Lake Superior district in the United States and in the Labrador-Quebec geosyncline in Canada. Iron formation occurs in a broad range of depositional environments as illustrated in Fig. 1, varying from the shallow water shelf environments common for the Lake Superior type to the highly varied depositional conditions classified as Algoma type, as well as in grabens, trenches, and along fault scarps as shown by the Snake River occurrences in the Mackenzie Mountains in the Yukon Territory in Canada. Manganese-rich beds occur with banded cherty iron formation in the various depositional environments outlined in this diagram, and are part of a very large group of chemical sediments in which the constituents are believed to have been derived mainly from volcanogenic or hydrothermal effusive rather than purely sedimentary sources. Park and MacDiarmid (1970) pointed out that the association of manganese oxides with iron formation is common enough that an explanation of the origin of iron formation must also explain the origin of the manganese oxides.

The volcanogenic model originally conceived in very general terms to account for the source and mode of deposition of iron and silica and other components in iron formations has been used effectively throughout the world to interpret the origin of the highly varied group of chemical sediments that occur within or closely associated with iron formations. This whole group of chemical sediments with many different sedimentary facies is recognized as the iron formation group of rocks and formed by the same genetic processes. It is unnecessary to elaborate

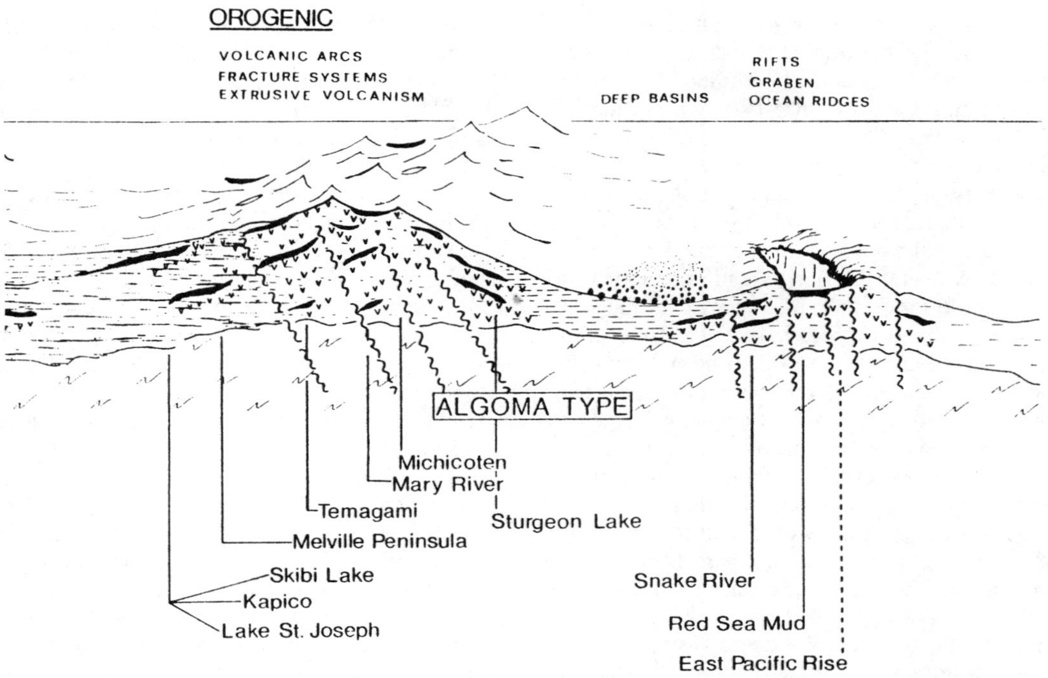

FIG. 1, continued.

in detail on the volcanogenic model. It is recognized that tectonic settings and depositional environments have varied greatly in the time and circumstances in which effusive hydrothermal processes formed iron formations and the great variety of chemically precipitated facies genetically related to them. In essence the volcanic model involves an outpouring of hydrothermal solutions charged with various metallic ions and chemical components which are precipitated in sedimentary basins under varying conditions of Eh and pH. The hydrothermal solutions are mobilized by convection in areas of higher thermal gradient in active tectonic belts and acquire metallic ions by leaching and solution of chemical components from the rocks through which they pass or by addition of volatiles and soluble constituents from magmatic sources. The hydrothermal solutions are mobilized and circulate through fault systems and porous rocks and are discharged into sedimentary basins through fumaroles, volcanic vents, and by various forms of effusive activity.

The physical characteristics, composition, and mineralogy of the chemical sediment thus formed are dependent on and controlled by a great number of factors that influence the depositional environment of the sedimentary basin. The various facies of chemical sediment formed by hydrothermal-effusive processes may be defined in terms of Eh and pH conditions in the basin, as reflected in the primary sedimentary minerals, and form oxide, silicate, carbonate, or sulfide facies as defined by James (1954) for iron formations; in terms of compositional variations such as banded carbonate iron formation, manganese-bearing iron formation, banded manganese silica, or jasper facies, as sulfide or polymetallic sulfide facies, or as gold-bearing carbonate facies; and in terms of physical and organic factors affecting texture, bedding, and lithology as seen in microbanding and in oolitic, granular, pisolitic, or framboidal textures.

The marked separation of manganese, iron, and sulfide facies in depositional basins, and in many cases the predominance of one facies to the exclusion of others, is a characteristic feature of many Algoma-type iron formations and the chemically precipitated facies associated with them. The group studying deposition of metalliferous sediments by hydrothermal-effusive processes along the East Pacific Rise points out that the degree of subsurface dilution is a major determinant of the final composition of the host springs and the chemical sediments associated with them (Ballard et al., 1981). Low dilution permits sulfide deposition, intermediate dilution leads to iron and sulfide systems, and manganese is unaffected because of its slow oxidation kinetics. Since oxidation of metals in the hydrothermal solutions is controlled by the degree of seawater dilution, a wide range of iron-manganese ratios may be expected in the sediments. Iron and sulfides may be oxidized in the subsurface parts of the system permitting effusion and deposition of very pure manganese precipitates, and pure sulfide facies are formed in effusive areas where there is little dilution.

Primary manganiferous facies are associated with cherty iron formation in most of the iron ranges throughout the world with the exception of those in North America where they are less common and inconspicuous if not absent. Banded siliceous manganese facies occur with all of the mineralogical facies of iron formation but are most commonly intermixed with oxide and carbonate facies or with fine grained clastic sediments in which their possible genetic relationship with iron formation is less apparent. Many of the manganiferous facies are highly metamorphosed with little evidence of their primary nature remaining and form gondite, the protore from which many high-grade manganese ores are derived by secondary enrichment processes (Roy, 1966). A review of the literature indicates that primary manganese-rich sedimentary facies are more commonly associated with Algoma-type iron formation and occur with shale, turbidites, greywacke, tuff, and sedimentary-volcanic assemblages of rocks in depositional environments more characteristic of Algoma than Lake Superior-type iron formation. They occur more consistently in Phanerozoic and Mesozoic than in Precambrian basins.

Manganiferous facies in Lake Superior-type iron formation are commonly very low grade and occur in carbonate or clastic facies of iron formation. Many occurrences were formed at the margins between continental shelf and deeper water sedimentary basins where quartzites, dolomites, and black shales of shelf environments are intermixed with argillites, turbidites, and various fine-grained clastic sediments. In nearly all cases where manganiferous facies are developed, tuff beds and tuffaceous materials are reported in the clastic sediments, indicating volcanic activity in the vicinity of the depositional basin and contemporaneous with the deposition of the manganese.

Information has been summarized in Table 1 from the references listed for a number of selected areas to indicate the common association of primary manganese-bearing facies with other iron-formation facies, to show the analogous development of manganese and iron-rich facies in the large group of chemical sediments, and the various kinds of sedimentary and volcanic rock associated with the iron-manganese facies which give an indication of the type of depositional environment in which they were precipitated.

The common association of sedimentary manganese deposits with iron formation was reviewed by Roy (1969) in his classification of manganese deposits. His distinction between manganese deposits associated with iron formations of volcanogenetic or nonvolcanogenetic source appears to be arbitrary. These data indicate that many important manganese deposits of the world are associated with iron-formation protores or constitute manganiferous facies of iron formation which were formed by the same genetic processes as the iron-formation facies. Some of the characteristic relationships of manganese and iron-rich facies of iron formation are illustrated in the following selection of deposits.

Karadzhal Iron-Manganese Deposits

The Karadzhal iron-manganese deposits in the Dzhail'min syncline in central Kazakhstan occur in a succession of quartz-magnetite-hematite and carbonate facies of iron formation in a thick sequence of Devonian sandstones, conglomerates, reddish-gray limestones, cherty calcareous shales, and volcanic rocks. The iron formation is closely associated with reddish limestone and ranges in thickness from 1 to 24 m over a distance of 15 to 20 km. Manganese oxide and carbonate beds are intermixed with the jasper and chert-carbonate facies of iron formation. Jasper facies of iron formation are associated with spilitic rocks in the northwest part of the syncline where they achieve their greatest thickness and contain up to 60% iron and probably average about 30%, while in the eastern part of the area the iron-formation beds are 1 to 5 m thick and contain up to 40% manganese and 6 to 10% iron. The iron to manganese ratios in

Table 1. Iron-Manganese Facies Associated with Iron-Formation

Part 1

Country:	**Australia**			**Africa**	**Zaire**	**South Africa**	**Botswana**	**Gabon**	**Sinai**
Region	Pilbara Goldfield	Phillips River	Yilgarn Goldfield	West	Katanga	Transvaal	Kalahari		
Deposit, Rock Group				Mokta, Nsuta, Tambao	Kisenge	Kuruman	Palapye	Moanda	Um Bogma
Age	Prot.	Archean		PreC	PreC	Proterozoic	Prot.	Prot.	Mesozoic
Manganese									
Content, %					<20–40	25		10–20	
Ratio Fe/Mn in MN-facies									
Average*						1/3		1/8	
Range*						4/1–1/45	4/1–1/4	1/40–30/1	
Facies									
MnO—Mn Oxide	MnO	MnO	MnO	MnO		MnO	MnO	MnO	MnO
MnC—Mn Carbonate				MnC	MnC	MnC		MnC	
MnSi—Mn Silicate				MnSi	MnSi			MnSi	
FeO—Iron Oxide	FeO	FeO	FeO	FeO		FeO	FeO	FeO	FeO
FeC—Iron Carbonate						FeC			
FeSi—Iron Silicate								FeSi	
FeS—Iron Sulfide								FeS	
Ch—Chert Silica	Ch	Ch	Ch	Ch		Ch	Ch	Ch	
Rock Association									
Bandarito									
Gondite				G	G				
Carbon Shale				CSh	CSh			CSh	
Limestone-Dolomite						LD		LD	LD
Shale	Sh			Sh		Sh	Sh	Sh	Sh
Sandstone-Quartzite				SQ	SQ	SQ	SQ	SQ	SQ
Turbidites						Tb	Tb		
Tuff			Tf		Tf	Tf	Tf		
Tuff				Tf		Tf	Tf	Tf	
Volcanics BV-basic				V		V	V	V	
Amphibolite				Amp	Amp				
Schists				Sch					
Gneisses				Gn	Gn				

the iron-formation change from 10/1 to 7/1 in the west to 1/1 and 1/1.5 in the east. The siliceous iron formations in the west of the area were deposited closer to the effusive centers while deposition of the thinner manganese-rich facies appears to have been distal from the effusive hydrothermal source of the metals. Occurrences of manganese facies in iron formation similar to those described at Karadzhal are reported in many other Algoma-type iron formations in Kazakhstan and the southern Ural Mountains (Sapozhnikov, 1963; Kalinin, 1965).

San Francisco Manganese Deposit

A transition from iron-oxide to manganese-oxide facies of iron formation occurs in the San Francisco manganese deposit at Jalisco, Mexico, described by Zantop (1978, 1981). Banded cherty iron and manganese-oxide facies of iron formation developed in a Tertiary lacustrine basin

Table continued

Part 2 Country:	**Brazil**					**Canada**		**Mexico**	**United States**
Region	Bahia	Minas Gerais	Para	Amapa	Mato Grosso	Appalachian	Nastapoka Islands	Jalisco	Western
Deposit, Rock Group	Bandarra			Serra de Navio	Morro do Urucum	Woodstock		San Francisco	Nevada California
Age	PreC.	Prot.	Prot.	PreC.	Cambrian	Silurian	Prot.	Tertiary	
Manganese									
Content, %			32–47						
Ratio Fe/Mn in MN-facies									
Average*			1/10	1/10	1/4	1/1		2/1	
Range*	4/1–15/1	1/5–1/20				2.4/1–1/12		40/1–1/50	
Facies									
MnO—Mn Oxide	MnO	MnO	MnO		MnO	MnO		MnO	MnO
MnC—Mn Carbonate	MnC	MnC	MnC	MnC		MnC	MnC		MnC
MnSi—Mn Silicate	MnSi		MnSi	MnSi					MnSi
FeO—Iron Oxide	FeO	FeO	FeO		FeO	FeO	FeO	FeO	FeO
FeC—Iron Carbonate		FeC							
FeSi—Iron Silicate		FeSi							
FeS—Iron Sulfide			FeS						
Ch—Chert Silica	Ch	Ch	Ch	Ch	Ch	Ch	Ch	Ch	Ch
Rock Association									
Bandarito	B								
Gondite									
Carbon Shale			CSh	CSh					
Limestone-Dolomite		LD	LD	LD		LD			
Shale		Sh	Sh		Sh	Sh	Sh	Sh	Sh
Sandstone-Quartzite	SQ	SQ	SQ	SQ	SQ		SQ		
Turbidites			Tb			Tb			Tb
Tuff							Tf	Tf	Tf
Volcanics BV-basic			V					V	V
Amphibolite	Amp		Amp						
Schists	Sch		Sch	Sch					
Gneisses	Gn		Gn						

in association with tuff, andesite flows, red mudstones and siltstones, conglomeratic tuff, and shales. The iron formation forms a stratified lens up to 3 m thick and 1.6×1 km in extent. Iron-manganese ratios range from 40/1 in the iron-oxide-rich facies at one side of the lens to 1/50 in the manganese-rich part with an overall ratio of 2/1. Zantop (1981) concluded that hydrothermal solutions from a volcanic source supplied the iron and manganese and has supported this view with data on the trace element content of the iron and manganese facies.

Postmasburg and Kuruman Manganese Deposits

Stratabound and stratiform manganese deposits of Postmasburg and Kuruman in the Transvaal of South Africa are developed in a thick sequence of jasper oxide and carbonate facies of iron formation and probably contain the larg-

Table continued

Part 3

Country:	India				USSR			
Region	Karnataka Dharwar	Orissa	Andhra Pradesh	Madhya Pradesh	Kazakhstan	Urals	Yenisey Ridge	Transcaucasia
Deposit, Rock Group		Iron Ore Group	Pakhal	Aravalli Supergroup	Karadhzal			Armenia
Age	Prot.	Archean	Prot.	Prot.	Devonian	Permian	Prot.	Cretaceous Eocene
Manganese								
Content, %		28–42	14–32				10	
Ratio Fe/Mn in Mn-facies								
Average*						2/1		1/2
Range*		1/1.5–1/1.3	1/1.8–1/4	1/25	10/1–1/1.5			
Facies								
MnO—Mn Oxide	MnO	MnO	MnO	MnO	MnO	MnO		
MnC—Mn Carbonate			MnC	MnC	MnC		MnC	
MnSi—Mn Silicate			MnSi	MnSi				
FeO—Iron Oxide	FeO	FeO	FeO	FeO	FeO	FeO	FeO	
FeC—Iron Carbonate					FeC		FeC	
FeSi—Iron Silicate								
FeS—Iron Sulfide				FeS				
Ch—Chert Silica	Ch	Ch	Ch	Ch	Ch	Ch	Ch	Ch
Rock Association								
Bandarito								
Gondite		G	G	G				
Carbon Shale		CSh						
Limestone-Dolomite	LD		LD	LD	LD	LD	LD	LD
Shale	Sh	Sh	Sh		Sh	Sh	Sh	Sh
Sandstone-Quartzite	SQ	SQ	SQ		SQ	SQ	SQ	SQ
Turbidites								
Tuff		Tf			Tf		Tf	Tf
Volcanics BV-basic	V	V			V		V	V
Amphibolite	Amp		Amp					
Schists	Sch		Sch					
Gneisses	Gn		Gn					

est resources of manganese known in any single region. The iron formation and cherty manganese facies form part of a thick group of rocks consisting of dolomite, limestone, shale, siltstone, quartzite, greywacke, andesite, lava, and tuff in which both syngenetic manganese-rich sediments and other deposits formed by secondary enrichment processes are present. Manganese facies in the jasper iron formation form two prominent lenses, one from 3 to 25 m and the other 4 to 6 m in thickness, which have an average manganese content of 20 to 30% and are separated by 5 to 22 m of banded iron formation. The banded jasper iron formation as well as the manganese carbonate facies transitional into manganese ore represent chemical sediments derived from a volcanic source (Sohnge, 1977).

Table continued

Part 4 Country:	Japan			Germany	Hungary	Finland	Portugal-Spain	New Zealand
Region	Ashlo Mts.	Kitakami Mts.	Tamba Mts.	Apls. Mts.	Bakony Mts.		South-	Northland
Deposit, Rock Group					Epleny	Kittila	Pyrite Belt	Auckland
Age	Paleozoic	Permian Jurassic	Miocene	Jurassic	Mesozoic	Prot.	Silurian	Mesozoic
Manganese								
Content, %						0.4–15		
Ratio Fe/Mn in Mn-facies								
Average*	1/1			1/2	1/2			1/20
Range*					1/20–1/5			
Facies								
MnO—Mn Oxide	MnO	MnO	MnO	MnO			MnO	MnO
MnC—Mn Carbonate	MnC	MnC	MnC	MnC	MnC	MnC	MnC	
MnSi—Mn Silicate	MnSi			MnSi		MnSi	MnSi	
FeO—Iron Oxide	FeO			FeO	FeO	FeO		FeO
FeC—Iron Carbonate		FeC	FeC	FeC		FeC		
FeSi—Iron Silicate						FeSi		
FeS—Iron Sulfide					FeS	FeS	FeS	
Ch—Chert Silica	Ch	Ch	Ch	Ch	Ch	Ch	Ch	Ch
Rock Association								
Bandarito								
Gondite								
Carbon Shale								
Limestone-Dolomite	LD			LD	LD	LD		
Shale	Sh	Sh	Sh	Sh	Sh	Sh	Sh	Sh
Sandstone-Quartzite	SQ	SQ		SQ	SQ			SQ
Turbidites	Tb	Tb				Tb		Tb
Tuff	Tf	Tf	Tf	Tf		Tf	Tf	Tf
Volcanics BV-basic	BV	BV	V			V	V	V
Amphibolite								
Schists								
Gneisses								

Woodstock Deposit Area

Banded cherty manganiferous iron formation near Woodstock, N.B., Canada forms part of a succession of thinly bedded Silurain gray, gray-green, and red slate; sandstone; greywacke; and limestone. The manganiferous jasper-hematite facies of the iron formation are up to 30 m thick with an iron content ranging from 11 to 30%, a manganese content from 12 to 25%, and an overall iron to manganese ratio of about 1.5. Several hundred million tons of potential manganiferous resource material have been outlined in the Woodstock area, and manganiferous facies are present in this group of rocks where they extend westward into the state of Maine. (Gross, 1967).

Manganese in Lake Superior-Type Iron Formation

Manganese facies in Lake Superior-type iron formation are uncommon in North America. Beds rich in manganese in carbonate facies of iron formation have been traced for several kilometers on Belanger and Flint Islands of the Nastapoka chain on the east side of Hudson Bay (Bell, 1878). A bed of rhodonite up to 20 cm thick was observed in the Mount Reed iron formation in northern Quebec and some beds in magnetite-hematite facies of iron formation near Wabush Lake in southern Labrador, Newfoundland contain up to 2% manganese. The manganiferous facies of iron formation in the Cayuna Range in Minnesota have been investigated more thoroughly as a source of manganese. Siderite facies of Algoma-type iron formation in the Michipicoten District, Ont., also contain up to 2% manganese.

Japanese Manganese Deposits

Bedded cherty manganese and manganese-iron sediments are widely distributed in Japan in association with basic tuffs, lavas, greywacke, sandstone, shale, and limestone ranging in age from Paleozoic to Miocene in the Kitakama, Ashio, and Tambo mountain belts. Some of the cherty manganese beds were regionally metamorphosed and others near Mesozoic granite intrusions are highly metamorphosed by contact metasomatic processes. The chert sequences, cherty manganese facies, and jasper iron facies are considered by Watanabe et al. (1970) to be syngenetic sediments genetically related to hot spring activity in the volcanic belts and their minor element content would suggest that they formed by submarine hydrothermal-effusive processes.

Australian Manganese Deposits

Manganese occurs with Precambrian banded iron formation in Western Australia in several areas mentioned by MacLeod (1965) and de la Hunty (1965). In the southern part of the Pilbara Goldfield manganese forms a capping on ferruginous chert and shale beds which are correlated with the Marra Mamba iron formation of the Hamersley iron province. The Marra Mamba iron formation also contains manganiferous facies containing 40% manganese that are protore for manganese deposits. Manganiferous facies distributed in Archean iron formation in the Phillips River Goldfield extend for a distance of 100 km along the south coast where large tonnages of low-grade material containing 30 to 40% manganese are present below the upper zone of enriched manganese ore. Manganiferous facies also occur in iron formation in the Yilgarn Goldfield and in the jaspilite at Mount Stuart in the Ashburton Goldfield.

Spanish-Portuguese Pyrite Belt

Manganese silicate and carbonate facies in red tuffaceous shales, red ferruginous jasper facies, and polymetallic sulfide facies are interbedded in the complex assemblage of volcanic and sedimentary rocks of Silurian age that form the Spanish-Portuguese pyrite belt (Strauss et al., 1977). The manganese, iron, and polymetallic sulfide facies in this belt have been compared to other Precambrian and Paleozoic iron formations of the Algoma type, and the association of manganiferous iron-oxide and carbonate facies with sulfide facies is not uncommon.

Other Deposits

Iron formations in many other parts of the world contain cherty manganese-oxide or carbonate facies including those in Minas Gerais, Bahia, and Morro do Urucum in Brazil, the Kalahari occurrences in Botswana, Africa, the Maliy Khingan area of the USSR, the Franciscan cherts in California, others in West Africa, and in the Guyana shield, in the Alps, and in southern Europe.

Red Sea Basins

The writer has no doubt that the recently deposited banded silica and iron-rich muds in the Atlantis II and other basins of the Red Sea, which have extensive manganese-oxide and polymetallic sulfide facies associated with them, are the prototype facies of an Algoma-type iron formation (Bischoff, 1969; James, 1969). The lithified equivalent of these muds and siliceous ooze would compare closely to some of the well-known occurrences of sulfide and manganese facies of iron formation listed in Table I that range in age from early Precambrian to Recent. They show very clearly that many different facies of chemical sediment that are directly comparable to older facies of iron formation may be closely related and precipitated contemporaneously by hydrothermal-effusive processes.

Identifying and Exploring for Manganese Resources

A review of the data from the wide range of manganese-iron-rich facies in iron formation listed in Table I leads to a number of generalizations that may be useful in identifying and in exploring for manganese resources of potential economic significance.

Manganese oxide and carbonate facies associated with sulfide facies appear to have formed in close proximity to hydrothermal-effusive centers while the banded cherty iron formations and siliceous iron-manganese facies were deposited some distance from these centres.

Banded cherty iron-oxide facies are subsidiary in quantity in iron-formation members where manganese oxide and carbonate facies are associated with polymetallic sulfide facies.

Laminated cherty beds are characteristic of and abundantly developed in most iron-rich facies with the exception of the massive sulfide facies, but chert beds and the silica content in these facies diminish in lateral transitions from iron-rich to manganese-rich facies. Many massive manganese carbonate and oxide facies that lack interbanded chert layers may be distal from hydrothermal-effusive centers but nevertheless genetically related to iron formation.

The common types of sedimentary manganese beds consisting of oxide or carbonate facies associated with black shales, phyllite, limestone, dolomite and greywacke, tuff, and volcanic rocks are considered by many to have formed by volcanogenic processes and can be related directly to the effusive hydrothermal-genetic model that is widely accepted for iron formation.

Probably iron oxide-rich facies were not deposited with manganese in some of the basins; they may have been lost by erosion or not recognized because of later metamorphic transformation to various silicate mineral assemblages in schists.

The primary nature of the protore for many of the secondary manganese occurrences is not well known or clearly defined so that the number of manganese occurrences related to iron formation may be even more extensive than suggested here. The gondite protores are generally considered to be highly metamorphosed siliceous manganese facies and the primary features of many manganese carbonate facies are destroyed in early stages of metamorphism and recrystallization (Roy, 1966; Krishna Rao et al., 1981).

The primary cherty manganiferous facies containing variable amounts of iron and manganese constitute significant low-grade resources of manganese. The iron formations with their associated manganese facies are part of a very large group of chemically precipitated sediments that were deposited by hydrothermal-effusive processes under a broad range of environmental conditions. Siliceous manganese facies are widely distributed throughout the world and conceivably can provide a source of manganese for local use where the ferromanganese and steel industries can accept a lower grade and quality of manganese raw materials.

Identification of specific regions where lower grade manganese-iron resources may be developed and assessment of the different kinds of material for future use are beyond the scope and the purpose of this chapter. It is noted that secondary manganese oxide materials that are enriched in part by natural processes and the banded siliceous or cherty manganese-iron facies in iron formation appear to be more amenable to upgrading and beneficiation treatment than the fine-grained manganese-bearing shales and clastic facies. If material with a higher iron to manganese ratio than that presently being used in industry could be accepted, the manganese facies material so widely distributed in association with the iron formations might enable decentralization of manganese supply and procurement from sources closer to the industrial sites where it is used. Low-grade manganese facies have been used in some areas in the past without beneficiation and many of them probably present no greater challenge for processing and beneficiation than taconite iron ore or some of the other facies of iron formation being concentrated.

The assessment of primary manganese facies as a resource is rendered difficult because of alteration of the primary beds by secondary oxidation and leaching processes which may leave complex assemblages of fine-grained manganese oxide minerals with only minor enrichment of manganese content. In other cases complex manganese silicate mineral assemblages are developed by metamorphic processes and the nature of the primary facies material is obscure. As study and research on iron formation proceeds throughout the world with the benefit of more detailed study of the mineralogy and stratigraphy and analysis of sedimentary basins, a clearer understanding of the origin of chemically precipitated sediments is emerging, and the extensive distribution and favorable geological environments for locating manganese resources are being revealed.

REFERENCES

Anderson, W.L., Dyer, R.C., and Torres, D.D., 1974, "Manganese Occurrences in the Itacaiunas River Basin, Southern Amazon Region, Brazil," *Problems of Ore Deposition,* Varna, Fourth IAGOD Symposium, 1974, Vol. 3, pp. 114–121.

Ballard, R.D., et al., 1981, "East Pacific Rise Study Group, Crustal Processes of the Mid-Ocean Ridge," *Science,* Vol. 213, pp. 31–40.

Banerji, A.K., 1977, "On the Precambrian Banded Iron-Formations and the Manganese Ores of the Singhbhum Region, Eastern India," *Economic Geology,* Vol. 72, pp. 90–98.

Bell, R., 1878, "Nastapoka Islands, Northwest Territories," Rept. of Progress, 1877–78, Canada Geological Survey, pp. 15C–18C.

Bischoff, J.L., 1969, "Red Sea Geothermal Brine Deposits: Their Mineralogy, Chemistry, and Genesis," *Hot Brines and Recent Heavy Metal Deposits in the Red Sea,* E.T. Degens and D.A. Ross, eds., Springer Verlag, New York, pp. 368–401.

de la Hunty, L.E., 1965, "Manganese Ore Deposits of Western Australia," *Geology of Australian Ore Deposits,* J. McAndrew and R.T. Madigan, eds., 8th Commonwealth Mining and Metallurgical Congress, Australia and New Zealand, 1965, Vol. 1, pp. 140–146.

de Villiers, P.R., 1970, "The Geology and Mineralogy of the Kalahari Manganese-Field North of Sishen, Cape Province," Memoir 59, South Africa Geological Survey, p. 84.

Dorr, J.V.N., II, 1973, "Iron-Formation and Associated Manganese in Brazil," *Genesis of Precambrian Iron and Manganese Deposits,* Earth Science 9, UNESCO, Paris, pp. 105–113.

Doyen, L., 1973, "The Manganese Ore Deposit of Kisenge-Kamata (Western Katanga) Mineralogical and Sedimentological Aspects of the Primary Ore," *Ores in Sediment,* G.C. Amstutz and A.J. Bernard, eds., International Union Geological Science, Series A, No. 3, Springer Verlag, New York, pp. 93–100.

Germann, K., 1973, "Deposition of Manganese and Iron Carbonates and Silicates in Liassic Marls of the Northern Limestone Alps (Kalkalpen)," *Ores in Sediments,* G.C. Amstutz and A.J. Bernard, eds., International Union of Geological Science, Series A, No. 3, Springer Verlag, New York, pp. 129–138.

Ghosh, D.B., 1972, "Iron and Manganese," *Records,* Geological Survey of India, Vol. 102, Pt. 2, pp. 61–78.

Grasselly, Gy., et al., 1969, "Data on the Geology and Mineralogy of the Epleny Manganese Ore Deposit," *Acta Mineralogica-Petrographica,* Vol. 14, Fasc. 1, p. 15–43.

Gross, G.A., 1965, *Geology of Iron Deposits in Canada,* Vol. 1, *General Geology and Evaluation of Iron Deposits,* Canada Geological Survey, 181 pp.

Gross, G.A., 1967, Vol. 2, *Iron Deposits in the Appalachian and Grenville Regions of Canada,* Canada Geological Survey, 111 pp.

Gross, G.A., 1968, Vol. 3, *Iron Ranges of the Labrador Geosyncline,* Economic Geology Report No. 22, Canada Geological Survey, 179 pp.

Gross, G.A., 1980, "A Classification of Iron Formations Based on Depositional Environments," *Canadian Mineralogist,* Vol. 18, pp. 215–222.

Guild, P.W., 1957, "Geology and Mineral Resources of the Congonhas District, Minas Gerais, Brazil," Professional Paper 290, US Geological Survey, 90 pp.

Gurvich, Ye.M., 1981, "The Carbonaceous Manganiferous Associations," *International Geological Review,* Vol. 23, No. 5, pp. 584–590.

James, H.L., 1954, "Sedimentary Facies of Iron-Formation," *Economic Geology,* Vol. 49, pp. 235–293.

James, H.L., 1969, "Comparison between Red Sea Deposits and Older Ironstone and Iron-Formation," *Hot Brines and Recent Heavy Metal Deposits in the Red Sea,* E.T. Degens and E.A. Ross, eds., Springer Verlag, New York, pp. 525–532.

Johnston, A.G., and McCartney, W.D., 1965, "Manganese Occurrences in Canada," Paper 64–37, Canada Geological Survey, 63 pp.

Kalinin, V.V., 1965, "The Iron-Manganese Ores of the Karadzhal Deposit," Institute of the Geology of Ore Deposits, Petrography, Mineralogy and Geochemistry, USSR Academy of Sciences, Moscow, translation of Russian, 123 pp.

Key, R.M., 1980, "Stratiform Manganese Mineralization in the Palapye Group Central-Eastern Botswana," *Proceedings,* 5th Quadrennial IAGOD Symposium, Stuttgart, E. Schweizerbart'sche Verlagsbuchhandlung (Nagele u. Obermiller), pp. 607–617.

Krishna Rao, S.V.G., Bhattacharyya, S., and Sarma, K.J., 1981, "Manganese Silicate-Carbonate Protore (Queluzite) from the Srikakulam Manganese Belt, Andhra Pradesh," *Journal,* Geological Society of India, Vol. 22, pp. 197–200.

Machado, R., and Ribeiro, F.E., 1980, "Geology and Genesis of the Manganese Deposit of Bandarra, Urandi District, Brazil," *Abstracts,* 26th International Geological Congress, Vol. 3, Paris, p. 962.

MacLeod, W.N., 1965, "Banded Iron Formations of Western Australia," *Geology of Australian Ore Deposits,* J. McAndrew and R.T. Madigan, eds., 8th Commonwealth Mining and Metallurgical Congress, Australia and New Zealand, 1965, Vol. 1, pp. 113–117.

Magaritz, M., and Brenner, I.B., 1979, "The Geochemistry of a Lenticular Manganese-Ore Deposit (UmBogma, Southern Sinai)," *Mineralium Deposita,* Vol. 14, pp. 1–13.

Natarajan, V., and Reddy, B.S.R., 1978, "Stratiform Manganese-Ore Deposits of Adelabad District, Andhra Pradesh," *Indian Minerals,* Vol. 32, No. 2, pp. 1–9.

Paakkola, J., Airas, K., and Hugg, R., 1980, "Manganese in the Iron Formations of the Kittila Greenstone Belt in Northern Finland," *Abstracts,* 26th International Geological Congress, Vol. 3, Paris, p. 983.

Park, C.F., Jr., and MacDiarmid, R.A., 1970, *Ore Deposits,* W.H. Freeman and Co., San Francisco, 421 pp.

Perseil, E.A., and Grandin, G., 1978, "Evolution Minéralogique du Manganèse dans Trois Gisements d'Afrique de l'Ouest: Mokta, Tambao, Nsuta," *Mineralium Deposita,* Vol. 13, No. 3, pp. 295–311.

Rakhmanov, V.P., and Tchaikovsky, V.K., 1980, "Manganese Ores in Activated Platforms," *Proceedings,* 5th Quadrennial IAGOD Symposium: Stuttgart, E. Schweizerbat'sche Verlagsbuchhandlung (Nagele u. Obermiller), pp. 125–141.

Roy, S., 1966, "Syngenetic Manganese Formations of India," Jadavpur University, Calcutta, 219 pp.

Roy, S., 1969, "Classification of Manganese Deposits," *Acta Mineralogical-Petrographica,* Vol. 14, Fasc. 1, pp. 67–83.

Roy, S., 1980, "Manganese Ore Deposits of India," *Geology and Geochemistry of Manganese,* I.M. Varentsov and Gy. Grasselly, eds., Vol. 2, *Manganese Deposits on Continents,* Stuttgart, E. Schweizerbart'sche Verlagsbuchhandlung (Nagele u. Obermiller), pp. 237–264.

Sapozhnikov, D.G., 1963, "The Karadzhal Iron-Manganese Deposit," *Transactions,* Institute of the Geology of Ore Deposits, Petrography, Mineralogy and Geochemistry, No. 89, USSR Academy of Sciences, Moscow, Chap. 3, No. 89, translation of Russian, 193 pp.

Sawkar, R.H., 1980, "Geology of the Manganese Ore Deposits of North Kanara District, Karnataka State, India," *Geology and Geochemistry of Manganese,* I.M. Varentsov and Gy. Grasselly, eds., Vol. 2, *Manganese Deposits on Continents* Stuttgart, E. Schweizerbart'sche Verlagsbuchhandlung (Nagele u. Obermiller), pp. 279–295.

Scarpelli, W., 1973, "The Serra do Navio Manganese Deposit (Brazil)," *Genesis of Precambrian Iron and Manganese Deposits,* Earth Science 9, UNESCO, Paris, pp. 217–227.

Sohnge, P.G., 1977, "Timing Aspects of the Manganese Deposits of the Northern Cape Province (South Africa)," *Time-and Strata-Bound Ore Deposits,"* D.D. Klemm and H.-J. Schneider, eds., Springer Verlag, New York, pp. 115–122.

Stanaway, K.J., Kobe, H.W., and Sekula, J., 1978, "Manganese Deposits and the Associated Rocks of Northland and Auckland, New Zealand," *New Zealand Journal of Geology & Geophysics,* Vol. 21, No. 4, p. 21–32.

Strauss, G.K., Madel, T.J., and Edez, F., 1977, "Exploration Practice for Strata-Bound Volcanogenic Sulphide Deposits in the Spanish-Portuguese Pyrite Belt: Geology, Geophysics, and Geochemistry," *Time- and Strata-Bound Ore Deposits,* D.D. Klemm and H.-J. Schneider, eds., Springer Verlag, New York, pp. 55–93.

Turnbull, I.M., 1972, "New Manganese Mineral Localities in Otago," *Industrial Minerals and Rocks, 1972,* New Zealand Geological Survey, pp. 41–46.

Valarelli, J.V., et al., 1977, "The Manganese Deposits of Buritirama (Para, Brazil)," *Schweiz. Mineral. Petrogr.,* Vol. 57, pp. 313–327.

Varentsov, I.M., and Rakhmanov, V.P., 1980, "Manganese Deposits of the USSR (A Review)," *Geology and Geochemistry of Manganese,* Vol. 2, I.M. Varentsov and Gy. Graselly, eds., *Manganese Deposits on Continents,* Stuttgart, E. Schweizerbart'sche Verlagsbuchhandlung (Nagele u. Obermiller), pp. 320–391.

Watanabe, T., 1970, "Bedded Manganese Deposits in Japan, a Review," T. Tatsumi, ed., *Volcanism and Ore Genesis,* University of Tokyo Press, pp. 119–142.

Weber, F., 1973, "Genesis and Supergene Evolution of the Precambrian Sedimentary Manganese Deposit at Moanda (Gabon)," *Genesis of Precambrian Iron and Manganese Deposits,* Earth Science 9, UNESCO, Paris, pp. 307–322.

Zantop, H., 1978, "Geologic Setting and Genesis of Iron Oxides and Manganese Oxides in the San Francisco Manganese Deposit, Jalisco, Mexico," *Economic Geology,* Vol. 73, pp. 1137–1149.

Zantop, H., 1981, "Trace Elements in Volcanogenic Manganese Oxides and Iron Oxides: The San Francisco Manganese Deposit, Jalisco, Mexico," *Economic Geology,* Vol. 76, pp. 545–555.

5

Low-Grade Chromium Resources

Bruce R. Lipin

US Geological Survey, Reston, VA

ABSTRACT

Chromite, the only ore of chromium, occurs almost exclusively in mafic-ultramafic rocks and their weathering products. However, not all kinds of mafic-ultramafic rocks are favorable hosts for possible low-grade chromite deposits. For instance, chromium-bearing basalts are not favorable because they commonly contain <0.3% Cr_2O_3, and chromite is almost always less than 50 mμ in diameter, if it is present at all. Chromite-bearing ultramafic rocks in stratiform and alpine complexes tend to be either ore grade (>25% chromite) or very low grade (<2% chromite); intermediate rocks are scarce. In most complexes, large tonnages exist only in the <2% chromite rocks and the composition of the chromite in those rocks may be unsuitable for metallurgical use. Some nickel laterite districts contain more than 10^8 tons of material with 2.0 to 5.5% Cr_2O_3, and hence, chromium could be a byproduct of nickel in some laterite districts. In addition, some chromiferous sand placers may contain tens to hundreds of millions of tons of material between 2 and 5% Cr_2O_3.

INTRODUCTION

Since 1797, when chromium was discovered, the mineral chromite has been the only natural source of chromium. Conventional chromite deposits contain sufficient tonnage to satisfy world demand for centuries. However, as with all natural resources, chromite is not distributed evenly worldwide. South Africa and Zimbabwe possess about 99% of the world's known chromite, and a third country, the USSR, has about a third of the remaining 1% (Fig. 1). This extreme concentration of chromite in two adjacent countries raises the possibility of future severe disruptions in supply (see Thayer and Lipin, 1978; and Lipin and Thayer, 1980, for a more extended discussion on the outlook for chromium supply from conventional deposits). Like most other major consumers of chromium, the United States does not have sufficient conventional chromite resources to withstand a long-term disruption of supply. Therefore, it is the purpose of this report to examine possible unconventional, low-grade sources of chromium.

The chromium industry recognizes three classes of chromite, depending upon its composition (Table 1). High-Cr chromite is used mostly to make steel alloys, principally stainless steel. High-Fe chromite traditionally has been used in the chemical industry for plating and pigment. In the last 10 to 15 years, however, changes in technology have made high-Fe chromite usable for producing steel alloys. High-Al chromite is used in its mineral form as refractory bricks for steelmaking furnaces. By far the most important use of chromium is in the metallurgical industry for producing oxidation- and corrosion-resistant alloys; over 70% of yearly consumption goes for this purpose. In addition, there is no substitute for chromite in these alloys. Chromite used for metallurgical purposes, especially high-

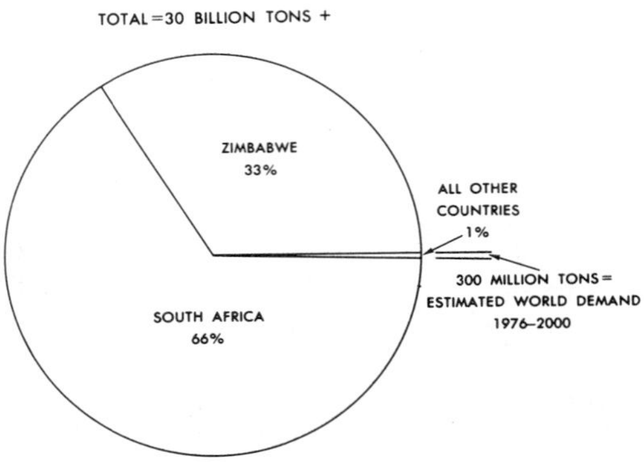

Fig. 1. Estimated chromite resources in South Africa, Zimbabwe, and the rest of the world compared to projected world demand from 1976 to 2000 (from Thayer and Lipin, 1978).

Cr chromite, commands the highest price (see Table 1).

UNCONVENTIONAL SOURCES OF CHROMIUM

Barton (this volume) defines two kinds of unconventional deposits; (1) an unconventional mineralogy, that is, a mineral or minerals not previously considered to be a source of a certain element, and (2) familiar ore minerals in an unconventional (not previously recognized) geological setting. This definition will be generally followed in this chapter.

Unconventional Mineralogy

At present, chromite is the only natural source of chromium, but many minerals contain chromium as a major constituent (Table 2, from Fleischer, 1980). In addition, about six chromium-bearing minerals have been found exclusively in meteorites. Except for chromite, fuchsite appears to be the only chrome-bearing mineral known to occur anywhere in large amounts. Fuchiste commonly occurs in Precambrian quartzites, sometimes called green quartzite, as in the Facer formation in Utah (Crittenden and Sorensen, 1980). Fuchsite-bearing quarzite occurs in layers 1 to 2 m thick and a few hundred meters at most along strike (Crittenden, 1981, oral communication) and contains about 150 ppm chromium. Compton (1972) described local green quartzite in the Raft River Mountains of northwest Utah. Some of these patches are large enough to quarry (Crittenden, 1981, oral communication). Fuchsite is also present in corundum-rich rocks in the Precambrian metavolcanics of southern Africa (Schreyer et al., 1981) where one fuchsite-bearing lense in Zimbabwe is 16 m thick and 50 m along strike.

Even these occurrences, however, are inadequate to be considered a potential source of chromium. Because yearly world demand for chromium is measured in millions of tons, any possible ore mineral must occur in large quantities, at least a few thousand tons of chromium equivalent. Among the known chromium-bearing minerals only chromite occurs in sufficient quantity to be a source of chromium.

Unconventional Geological Settings

Before considering unconventional settings for chromite, we must know where it has been mined traditionally. Chromite has been mined from three geologic settings: (1) stratiform intrusions such as the Stillwater complex, Montana; Bushveld Complex, South Africa; and Great Dyke, Zimbabwe; (2) ophiolites (or alpine peridotites) such as the Zambales Complex in the Philippines; the Guleman-Elazig region in Turkey; Tiebaghi in New Caledonia; and the Troodos Complex, Cyprus; these will hereafter be called podiform deposits; and (3) placer deposits in areas that drain chromite-bearing ultramafic rocks. Because ophiolite complexes usually occupy greater

TABLE 1. Classification of Chromite by Composition, Type of Deposit, and Principal Uses

Class of ore	Composition (by wt)	Type of deposit	Major use	1980 year-end price
High-Cr	Cr_2O_3 46–55% Cr:Fe >2:1	Podiform and stratiform	Metallurgical	\$110/metric ton
High-Fe	Cr_2O_3 42–64% Cr:Fe <2:1	Stratiform	Metallurgical Chemical	\$51–55/metric ton
High-Al	Cr_2O_3 33–38% Al_2O_3 22–34%	Podiform	Refractory	\$96/metric ton

TABLE 2. Terrestrial Minerals Containing Chromium as a Major Constituent

Name	General Forumla	Wt % Cr
Barbertonite	$Mg_6Cr_2(CO_3)(OH)_{16}\cdot 4H_2O$	16
Bentorite	$Ca_6(Cr,Al)_2(SO_4)_3(OH)_{12}\cdot 6H_2O$	5
Bracewellite*	$CrO(OH)$	61
Brezinaite	Cr_3S_4	47–50
Chromian diopside	$Ca(Mg,Fe,Cr)Si_2O_6$	0.1–8
Chromian geikielite	$(Mg,Fe^{2+},Cr, Fe^{3+})(Ti,Cr,Fe^{3+})O_3$	0.5–8.5
Chromian garnet	$(Cr,Mg)_3(Al,Cr)_2(SiO_4)_3$	0.1–13
Chromite	$(Mg,Fe^{2+})(Cr,Al,Fe^{3+})_2O_4$	10–44
Chromatite	$CaCrO_4$	33
Chromian clinochlore[+]	$(Mg,Fe^{2+})(Al,Cr)_2(Al_2,Si_2)O_{10}(OH)_8$	0.5–12
Cochromite	$(Co,Ni,Fe^{2+})(Al,Cr)_2O_4$	34–37
Crocoite	$PbCrO_4$	16
Dietzeite	$Ca_2(IO_3)_2(CrO_4)$	10
Donathite[≠]	$(Fe^{2+},Mg)(Cr,Fe^{3+})_2O_4$	28–30
Embreyite	$Pb_5(CrO_4)_2(PO_4)_2\cdot H_2O$	7
Eskolaite	Cr_2O_3	64–68
Fornacite	$(Pb,Cu)_3[(Cr,As)O_4]_2(OH)$	6
Fuchsite	$K(Al,Cr)_2(AlSi_3)O_{10}(OH)_2$	0.5–6
Grimaldiite[§]	$CrO(OH)$	61
Guyanaite[¶]	$CrO(OH)$	61
Hemihedrite	$Pb_{10}Zn(CrO_4)_6(SiO_4)_2(F,OH)_2$	13–14
Iranite	$Pb_{10}Cu(CrO_4)_6(SiO_4)_2(F,O)$	10
Knorringite	$Mg_3Cr_2(SiO_4)_3$	12–23
Lopezite	$K_2Cr_2O_7$	35
Loveringite	$(Ca,Ce)(Ti,Fe^{3+},Cr,Mg)_{21}O_{38}$	0.5–10
Macquartite	$Pb_3Cu(CrO_4)SiO_3(OH)_4\cdot 2H_2O$	6
Manganochromite	$(Mn,Fe^{2+})(Cr,V)_2O_4$	41–62
Mariposite[#]	$K(Al,Cr)_2(Si_{3+x}Al_{1-y})O_{10}(OH)_2$	0.5–6
McConnellite	$CrOOCu$	35
Mountkeithite	$[(Mg,Ni)_9(Fe^{3+},Cr,Al)_3(OH)_{24}]^{3+}[(CO_3,SO_4)_{1.5}(Mg,Ni)_2(SO_4)_2(H_2O)_{11}]^{3-}$	2.2–6
Nichromite	$(Ni,Co,Fe^{2+})(Cr,Fe^{3+},Al)_2O_4$	31–37
Philipsbornite	$PbAl_3H(As,CrO_4)_2(OH)_6$ (?)	2.–2.5
Phoenicochroite	$Pb_2(CrO_4)O$	8–10
Redingtonite	$(Fe^{2+},Mg,Ni)(Cr,Al)_2(SO_4)_4\cdot 22H_2O$	0.5–3
Redledgeite	$Mg_4Cr_6Ti_{23}Si_2O_{61}(OH)_4$	11
Rilandite	$(Cr,Al)_6SiO_{11}\cdot 5H_2O$	33
Santanaite	$9PbO \bullet 2PbO_2\cdot CrO_3$	2
Schreyerite	$(V,Cr,Al)_2Ti_3O_9$	0.7–3.6
Shuiskite	$Ca_2(Mg,Al,Fe)(Cr,Al)_2[(Si,Al)O_4](Si_2O_7)(OH)_2\cdot H_2O$	10–17
Stichtite	$Mg_6Cr_2(CO_3)(OH)_{16}\cdot 4H_2O$	6–19
Tarapacaite	K_2CrO_4	27
Uvarovite	$Ca_3Cr_2(SiO_4)_3$	21
Vauquelinite	$Pb_2Cu(CrO_4)(PO_4)(OH)$	7
Vuorelaineite	$(Mn,Fe,^{2+})(V,Cr)_2O_4$	3.2–21
Yedlinite	$Pb_6CrCl_6(O,OH)_8$	4

*Orthorhombic.

[+]This has been called the now discredited name Kaemmererite.

[≠]Tetragonal AB_2O_4 structure.

[§]Trigonal.

[¶]Orthorhombic, different space group than bracewellite.

[#]Chromian phengite, $Y = 4/3x$ in the formula.

areas than stratiform intrusions, placer deposits are more common in ophiolite terranes.

We must also consider another factor of the conventional chromite deposits—grade. Stratiform and podiform deposits that contain less than 25% chromite (about 13% Cr_2O_3) are not mined. Placer deposits containing smaller amounts of chromite (about 3%) have been mined, because simple and inexpensive separation of chromite is possible.

Where then do we look for unconventional settings? Remembering that any potentially exploitable chromite deposit should contain at least a few thousand tons of chromium, we will examine places where chromite is a minor or accessory, but common mineral. Present knowledge indicates that mafic and ultramafic rocks have the highest average chromium contents of all major rock types and are the only rocks that contain chromite in large quantities. We will consider five environments in which chromite has never been mined: (1) continental flood basalts, (2) oceanic basalts, (3) ultramafic rocks in ophiolites which contain only accessory chromite (<5%), (4) ultramafic rocks in stratiform complexes which contain only accessory chromite, (5) laterites that are formed by weathering of ultramafic rocks. In addition, we will consider one environment from which chromite has been mined, placer-type deposits.

Continental Flood Basalts: Flood basalts represent very large volumes of mafic rocks. Unfortunately, as typified by the Columbia River basalt group in the northwestern United States, flood basalts are rather poor in chromium compared with other types of basalt (Fig. 2). According to the data of Swanson and Wright (1981), Columbia River basalts do not contain more than about 250 ppm Cr and average well below 100 ppm Cr. Furthermore, chromite is almost never observed as a phase in the Columbia River basalts (T. L. Wright, 1981, oral communication), and presumably chromimum occurs in pyroxenes and glass.

Oceanic and Precambrian Basalts: Oceanic basalts are another class of mafic rock that cover very large areas. Their compositions are more primitive, that is, higher $Mg/(Mg+Fe^{2+})$ and higher chromium than Columbia River basalts (Fig. 2). Highest chromium concentrations are about 1000 ppm and average about 400 ppm. In addition, most studies of oceanic basalts report chromite as an accessory phase. We will

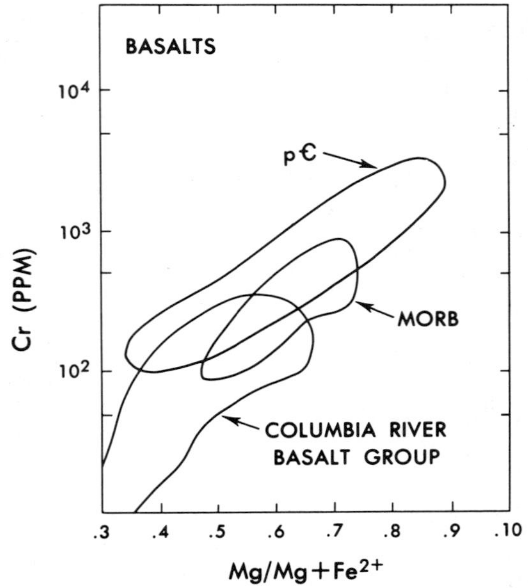

Fig. 2. Cr (ppm) vs. $Mg/(Mg + Fe^{2+})$ for Precambrian basalts (P €), mid-ocean ridge basalts (MORB), and Columbia River basalts. Modified from Bence et al. (1980).

more closely examine Kilauea and Makaopuhi lavas as examples of oceanic basalts. These Hawaiian basalts have similar Cr vs. $Mg/(Mg+Fe^{2+})$ values to midocean ridge basalts (MORB), and chromite is a common accessory mineral in the Hawaiian basalts.

Evans and Wright (1972, p. 220) reported average analyses of 70 chromite grains in six samples from Kilauea and Makaopuhi (Fig. 3). Compared with most chromite being mined, chromite from Hawaii is low in Cr_2O_3 and Cr/Fe. Chromite containing less than 40% Cr_2O_3 and with Cr/Fe less than 1.2 is not usable for metallurgical purposes (G. Watson, 1981, oral communication.). Hawaiian chromite is either very close to, or below, these minimums. It is interesting that the Kilauea chromites continue the compositional trend of the chromite seams in the Bushveld Complex.

Although some chromite in Hawaiian basalts meets minimum compositional standards, grade and recoverability are still practical problems. The average grain size of the Hawaiian chromites studied by Evans and Wright is 29 mμ, which would require crushing to at least 600 mesh to free the chromite. In addition, the Ki-

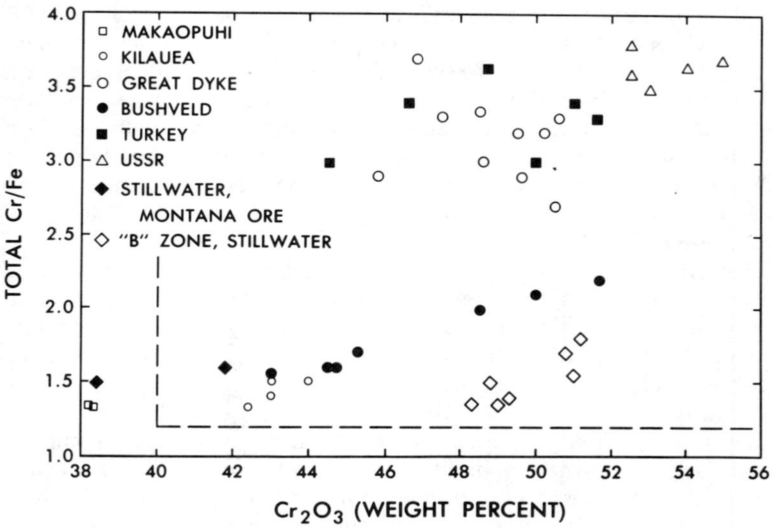

Fig. 3 Cr/Fe vs. Cr_2O_3 of chromite ore from various geologic environments compared to chromite compositions from Hawaii. Dashed lines represent the minimum Cr/Fe (1.2) and Cr_2O_3 (40%) acceptable for metallurgical use with present technology.

lauea samples average 0.17% Cr_2O_3 which, if it were in chromite, would represent about 0.4% chromite. However, Cr_2O_3 will also partition into other minerals and glass, reducing the amount of chromite in oceanic basalts to less than 0.1%.

Precambrian basalt can reach 2000 ppm chromium (Fig. 2), which is equivalent to about 0.6% chromite if all the chromium is in the chromite. However, most Precambrian basalt is altered and metamorphosed so that the originial mineralology has been obliterated.

In summary, basalts are not suitable as possible low-grade sources of chromite.

Ultramafic Rocks in Ophiolites: In order to understand guides for low-grade chromite deposits in ultramafic rocks in ophiolites, it is necessary to describe the occurrence of conventional podiform chromite deposits in this environment. The stratigraphically lowest unit in most ophiolites is a tectonite peridotite, which is dominantly harzburgite, ranging in thickness from a few hundred meters to 7 km (Fig. 4). This tectonite unit commonly grades upward into a cumulate unit. The cumulate unit is composed of dunite at its base which grades upward into wehrlite or troctolite, layered olivine gabbro, and then into massive gabbro in which there is no evidence for movement of crystals. In most ophiolites, the tectonite peridotite contains many lenses of dunite, most of which are concordant with the foliation in the harzburgite. Near the base of the tectonic harzburgite the dunite lenses are small, commonly less than 10 m^2 in extent, and widely spaced. The dunite lenses become larger and more numerous upward. Near the stratigraphic top of the harzburgite these lenses coalesce into a continuous dunite which marks the bottom of the cumulate unit. In this transition zone between harzburgite and dunite, most large podiform chromite deposits occur. Some chromite deposits are in the dunite lenses surrounded by harzburgite and some are in the overlying, continuous dunite. All podiform chromite deposits

IDEALIZED OPHIOLITE SECTION

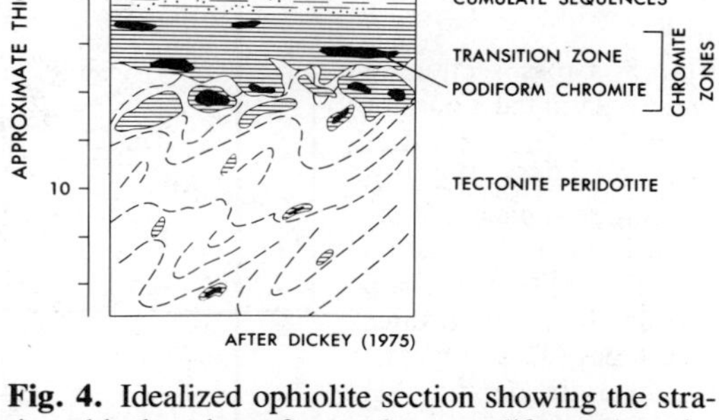

AFTER DICKEY (1975)

Fig. 4. Idealized ophiolite section showing the stratigraphic location of most large podiform chromite deposits (from Dickey, 1975).

in harzburgite are in a dunite envelope; the envelope may be less than 1 m thick or tens of meters thick. Therefore, two main rock types occur in ophiolites in which low-grade chromite deposits may be sought, dunite and harzburgite.

Tonnage and grade distribution of conventional chromite deposits in ophiolites are influenced by the sharp boundaries that are typical of podiform chromite deposits. A deposit in the Vourinos Complex, Greece (Fig. 5), is a good example. An ore body may range from 25 to 100% chromite, but there is very little tapering off of grade into host rock; instead there is an abrupt change from ore grade to background values. Assay data and logs of drill core from the Chambers and Iron King mines, Grant County, OR (Hundhausan et al., 1956, p. 46–57), show that the chromite-dunite contact is seldom more than 1.5 m thick and is usually sharp. Thus, in some deposits the contact is knife-sharp, whereas in others that contact is gradational but almost always less than 2 m wide. As a result, ultramafic rocks in ophiolites contain either background concentrations of chromite (0–5%, average 1–2%) or ore-grade segregations (>25%) with a paucity of rocks containing chromite concentrations between these values (Arai, 1980, p. 146).

A number of ophiolite localities have acquired the reputation of containing large tonnages averaging 4–10% Cr_2O_3, or about 8–20% chromite. These places include the Seiad area and the Pilliken area in California, Red Mountain, Alaska, and some areas in Turkey. This is an apparent contradiction to the statement previously that there is a paucity of rocks with less than 25% chromite and more than 2% chromite. However, field observations in Turkey (T. P. Thayer, oral communication, 1982), Red Mountain (P. W. Guild, oral communication, 1982), and both areas in California (Lipin, unpublished data) confirm that the gap in chromite abundance exists. In these areas the ultramafic rock with less than 2% chromite contains some thin lenses of chromite, bringing the average chromite content of a large volume of rock into the 5–10% chromite range.

In addition, one cannot predict the grade of chromite concentrate that may be obtained from accessory chromite in the dunite and harzburgite host rocks by the composition of segregated chromite in nearby ore bodies. Figs. 6 and 7 show compositional range of segregated and accessory chromite from two rather complete ophiolite suites, New Caledonia (LeBlanc et al., 1980), and Semail, Oman (Brown, 1980), and one area of scattered dunites in North Carolina that are probably remnants of a disrupted ophiolite (Lipin, in press). In all these areas accessory chromite composition differs from that of segregated chromite in that, for a given chromium value, accessory chromites are generally richer in iron than segregated chromite.

Two different mechanisms tend to enrich accessory chromite in iron. Irvine (1967, p. 83–85) demonstrated that accessory chromite will gain more iron than segregated chromite during subsolidus cooling than segregated chromite. This is because the distribution coefficient, $K = (Mg/Fe^{2+})_{silicate} \cdot (Fe^{2+}/Mg)_{chromite}$ increases with decreasing temperature. Therefore, when chromite is the dominant mineral (Fig. 8) its final composition will be close to its initial composition. But when olivine or pyroxene is dominant the final chromite composition will be much more iron rich than the initial composition. The

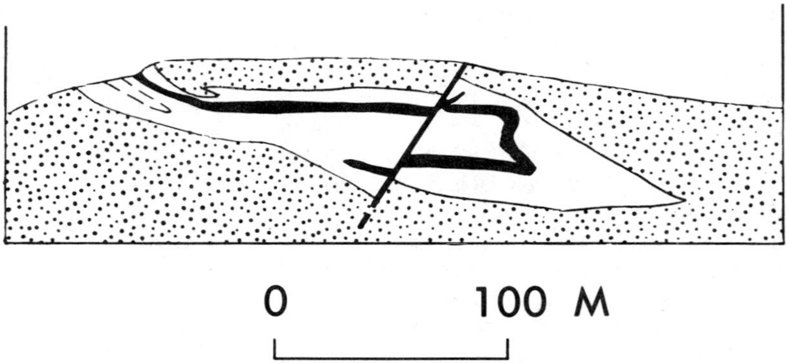

Fig. 5. Cross section of mine P-25 in the Vourinos Complex, Greece (after Moores, 1969, Fig. 6). Chromitite is black, dunite is clear, and harzburgite is stippled. Note the sharp contacts between all three rock types. This is typical of chromite deposits in ophiolites.

Fig. 6. Compositions of segregated and accessory chromite.

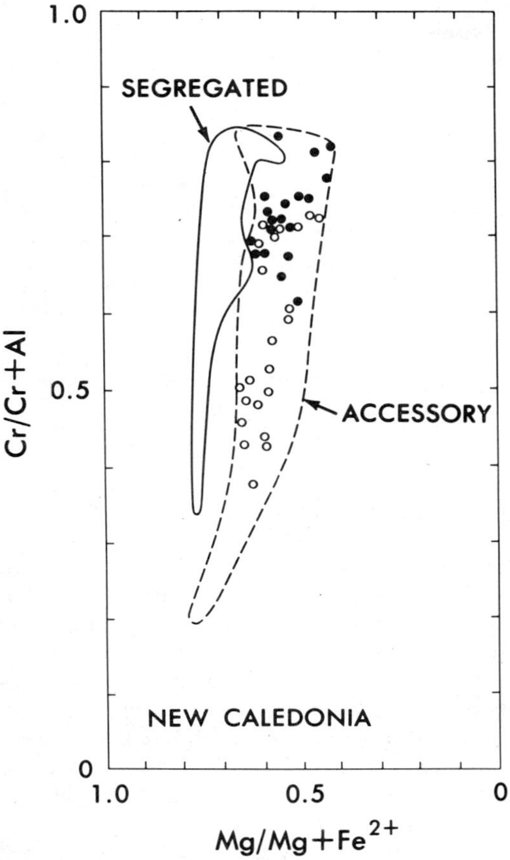

A. New Caledonia (Leblanc et al., 1980). The dots are accessory chromite in dunite, the open circles are accessory chromite in harzburgite.

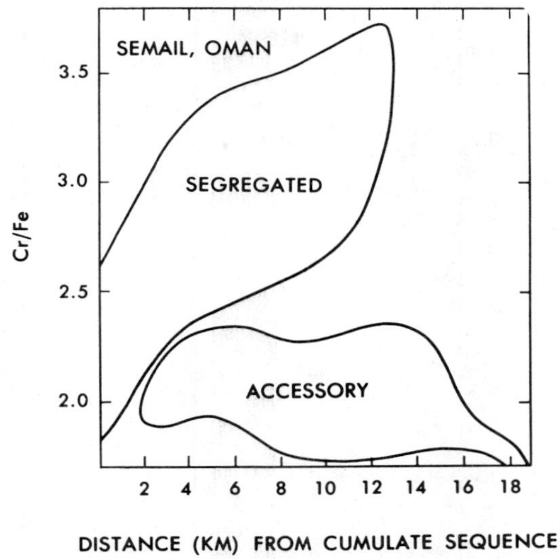

B. Semail, Oman (Brown, 1980).

other mechanism is alteration or metamorphism of the ultramafic rock. During alteration of ultramafic rocks, chromite may gain iron through reactions such as

$$\text{chromite} + \text{serpentine} + \text{talc} + \text{fluid} \rightleftharpoons \text{Fe-rich chromite} + \text{chlorite}$$

(Evans and Frost, 1975).

The fact that accessory chromite is more Fe-rich (i.e. has a lower Cr/Fe ratio) than segregated chromite is disappointing in terms of possible exploitation because chromite with a high Cr/Fe ratio (>2) and high Cr_2O_3 content (>46%) commands the highest price of all grades of chromite.

In addition to the compositional difference between segregated and accessory chromite is the difference in composition between accessory chromite in dunite and harzburgite. In the Canyon Mountain complex, OR, Himmelberg and Loney (1980) show that chromite in harzburgite is much more aluminous than chromite from dunite (Fig. 9). Leblanc et al. (1980) and Arai (1980), in New Caledonia and Japan, respectively, also show that chromite in harzburgite contains more Al_2O_3 than chromite in dunite (see Fig. 6a for the data of Leblanc et al.). Indeed, much of the chromite in harzburgite contains well over 20% Al_2O_3 and less than 40% Cr_2O_3, which is unsuitable for metallurgical use. However, accessory chromite in dunite is usuable. Dickey and Yoder (1972) explain the higher Al in chromite associated with harzburgite by subsolidus reaction of chromite with Al-bearing pyroxene which is, of course, absent in dunite. Their experimental results are compatible with such a mechanism. Thus, accessory chromite in dunite is the best exploration target for low-grade chromite in ophiolites. Some work on large dunite areas is being done in Turkey (T. Engin, 1978, oral communication) and in the United States by the US Geological Survey in North Carolina (Lipin, in press), Oregon (Lipin and Page, unpublished data), in the Seiad area, California, the Pilliken area of California (Nicholson and Lipin, 1982), Washington, and Alaska.

An exception to the pattern of high-Cr chrom-

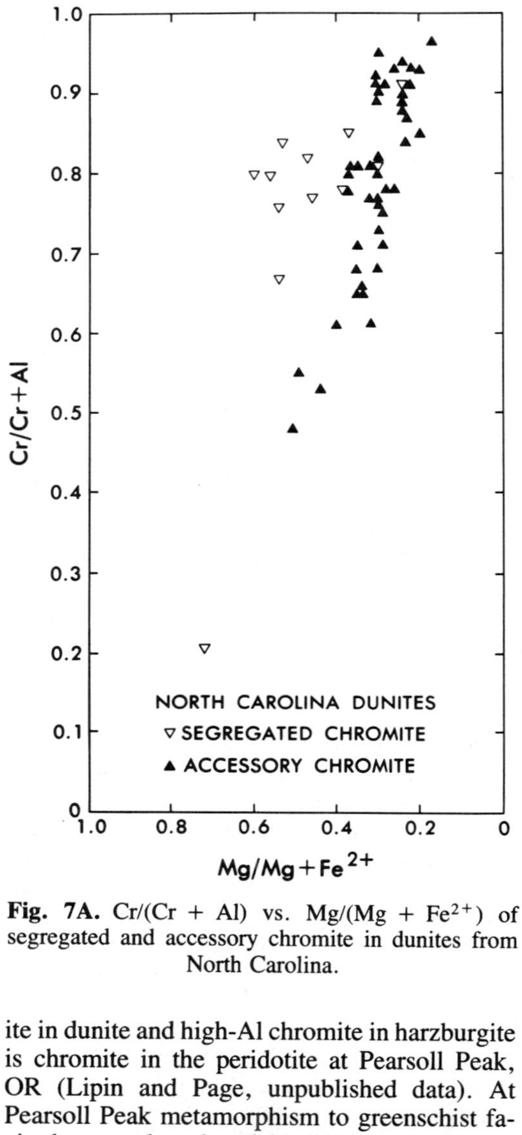

Fig. 7A. Cr/(Cr + Al) vs. Mg/(Mg + Fe^{2+}) of segregated and accessory chromite in dunites from North Carolina.

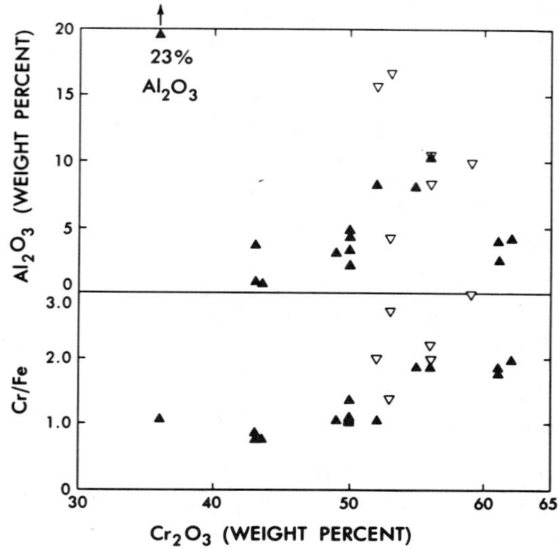

Fig. 7B. Cr_2O_3 vs. Cr/Fe and Al_2O_3 in North Carolina chromites.

ite in dunite and high-Al chromite in harzburgite is chromite in the peridotite at Pearsoll Peak, OR (Lipin and Page, unpublished data). At Pearsoll Peak metamorphism to greenschist facies has greatly reduced the difference in chromite composition in the two rock types.

In the absence of technology to use high-Al chromite in metallurgical processes, dunite in ophiolites is the preferred rock type in which to search for low-grade chromite. However, if high-Al chromite becomes metallurgically usable, huge areas of harzburgite containing 0.5–2.0% chromite would beome more attractive. One example of this is the Josephine Peridotite in Oregon and California. This body is about 640 km^2 in area and about 95% harzburgite averaging 0.55 vol % chromite (Dick, 1976).

Ultramafic Rocks in Stratiform Intrusions: Most of the information on chromite in stratiform intrusions in this report is based upon data from the Stillwater and Bushveld complexes. This is not because I believe that low-grade chromite will be mined in either of these complexes in the foreseeable future; but because the work of E. D. Jackson in the Stillwater and E. N. Cameron in the Bushveld has resulted in a more thorough knowledge of accessory chromite in these two complexes than in any other stratiform intrusion. Therefore, data concerning the accessory chromite in the Stillwater and Bushveld Complexes will be cited as examples of what we may expect to discover about other stratiform intrusions that are not so well known.

Stratiform intrusions commonly contain a significant proportion of ultramafic rock. In the Stillwater Complex, for example, the average thickness of the ultramafic zone is nearly 1.1 km over a maximum exposed stratigraphic thickness of about 5.5 km (Jackson, 1961, pp. 2–4). If we use Jackson's data on proportions of rock types and average volume percent of chromite in these rocks (1961, pp. 5, 86), the amount of chromite in a cubic kilometer of ultramafic zone exclusive of the chromite layers may be calculated; it is nearly 200 million tons per cubic kilometer. When one considers that the strike length of the Stillwater Complex is just over 48 km, the tonnage is staggering. Even

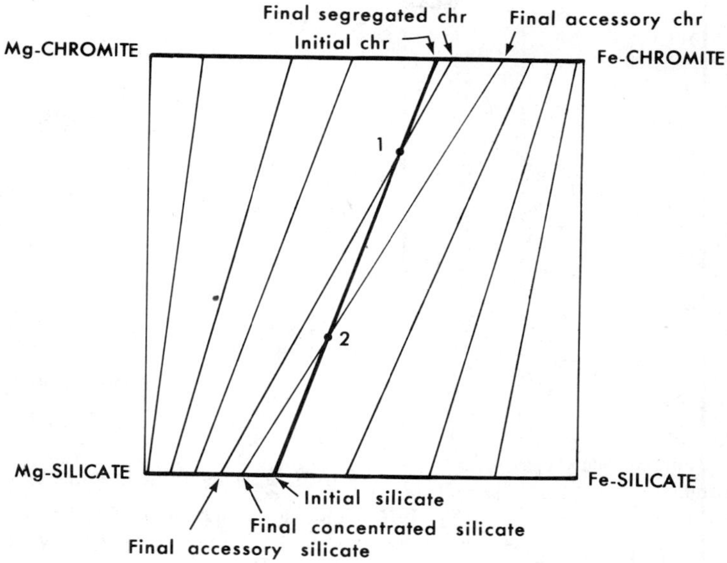

Fig. 8. Mg-Fe relations between chromite and olivine (or pyroxene). The horizontal axis is the Mg/Fe^{2+} ratio of chromite and silicate and the vertical axis is the relative proportion of the two minerals. Irvine (1967) assumed that $K = (Mg/Fe^{2+}{}_{\text{silicate}} \cdot Fe^{2+}/Mg_{\text{chromite}})$ increases with falling temperature. Therefore chromite in a rock with a high chromite/silicate ratio (point 1) will have a final composition close to its initial composition. However, chromite in a dominantly silicate unit (point 2) will have a final composition much more Fe-rich than its initial composition.

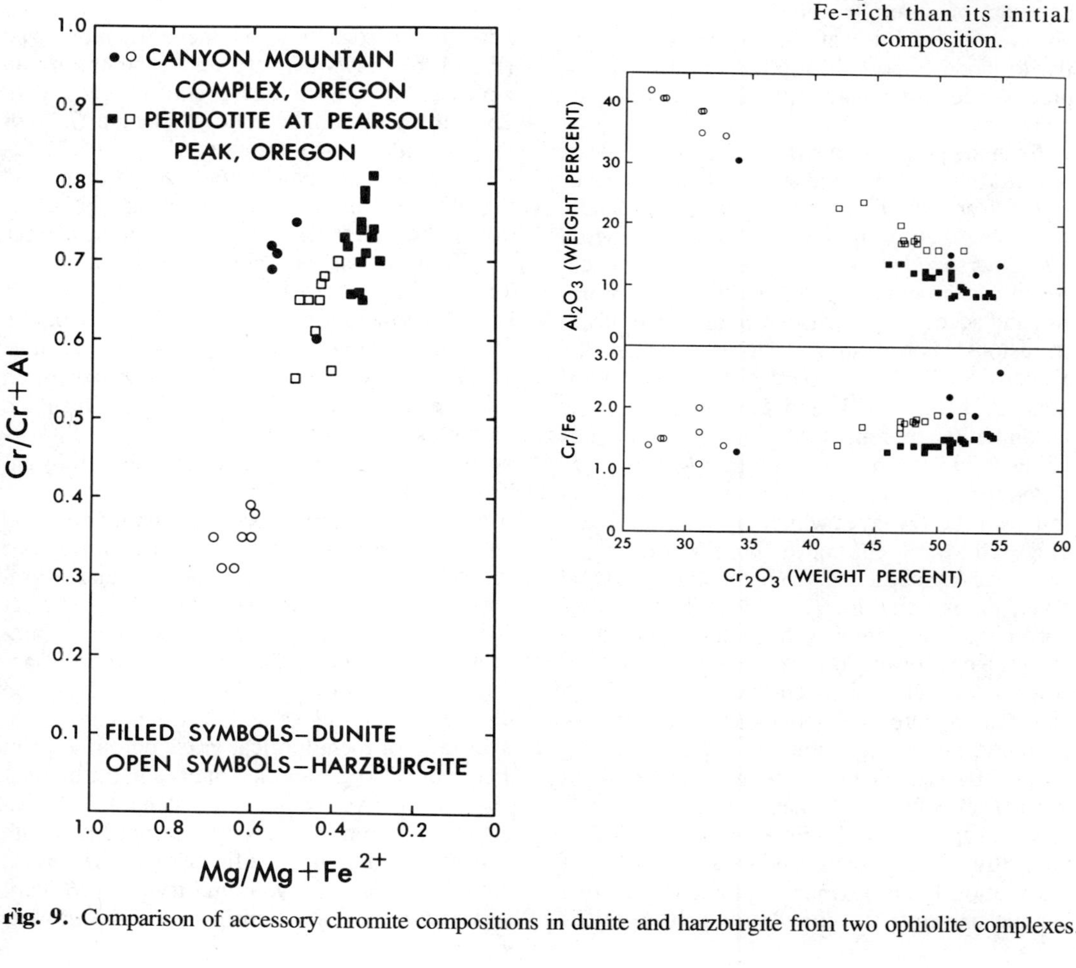

Fig. 9. Comparison of accessory chromite compositions in dunite and harzburgite from two ophiolite complexes.

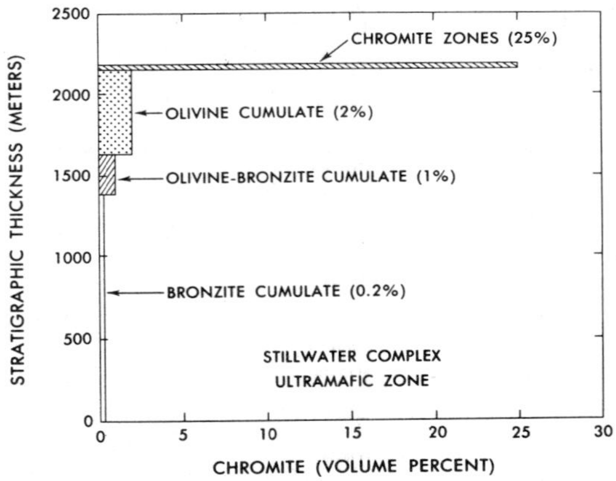

Fig. 10A. Stratigraphic thickness of rock types vs. volume percent chromite in the Mountain View area of the Stillwater Complex.

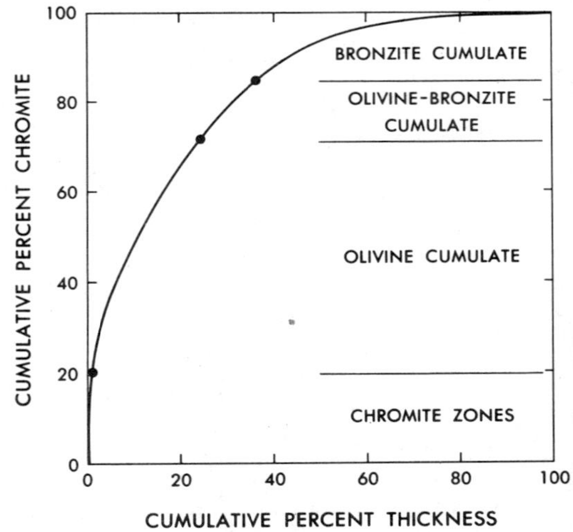

Fig. 10B. Cumulative amount of chromite vs. cumulative thickness of the rock types in the Mountain View area.

small stratiform complexes have potentially huge amounts of chromite. The average amount of chromite in the Stillwater Complex is very similar to modes reported for other stratiform complexes (see, for example, Irvine and Baragar, 1972).

As in the podiform deposits, contacts between stratiform chromite deposits and their host rocks range from sharp to rapidly grading, usually over less than 1 m (Jackson, 1968; Cameron and Emerson, 1959). Thus, there is also a paucity of rocks between ore-grade (<25% chromite) and accessory (>2% chromite) in stratiform intrusions. For example, let us consider the Eastern Bushveld Complex. The lower critical zone is 407 m thick and contains 3.4 wt % chromite (Cameron, 1980, p. 860), which is about 2.5% chromite by volume. Much of the chromite in the lower critical zone is concentrated in the D_2 layer which is 16 m thick (average 20 vol % chromite). If the 16 m of the D_2 layer is subtracted from the lower critical zone, the remaining 391 m average 1.8 vol % chromite. Cameron also shows that the remaining thickness of the lower critical zone contains several thin chromite-rich layers with abrupt changes in chromite abundance into the surrounding rocks (see Cameron, 1980).

Fig. 10a is a chart of stratigraphic thickness (related, of course, to tonnage) vs. average grade of rock types in the Mountain View section of the Stillwater Complex. The stepwise nature of the relationship is obvious. A plot of cumulative percent chromite vs. cumulative percent thickness of the rock units in the ultramafic zone (Fig. 10b) shows that two units, the chromitite zones and olivine cumulate, constitute less than 25% of the ultramafic zone yet contain 72% of the chromite in the ultramafic zone. The chromite zones contain about 20% of the chromite and the olivine cumulate contains 52%. The situation in the Stillwater Complex and in most other stratiform intrusions is similar to that in ophiolites; olivine-rich rocks are the most favorable hosts for low-grade chromite. Some exceptions to this general rule do exist. For instance, chromite is disseminated throughout orthopyroxene-rich units and some anorthosite units in the Bushveld Complex. Most of the chromite in the Fiskenaesset Complex, Greenland, is also within anorthosite units.

Another similarity between ophiolites and stratiform intrusions is the systematic difference in composition between accessory and segregated chromite. In a detailed section of the Lower Critical zone of the Bushveld Complex, Cameron (1980) showed that the Cr/Fe ratio of accessory chromite is lower than segregated chromite and is close to 1.2, the minimum usable ratio of metallurgical grade chromite (Fig. 11). Note also the abrupt changes in the amount of chromite over very small distances. The systematic difference in chromite composition with concentration is not confined to the Bushveld complex. Jackson (1969) and Irvine (1967) observed the same relationships in the Stillwater

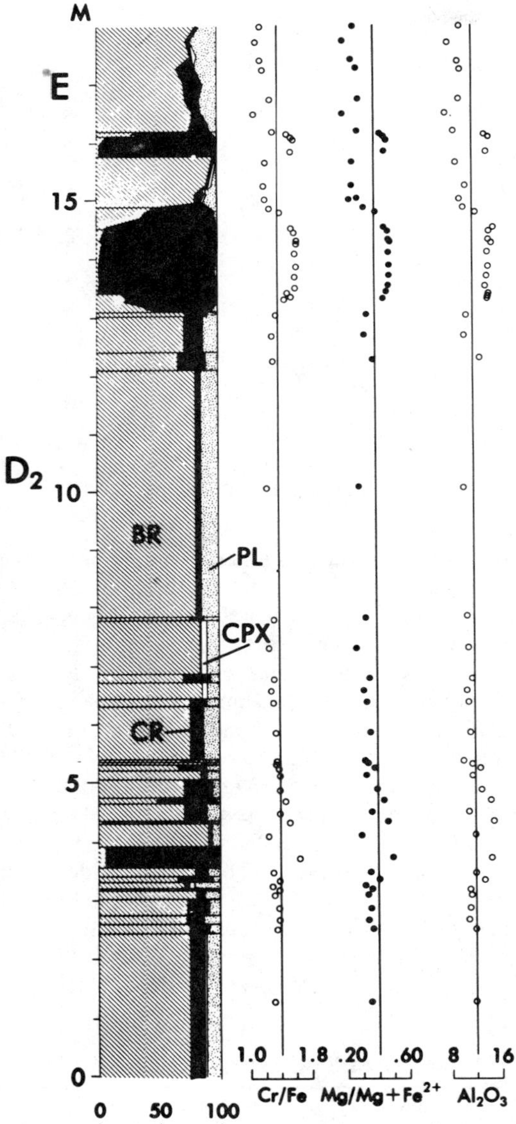

Fig. 11. Log, modes, and chemistry of the D_2 layer of the Lower Critical zone of the Eastern Bushveld Complex (Cameron, 1980). Note the positive correlation of Cr/Fe with chromite mode.

and Muskox Complexes, respectively. The mechanisms explaining the iron-rich nature of accessory chromite in ophiolites apply equally well to chromite in stratiform intrusions.

In summary, ophiolites and stratiform intrusions comprise large volumes of ultramafic rocks that contain huge tonnages of accessory chromite. The composition of much of this chromite is such that it would be of limited usefulness for metallurgical purposes with present technology. However, large areas of dunite in ophiolites and stratiform intrusions are possible sources of low-grade chromite.

Laterites: Ferruginous laterites form by weathering of peridotite in a humid tropical climate having a wet/dry seasonal variation. Other conditions for development of a laterite, according to Lawrence (1974, p. 523), include: (1) an extensive, low altitude peneplane which is well drained; (2) block faulting and some brecciation in the peridotite to assist weathering by percolating ground water; (3) minimum erosion in order to preserve the soil.

Of all the minerals in peridotite, chromite is least susceptible to alteration, and therefore it is concentrated in the residual soil as MgO and SiO_2 are leached out. A typical laterite profile from New Caledonia (Fig. 12, modified from Troley et al., 1979, pp. 102–103) shows that Cr_2O_3, and hence chromite, is concentrated in the upper part of the laterite whereas nickel is mined from "earthy ore," "soft saprolite," "ore with boulders," and "rocky ore" units. In New Caledonia, the nickel in these units is found in a poorly defined group of hydrated magnesian silicates, and thus the nickel ore in New Caledonia is called silicate ore. However, in other localities, such as Greenvale, Australia, and Moa Bay, Cuba, nickel is mined from higher units where it is found in solid solution with iron in goethite (Lawrence, 1974, p. 526). In this case, the nickel ore is called oxide ore. Thus, chromite occurs in the same units in which nickel oxide ore is mined.

Because of the intrinsic difference in chromite compositions in dunite and harzburgite, it comes as no surprise that chromite composition in laterites depends on the rock from which it was derived. Bulk Cr/Fe ratios are higher in laterites derived from dunites than in laterite derived from harzburgite (Fig. 13). Note that chromite concentrates from dunites have a Cr/Fe ratio of more than 2, which is suitable for metallurgical use.

Laterites also meet the requirement that a potential source of chromium must occur in large tonnages. Table 3 shows that, as of 1975, the worldwide reserves of lateritic nickel ore were more than 3 billion tons. If the average Cr_2O_3 content is 2.5% (corresponding to about 5%

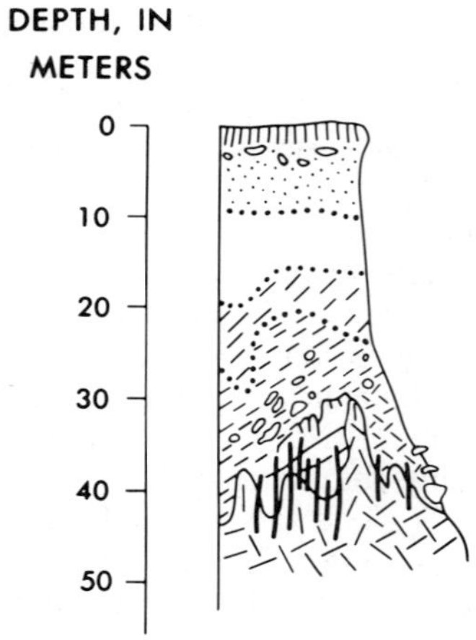

	WEIGHT PERCENT		
	Cr_2O_3	Fe_2O_3	Ni
IRON CRUST PISOLITES	5	75	0.3
RED LIMONITE	3.5	72	0.9
YELLOW LIMONITE	2	70	1.4
EARTHY ORE	2	32	2.0–3.0
SOFT SAPROLITE	1.2	23	2.3
ORE	1	17	2.5
ROCKY ORE	0.7	12	3
FRESH ROCK	0.4	8	0.3

Fig. 12. Typical profile and chemistry of laterite in New Caledonia (after Troley et al., 1979).

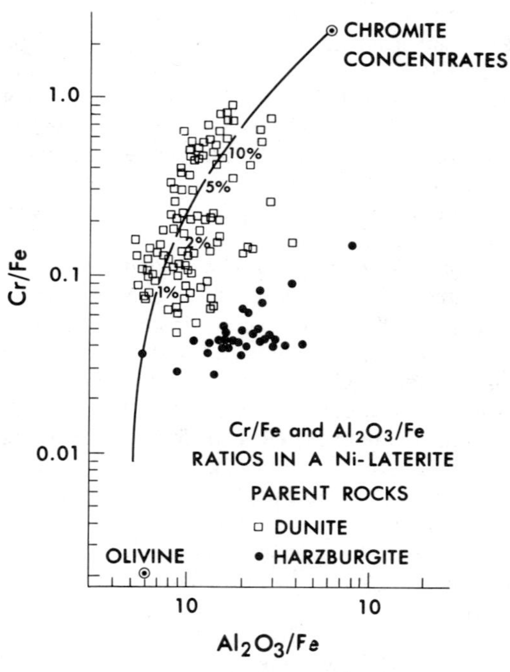

Fig. 13. Cr/Fe vs. Al_2O_3/Fe in a laterite in the Rockhampton district of Queensland, Australia (after Golightly, 1979). The curve represents a mixing line between a typical olivine composition and chromite composition. Breaks in the curve are the expected compositions for a parent rock containing 1, 2, 5, and 10 percent chromite.

chromite), known laterites contain more than 150 × 10^6 tons of chromite. If only a few percent of the known nickel-laterites occur over dunite, the chromite potential is significant.

Chrome Sand: Some stream placer deposits in North Carolina, Maryland, Pennsylvania, and California have been mined for chromite. Of much more potential importance, however, are areas in which several streams drain an area of peridotite near shore. The heavy mineral content of several streams is then reworked by wave action and extensive sand deposits may develop. The black sands of the raised beaches in Oregon (Griggs, 1945) are an example of such a combination of processes. About 500,000 tons of chromite are estimated in known deposits (Griggs,

TABLE 3. Laterite Reserves*

Country	Metric tons × 10^6
Albania	20
Australia	220
Brazil	90
Columbia	70
Cuba	300
Dominican Republic	65
Greece	182
Guatemala	81
Indonesia	555
New Caledonia	914
Philippines	802
United States	12
Venezuela	40
Yugoslavia	54

*From Peters (1975).

1945) and another 500,000 tons may be expected in undiscovered beach sand deposits.

Good potential exists elsewhere for this kind of chromite, for instance, the Semail ophiolite, Oman, and some islands in the southwest Pacific region such as the Philippines, New Caledonia, Papua-New Guinea, and Indonesia.

CONCLUSIONS

Chromite will continue to be the only ore of chromium because it is the only chromium-bearing mineral that is abundant enough to satisfy industrial demand. Some common silicates, especially micas, contain as much as 5% chromium, but the occurrence of these Cr-enriched phases is limited and extraction of chromium from the silicate lattice is not economically feasible. The most promising geologic environments for large low-grade chromite deposits are dunites associated with ophiolites, metamorphosed ophiolites, laterites developed over dunites, and chrome sand deposits. Because nickel is being mined from many laterites, the right economic conditions could spur a number of producers to seriously consider producing chromium as a byproduct of nickel in some lateritic ores.

ACKNOWLEDGMENTS

The author is indebted to P. B. Barton, M. P. Foose, and especially T. P. Thayer for many stimulating discussions that helped to crystallize the ideas presented in this paper. M. Fleischer provided the latest information on chromium-bearing minerals. G. Watson of the Ferroalloy Association provided information on minimum compositional requirements for metallurgical chromite. Perceptive reviews by J. H. De-Young, Jr., and T. P. Thayer greatly improved the manuscript. Any paper concerned with chromite and ultramafic rocks owes a large debt to E. N. Cameron, whose careful observations and innovative ideas are the building blocks for future work on stratiform intrusions and the origin of chromite deposits.

REFERENCES

Arai, S., 1980, "Dunite-Harzburgite-Chromitite Complexes as Refractory Residue in the Sangun-Yamaguchi Zone, Western Japan," *Journal of Petrology*, Vol. 21, Pt. 1, pp. 141–165.

Bence, A. E., Grove, T. L., and Papike, J. J., 1980, "Basalts as Probes of Planetary Interiors: Constraints on the Chemistry and Mineralogy of Their Source Regions," *Precambrian Research*, Vol. 10, pp. 249–279.

Bliss, N. W., and McLean, W. H., 1975, "The Paragenesis of Zoned Chromite from Central Manitoba," *Geochimica et Cosmochimica Acta*, Vol. 39, pp. 973–990.

Brown, M., 1980, "Textural and Geochemical Evidence for the Origin of Some Chromite Deposits in the Oman Ophiolite," *Ophiolites*, A. Panayiotou, ed., Geological Survey Department, Nicosia, Cyprus, pp. 714–721.

Cameron, E. N., 1980, "Evolution of the Lower Critical Zone, Central Sector, Eastern Bushveld Complex, and its Chromite Deposits," *Economic Geology*, Vol. 75, pp. 845–871.

Cameron, E. N., and Emerson, M. E., 1959, "The Origin of Certain Chromite Deposits of the Eastern Part of the Bushveld Complex," *Economic Geology*, Vol. 54, pp. 1151–1213.

Compton, R. R., 1972, "Geologic Map of the Yost Quadrangle, Box Elder County, Utah and Cassia County, Idaho," I-672, US Geological Survey, 1:31, 680.

Crittenden, M. D., and Sorensen, M L., 1980, "The Facer Formation, a New Early Proterozoic Unit in Northern Utah," Bulletin 1482-F, US Geological Survey, pp. F1–F28.

Dick, H. J. B., 1976, "The Origin and Emplacement of the Josephine Peridotite of Southwestern Oregon," Ph.D. Thesis, Yale University, New Haven, CT, 409 pp.

Dickey, J. S., Jr., 1975, "A Hypothesis of Origin for Podiform Chromite Deposits," *Geochimica et Cosmochimica Acta*, Vol. 39, pp. 1061–1074.

Dickey, J. S., Jr., and Yoder, H. S., Jr., 1972, "Partitioning of Chromium and Aluminum between Clinopyroxene and Spinel," Yearbook 71, Carnegie Institute Washington, pp. 384–392.

Evans, B. W., and Frost, B. R., 1975, "Chrome Spinel in Progressive Metamorphism—a Preliminary Analysis," *Geochimica et Cosmochimica Acta*, Vol. 39, pp. 959–972.

Fleischer, M., 1980, "Glossary of Mineral Species," *Bowie Mineralogical Record*, 145 pp.

Golightly, J. P., 1979, "Nickeliferous Laterites: A General Description," D. J. I. Evans, R. S. Shoemaker, and H. Vetman, eds., *International Laterite Symposium*, AIME, New York, pp. 3–23.

Griggs, A. B., 1945, "Chromite-Bearing Sands of the Southern Part of the Coast of Oregon," Bulletin 945-E, US Geological Survey, pp. 113–150.

Himmelberg, G. R., and Loney, R. A., 1980, "Petrology of Ultramafic and Gabbroic Rocks of the Canyon Mountain Ophiolite, Oregon," *American Journal of Science*, Vol. 280-A, pp. 232–268.

Irvine, T. N., 1967, "Chromian Spinel as a Petrogenetic Indicator: Part 2, Petrologicl Applications," *Canadian Journal of Earth Sciences*, Vol. 4, pp. 71–103.

Irvine, T. N., and Baragar, W. R. A., 1972, "Muskox Intrusion and Copper-Mine River Lavas, Northwest Territories, Canada," Excursion A29, 24th International Geological Congress, Montreal, Que., Canada, 70 pp.

Jackson, E. D., 1961, "Primary Textures and Mineral Associations in the Ultramafic Zone of the Stillwater Complex, Montana," Professional Paper 358, US Geological Survey, 106 pp.

Jackson, E. D., 1968, "The Chromite Deposits of the Stillwater Complex, Montana," *Ore Deposits of the United*

States, 1933–1967, J. D. Ridge, ed., AIME, New York, Vol. 2, pp. 1495–1510.

Jackson, E. D., 1969, "Chemical Variation in Coexisting Chromite and Olivine in Chromite Zones of the Stillwater Complex," *Economic Geology*, Monograph 4, pp. 41–71.

Lawrence, M. J., 1974, "Supergene Nickel Deposits—Origins and Features," *Proceedings*, The Austrailasian Institute of Mining & Metallurgy Conference, Southern and Central Queensland, pp. 523–542.

Leblanc, M., et al., 1980, "Essai sur la Genese des Corpo Podiforms de Chromitite dans les Perioditite Ophiolitiques: Etude des Chromites de Novelle-Caledonie et Comparison avec Alles de Mediterranei Oriental," *Ophiolites*, A. Pannayioutou, ed., Geological Survey Department, Nicosia, Cyprus, pp. 691–701.

Lipin, B. R., in press, "Chromite in Dunites from the Blue Ridge Province, North Carolina," *American Journal of Science*.

Lipin, B. R., and Thayer, T. P., 1980, "The Outlook for Small Chromite Mines," R. F., Meyer, and J. S. Carmen, eds., *The Future of Small Scale Mining*, UNITAR, New York, pp. 191–196.

Moores, E. M., 1969, "Petrology and Structure of the Vourinous Ophiolite Complex, Northern Greece," Special Paper 118, Geological Society of America, 74 pp.

Nicholson, S. W., and Lipin, B. R., 1982, "Low-Grade Chromite Deposits in Alpine-Type Peridotites: An example at the Pilliken Area, California," Abstracts with Programs 1982, Geological Society of America, Vol. 14, pp. 576.

Peters, H. D., 1975, "Nickel," *Verlag Gluckauf GMBH*, Essen, 111 pp.

Schreyer, W., Werding, G., and Abraham, K., 1981, "Corundum-Fuchsite Rocks in Greenstone Belts of Southern Africa: Petrology, Geochemistry, and Possible Origin," *Journal of Petrology*, Vol. 22, pp. 191–231.

Swanson, D. A., and Wright, T. L., 1981, "Guide to Geologic Field Trips between Lewiston, Idaho, and Kimberly, Oregon, Emphasizing the Columbia River Basalt Group," *Guides to Some Volcanic Terranes in Washington, Idaho, Oregon, and Northern California*, Circular 838, US Geological Survey, 14 pp.

Thayer, T. P., and Lipin, B. R., 1978, "A Geological Analysis of World Chromite Production to the Year 2000 A.D.," *Proceedings*, Council of Economics, AIME, New York, pp. 143–152.

Troley, G., et al., 1979, "Nickel Deposits in New Caledonia—Some Factors Influencing Their Formation," *International Laterite Symposium*, D. J. I. Evans, R. S. Shoemaker, and H. Veltman, eds., AIME, New York, pp. 85–120.

6

Uranium in Phosphate

Donald L. Everhart

Consulting Geologist, Grand Junction, CO

ABSTRACT

Beginning in 1906, data has been compiled on the distribution of uranium in phosphate deposits. Particular emphasis has been placed on uranium in phosphate by a number of geologists and geochemists in the US Geological Survey in research sponsored by the US Atomic Energy Commission (now Nuclear Regulatory Commission) and the US Department of Energy (DOE) during the 30-year period from 1951 to 1981. A comprehensive study of uraniferous phosphate resources the United States and the Free World was completed, under the auspices of DOE, in 1979.

Apatite is the main carrier of uranium in phosphate. Uranium generally replaces calcium in the mineral lattice. In marine phosphorite throughout the world, the uranium content is characteristically in the range of 0.005 to 0.02% by weight. Phosphorite takes in uranium by structural substitution in apatite or other mineral phases, by absorption, or by the formation of independent mineral phases. Apatite derived from the interaction of ground water, guano, and limestone is typically relatively low in uranium content—in the 0.001 to 0.005% range. In primary accessory apatite of igneous rocks, as well as magmatic and metasomatic apatite segregations of alkalic rocks and carbonatite, uranium content is generally between 0.001 and 0.01%. In all these phosphate deposit types, secondary enrichment of uranium through leaching and redeposition has produced higher grade uranium deposits in certain areas.

Bedded marine phosphorites comprise the vast majority of the uraniferous phosphate resources of the world. Most of these phosphorites have accumulated on low- to middle-latitude continental shelves during specific times in geologic history, namely the Cambrian, Mississippian, Permian, Cretaceous, Eocene, and Miocene periods. A smaller part of the world's phosphate resources are igneous apatite concentrations in carbonatite and other alkalic complexes. These are also a potential source of uranium in Brazil and Canada.

The total uranium contained in more than 290,000 million tons of the Free World's total estimated recoverable phosphate product is estimated to be nearly 30 million tons. This is "mineral inventory," not modified in any way by economic considerations.

In the relatively near-term projections, uranium from phosphate will most likely be produced only as a byproduct in the manufacture of phosphoric acid. Assuming favorable mineral economic conditions, a total of about 620,000 tons of uranium is projected to be recoverable between now and the year 2025, from known sources of phosphoric acid throughout the world.

As a current "unconventional mineral resource," uranium in phosphate can conceivably evolve into a "conventional" resource shortly after the year 2000, given an increase in the price of uranium of five- to tenfold, and contin-

uation in the phosphate industry of the historic 4–5% annual increase in demand, and attendant increase in price.

INTRODUCTION

From a review of the literature, it appears that R. J. Strutt (1906) was the first to publish data and certain conclusions on the distribution of radium (and therefore uranium) in phosphatic rocks and fossils. These were reported in the *Proceedings* of the Royal Society of London in 1906. Significant research in the United States of America began in earnest with the massive uranium investigations sponsored by the US Atomic Energy Commission (AEC; now Nuclear Regulatory Commission) in the late 1940s and throughout the 1950s. Scores of publications on the subject have continued to appear throughout the 1960s, 1970s, and into the 1980s. Several prominent leaders in this area of research have emerged, particularly in the US Geological Survey (USGS). Data have been gathered by V. E. McKelvey (1956) and his associates in their work on the Phosphoria formation of Idaho, Utah, Montana, and Wyoming. The uranium in the marine phosphorites of the southeastern United States has been documented by J. B. Cathcart (1956) and others. Since the early 1950s Z. S. Altschuler (1980) and his associates in the USGS have specialized in investigations into the geochemistry of uranium in apatite and phosphorite. As recently as April 1980, Altschuler published a summary on the bearing of geochemistry on the recovery of uranium in phosphorites.

F. T. Manheim (personal communication, 1982) has pointed out some extensive data and work by Russian scientists, in which uranium in phosphate is discussed in certain publications that discuss phosphate deposits in general. These publications include those by Baturin (1978), Gimmelfarb (1965), Bushinskii (1966 and 1969), Arsenyev (1980), Gilinskaya (1979), Frolov and Rudy (1979), Kazarrinov and Krasilnikov (1980), and by Zverev and Mikhailov (1980).

As part of the recent intensive National Uranium Resource Evaluation (NURE) program (launched by the AEC in 1974 and concluded by the Department of Energy in September 1981), an in-depth review of uraniferous phosphate resources in the United States and the Free World was commissioned. This 1400-page report, in three volumes, was completed, under contract, by Earth Sciences Inc., in June 1979. R. H. DeVoto and D. N. Stevens (1979) were the editors, and significant sections on uraniferous phosphate resources were contributed by consultants R. C. Fountain and Associates, and G. D. Emigh.

The author of this paper has had the privilege of being associated, since the late 1940s, in uranium and phosphate exploration, resource assessment, and research programs with the aforementioned researchers, and the following summary is taken largely from the published works of these colleagues, and others.

GEOCHEMICAL AND MINERALOGIC OCCURRENCE OF URANIMUM IN PHOSPHATE DEPOSITS

The mineral apatite is the main carrier of uranium in phosphate. Although it contains only traces (0.00x to 0.0y%) of uranium, the primary mineral apatite is a minor constituent in most rocks and the major concentration of many very large deposits. It is instrumental in both the dispersal and the concentration of uranium in the earth's crust. The widespread occurrence of apatite as an accessory mineral in igneous rocks makes it a disperser of uranium in primary differentiation. It is prominent in both felsic and mafic igneous rocks. Apatite is also common in shales and limestones in the form of phosphatic nodules, concretions, and fossil replacements. It thus plays a major role in the geochemical cycle of uranium. Apatite is also one of the few minerals causing strong localization of uranium because of huge accumulations of marine sedimentary phosphate. Such phosphate is a carbonate-fluorapatite mixture (Altschuler, 1980).

Another concentration factor is the association of uranium with phosphate in seas contributing to uraniferous black shale deposition (Adler, 1981). The increased phosphate content of upwelling sea systems provides the nutrition for rapid and extensive plankton development and the resultant enrichment of uranium in plankton and the degraded organic matter derived therefrom. The higher the rate of decomposed organic matter to remnant organic matter the higher the concentration of uranium in the carbonaceous residuum.

In considering the nature of uranium occurrence in apatite and phosphorite, a key feature is the wide variation in apatite composition (Table 1). Basically, apatite is a hexagonal network structure composed of PO_4 tetrahedra, a simple anion like flourine, and a divalent cation, usually Ca. Apatite is a host to many substitutions by cations, anions, and anionic radicals that approximate the normal constituents in size and charge. As a result, many apatites depart significantly from the composition of simple igneous fluorapatite, depending on their environment and genesis. The uranium present in the majority of apatites replaces calcium in the lattice, as noted by phosphate rock types in the following discussion. In sedimentary apatite, carbonate is omnipresent and influences the crystal chemistry (Manheim and Gulbransen, 1979).

OCCURRENCE OF URANIUM IN PHOSPHATE BY TYPES OF DEPOSITS

It has been established, from thousands of chemical analyses throughout the world, that marine phosphorites characteristically contain uranium in the range of 0.005 to 0.02% by weight, that most of this uranium is in the apatite, and that many individual deposits or isolated concretions may be enriched to as much as 0.*X*%. In general, uranium and P_2O_5 contents show a close correlation, but this relationship is commonly altered by the leaching, weathering, or supergene enrichment of uranium.

Representative analyses of uranium content in various kinds of apatite and phosphorite are shown in Table 2. Some generalities can be noted:

1) In marine phosphorite, uranium content is typically in the range of 0.005 to 0.02% except for reported values up to 0.05%, recently cited from the People's Republic of China. These marine phosphorites derived their uranium originally, according to conventional theory, from seawater, which has only 2–3 ppb uranium and 50–70 ppb of P_2O_5. The consensus of geochemists is that uranyl in seawater is reduced to U^{+4} before emplacement in marine apatite, mainly as U^{+4} substituting for calcium. Stanley R. Riggs (1974) has further suggested that additional uranium may have been introduced by suboceanic vulcanism, which may have played a role in the genesis of at least some marine phosphorite deposits throughout the world.

2) Apatite derived from the interaction of ground water, guano, and limestone is typically relatively low in uranium content—in the 0.001 to 0.005% range. Guano-derived phosphorite that has been subjected to sea spray and/or marine flooding tends to become slightly enriched in uranium.

3) The apatite of interior lacustrine basins, which are typically rich in igneous detritus and/or volcanic ash, may be enriched in uranium to as much as several tenths of a percent (0.*x*%) occurring in very thin phosphatic layers and lenses.

4) In primary accessory apatite of igneous rocks, as well as magmatic and metasomatic apatite segregations of alkalic rocks and carbonatites, uranium content is generally between 0.001 and 0.01%—unless there has been enrichment during weathering—as at Araxa, Brazil.

Researchers generally have concluded that phosphorite takes in uranium in three ways, that is, by: (1) structural substitution in apatite or other mineral phases, (2) absorption, or (3) the formation of independent mineral phases.

Uranium mineralogic research, as the basis for beneficiation studies, is required on individual phosphorite deposits throughout the world because where uranium occurs as an absorbed or separate mineral phase, it is amenable to selective extraction, and is potentially recoverable from the phosphate mining products. Where uranium occurs principally as a structural substitute, however, apatite must be chemically broken down and practical recovery can be realized only as a byproduct of phosphoric acid production.

Obviously, a fundamental necessity in any consideration of uranium in phosphate as a mineral resource, is to discern and define the areas, zones, or varieties of phosphorite that are characteristically more enriched in uranium than the average phosphate in any major field. To the author's knowledge, this has not yet been systematically initiated anywhere in the world. However, the work of the US Geological Survey

TABLE 1. Apatite Mineralogy

APATITE: A group of minerals having the general formula

$$A_5(XO_4)_3\ (F, OH, Cl)$$

where

$$A = \underline{Ca}, U, Sr, Ba, Pb, Na, Ce, \text{ or } Y$$

and

$$X = \underline{P}, As, V, \text{ or } Si$$

Fluorapatite: $Ca_5(PO_4)_3F$
Chlorapatite: $Ca_5(PO_4)_3Cl$

From Michael Fleischer, *Glossary of Mineral Species* (1975).

TABLE 2. Uranium in Representative Phosphorite and Apatite Materials*

Deposits	Wt % P_2O_5	Wt % U	Source
Marine Phosphorite Concentrates (Regional Averages)			
• Bone Valley formation, Florida/central district	30–35	0.014	Cathcart (1978)
• Phosphoria formation, northwestern USA	30–35	0.009	Gulbransen (1966)
• Pungo formation, North Carolina	30–35	0.007	Cathcart (1978)
• Oulad-Abdoun basin, Morocco	34	0.014	Altschuler (1952)
• Georgina basin, Australia	30–35	0.008	Cook (1972)
• Guizhou Province, China	25–28	0.04	Everhart⁺ (1981)
Seafloor Nodules			
• Off California	22–30	0.004–0.013	Altschuler (1980)
• Off Southwest Africa	20–33	0.003–0.009	Baturin et al. (1974)
• Off Southwest Africa (Pelletal Phosphate)	>35	0.012	Bremmer (1978)
Island Phosphorites (Guano-Derived)			
• Curacao Island	>50	0.002	Altschuler (1980)
• Ocean Island	>40	0.001–0.002	Altschuler (1980)
• Nauru Island	>40	0.005–0.007	Altschuler (1980)
Lacustrine Phosphorites			
• Green River basin, Utah and Wyoming	<19	0.001–0.29	Love (1964)
• Siberia, USSR	<25	0.01–0.4	Gavshin et al. (1974)
Aluminum Phosphate Deposits			
• Florida/Central District	<30	0.004–0.04	Cathcart (1956)
Igneous and Carbonatite Apatites			
• Kola Peninsula, USSR	>40	0.001–0.002	Altschuler (1980)
• Palabora, South Africa	>40	0.003–0.004	Altschuler (1980)
• Araxa, Brazil	>40	0.02	Altschuler (1980)
• Diorite, Idaho	>40	0.006	Altschuler (1980)
• Granodiorite, Colorado	>40	0.007	Altschuler (1980)
• Granite, Colorado	>40	0.002	Altschuler (1980)

*From Z. S. Altschuler, *Proceedings, 2nd International Congress on Phosphorus Compounds* (April 1980).
⁺Personal communication.

in two major phosphate districts of the United States has indicated some general patterns that should be further investigated.

By mapping, sampling, and chemical analysis, it is has been determined that the apatite in the Phosphoria formation of southeastern Idaho and western Wyoming that is richest in organic matter has more uranium than the average (Gulbransen, 1966; McKelvey and Carswell, 1956; Sheldon, 1959). This is *not* necessarily due to any appreciable uranium in the organic matter itself, but to the high organic productivity creating reducing conditions in the bottom waters and muds, thus increasing the reduction of uranyl for uptake by the apatite.

The distribution of uranium among the petrographic varieties of phosphate in the Bone Valley formation of Florida is another example of systematic primary uranium enrichment. From the analyses of hundreds of beneficiation composites, it has been shown that the "pebble fraction" is consistently and appreciably richer in uranium than the "concentrate" fraction although the latter is consistently higher in P_2O_5 (Cathcart, 1956). This pattern is geologically mappable because the pebble deposits are characteristic of the near-shore shallow water facies, where the pebbles have incorporated additional uranium during reworking, presumably by ion exchange with seawater. Fortunately, the industrial production of phosphoric acid in Florida has been based largely on the pebble fraction.

Finally, appreciable and significant enrichment of uranium occurs where weathering is intense and long enough to totally dissolve apatite in uraniferous phosphate rocks. Such alteration effect on contained uranium has been studied in detail in certain argillaceous terranes such as Florida, Senegal, Nigeria, and Siberia; in calcareous terranes like South Carolina, Tennessee, and North Florida; and in the carbonatite terrane at Araxa, Brazil. Zones of significant uranium enrichment have been modeled, identified, and mapped in selected areas, in all of these terranes (Altschuler, 1973; White and Warin, 1964; Zanin 1967, 1969; Filcoteaux et al., 1977; and Filcoteau, 1980).

WORLD DISTRIBUTION OF URANIFEROUS PHOSPHATE DEPOSITS

Phosphate mining, development, and exploration have been active in many parts of the world since the beginning of the 20th century. As a result of these 80 years of work, a pattern of phosphate distribution in the earth's crust is emerging. Since the early 1950s efforts have begun to determine the distribution of uranium within the phosphate resources of the world. Some generalities can be presented.

Bedded marine phosphorites comprise the vast majority of the uraniferous phosphate resources of the world. Fig. 1 shows the location and geologic age of the major marine phosphorite resources exclusive of those in the USSR and the Far East. In recent years, the International Geological Correlation Program-Project 156 has made great strides in developing data and research on Proterozoic-Cambrian phosphorites in Asia, including the USSR. Geoscientists from the USSR, Vietnam, India, Malaya, Indonesia, Korea, and Japan are all active in this effort. The Bureau of Uranium Geology in the People's Republic of China has recently furnished some information on uraniferous phosphorite of Lower Cambrian and Carbonaceous ages in Guizhou province. It is expected that the uranium resources in Russia and Far Eastern phosphorite deposits will become much more widely described through publications in the near future.

Most of the marine phosphorite deposits of the world have accumulated on low- to middle-latitude continental shelves during a few specific times in geologic history. Older presumably low-latitude continental shelf phosphorites were deposited during Cambrian time in Australia and Brazil, during Mississippian time in Alaska and Utah, and during Permian time to form the world's largest conterminous phosphate resource—the Phosphoria formation of Idaho, Wyoming, Utah, and Montana. There are many important phosphorites of Upper Cretaceous and Eocene age, which were deposited on the south side of the Tethys Sea (predecessor of the Mediterranean) and on both the east and west borders of the Atlantic Ocean. The prolific phosphorite sedimentation during Miocene time is represented by the low- to middle-latitude continental shelf deposits on both coasts of North America—particularly the important phosphate deposits of the southeastern United States, those of Baja, California, Mexico, and those of the west coast of southern Africa. The phosphorites of Recent age in Baja, California and offshore South Africa, and the deposits of Pliocene age in the southeastern United States are generally regarded as reworked phosphatic formation of Miocene age.

A smaller part of the world's phosphate resources are igneous apatite concentrations in carbonatites and other alkalic complexes. Such intrusives have developed along rift zone systems, for example in eastern Africa. These are locally important sources of phosphate in the USSR, South Africa, Finland, and Brazil, and could potentially be a source of uranium in Brazil and Canada.

URANIUM RESOURCES IN PHOSPHATE

The three-volume report published in June 1979, by Earth Sciences Inc. under contract to the US Department of Energy, entitled *Uraniferous Resources and Technology and Economics of Uranium Recovery from Phosphate Resources of the United States and the Free World*, summarizes data on the uraniferous phosphate resources of the United States in detail and of the rest of the non-Communist world more generally (Table 3). The author has added personally estimated figures from China, from information gathered on a field review in that country in 1981. The Earth Sciences report also evaluates the technical, economic, and environmental feasibility of production of uranium as a byproduct, coproduct, or single product from

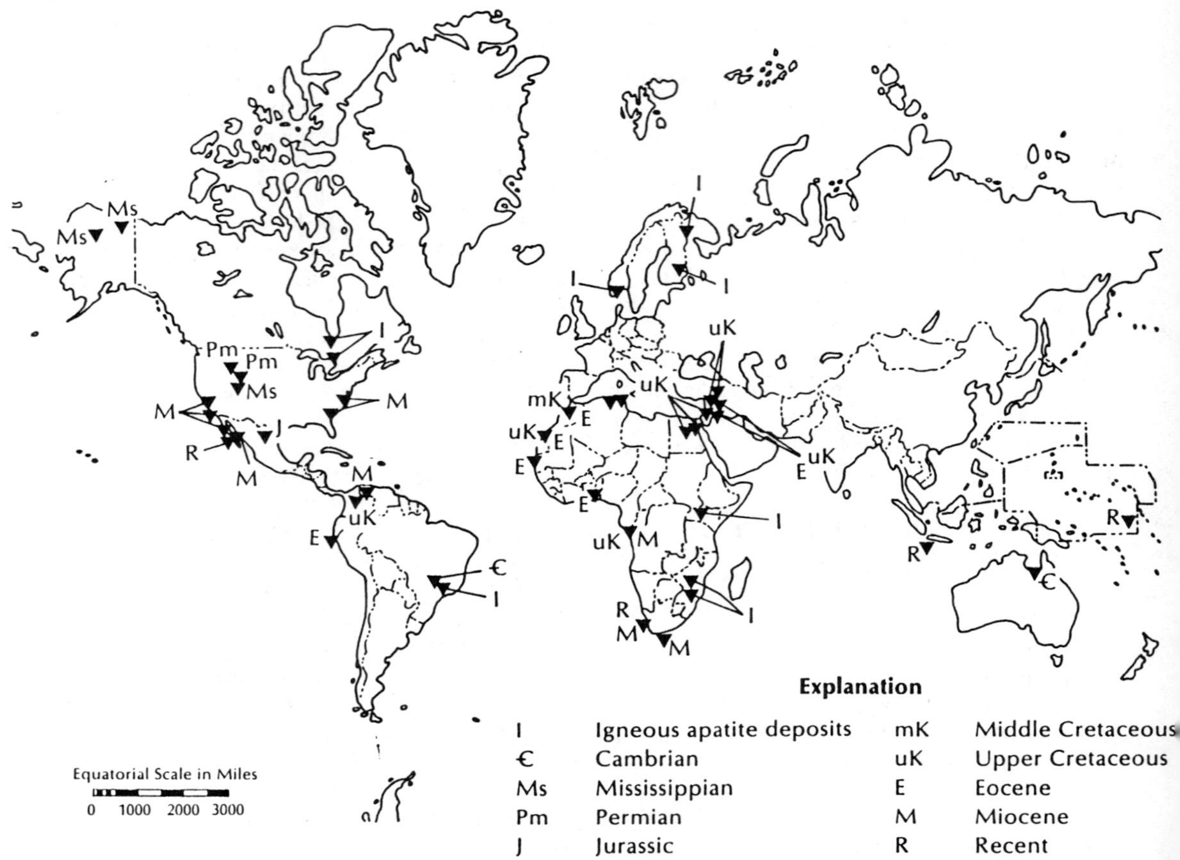

FIG. 1. Location and geologic type of the major phosphorite deposits in the Free World.

the uraniferous phosphate resources. The resource data include all information then available on the location, tonnage, and grade of the resources with respect to phosphate, uranium, and other potential byproducts. The work reflected in this report has confirmed that the uraniferous phosphate resources and the estimated recoverable phosphate product of the world are immense, enough to supply the required phosphate production of the world, at current levels, for several millenia. The total uranium contained in more than 290,000 million tons of the total estimated *recoverable* phosphate product is estimated to be nearly 30 million tons. It must be stressed that these are so-called *mineral inventory* figures and are in no way modified by economics or the fact that most of these resources cannot, and will not, be commercially processed under current circumstances.

Because most of the world's phosphate resources are marine phosphorites, the uranium contents of their phosphate products are relatively uniform, ranging from 60–130 ppm, with few exceptions. The phosphate-producing fields and the principal known resource areas of the world (central and southern Florida, Morocco, Idaho, Utah, Western Sahara, Mexico, Jordan, east Florida, and Peru) have a relatively uniform, high uranium concentration, from 100 to 120 ppm. Other major phosphate producing and resource areas (North Carolina, Wyoming, Montana, other Florida districts, Australia, Tunisia, and Algeria) have a phosphate product with a lower, but still significant, concentration of 60–90 ppm uranium. This is representative of the feed now being received, and projected to continue, at the world's phosphoric acid plants.

TABLE 3. Uraniferous Phosphate Resources of the Free World*

Area	Phosphate Resources, millions of tons	Potentially Minable Phosphate Product ~30% P_2O_5, millions of tons	Estimated Recoverable Phosphate Product, ~30% P_2O_5 millions of tons	Average Uranium Concentration of Phosphate Product, ppm U	Estimated Recoverable Uranium, thousands of tons
United States					
• Southeastern	170,526	33,187	19,890	110	1,565
• Western (Phosphoria Formation)	825,262	513,908	202,324	104	20,783
Other	9,412	2,255	1,128		111
Total	1,005,200	549,350	223,342	98	22,459
Rest of World					
• Australia	4,000	1,600	1,200	68	82
• Jordan	16,000	4,000	3,000	120	360
• Mexico	51,000	5,534	4,150	100	415
• Morocco	88,000	60,000	45,000	110	4,950
• Peru	10,000	3,000	2,250	100	225
• South Africa	105,300	50,770	3,600	12	43
• Western Sahara	11,000	7,400	5,550	100	555
• People's Republic of China†	2,930	1,170	600	400	60
Other	11,770	6,176	4,200	30–160	410
Total	300,000	139,650	69,550		7,100
Total World (Available Information)	1,305,200	689,000	292,892		

*From *Uraniferous Phosphate Resources: United States and Free World,* Vol. 1, US Department of Energy Report GJBX–110(79), Earth Resources Inc., June 1979.

†China data from D. L. Everhart, 1982.

MINERAL ECONOMICS OF URANIUM IN PHOSPHATE

The foregoing discussion summarizes the data available, and indicates the methods and rationale that have been used to estimate quantitatively the uranium resources contained in the Free World's phosphate deposits and the mined and processed products that are likely to be derived from them. Uranium in phosphate, as a mineral resource, however, can be dealt with only in the light of real-world mineral economics. The Earth Sciences Inc. study of 1979 included the technical and economic feasibility of the production of U_3O_8 as: (1) a byproduct of phosphate production, (2) a coproduct, or (3) a single product.

By far the most feasible route is the production of uranium as a byproduct, and the most attractive district for developing the process has been in central Florida. The pebble fraction of the phosphate rock there contains somewhat more uranium than the phosphate rock in the western United States, northern Florida, or Carolina. Hence most research and development effort to date has been directed toward uranium recovery from deposits in the central Florida area. No process has been discovered or developed that selectively extracts uranium directly from phosphate rock which, for the purposes of this discussion, is characterized simply as tricalcium phosphate (Ca_3 (PO_4)) [McGinley, 1980]. Alkaline leach methods have proved to be ineffective, and in acid leaching, the uranium extraction is proportional to the dissolution of the calcium phosphate. The only existing commercial processes for uranium recovery are therefore those based on complete digestion of the rock by acid solution. The standard current process is the digestion of rock by sulfuric acid to manufacture "wet process" phosphoric acid, which is an intermediate stage in the production of high-analysis chemical fertilizers and other phosphate chemicals. At the present time, about 80% of phosphate rock consumed in the United States is converted to wet-process phosphoric acid. The remainder is used in the production of other industrial and agricultural products, from which the uranium is not presently recoverable.

There are other limitations on the quantity of byproduct that can be recovered from phosphoric acid. Uranium recovery is not yet economically attractive for phosphoric acid plants producing less than 100,000 tons of P_2O_5 per year, or for plants digesting rock relatively low in uranium content. Nor has technology been fully developed for recovery uranium from concentrated acid (greater than 40% P_2O_5). With these constraints, there is consensus (Earth Sciences Inc. and F. McGinley) that by about 1985, no more than approximately 3000–3500 tons of uranium per year might be economically recoverable from United States phosphoric acid plants. Another 2100 tons in other non-Communist countries and possibly 400 tons in China might bring the world's known total to 6000 tons per year of uranium as a byproduct from phosphoric acid production. In 1980 about 800 tons of such uranium were produced from six plants in the United States and one in Canada. The production of this byproduct in 1981 was 1285 tons. The implied nearly threefold increase in the next three years certainly assumes relatively high prices and strong demand for both uranium and phosphate, neither of which existed in November 1982. Assuming favorable mineral economic conditions, a total of some 220,000 tons of uranium are projected as being recoverable as a byproduct from phosphoric acid plants in the United States between now and the year 2025 (Earth Sciences Inc., 1979); a total of approximately 620,000 tons of uranium is projected to be recoverable from known sources of phosphoric acid throughout the world.

The Earth Sciences Inc. analysis of mining the currently known oregrade phosphate resources (containing about 18% P_2O_5) in the United States by surface, underground, and heretofore untried deep underground methods, and processing the ore solely for its uranium content, indicates costs of $175 to $250 per lb uranium (in 1979 dollars). There is no present indication that uranium demand in the foreseeable future will command the prices necessary to accommodate such costs.

CONCLUSIONS

Uranium in phosphate as a byproduct of phosphoric acid is currently a "conventional" mineral resource.

The production of such uranium is beneficial because the process "cleans up" the uranium in phosphate being used in fertilizers and constitutes a source that requires no new mines, therefore eliminating certain environmental concerns and procedures.

The current state-of-the-art byproduct process has limitations and requires continued research and development to achieve substantial technical and economic improvement.

The world data base on the distribution of uranium in phosphate is in its infancy and requires major growth and development.

As a current "unconventional mineral resource," uranium in phosphate can conceivably evolve into a "conventional" resource in the early part of the next millenium, given an increase in the price of uranium of five- to tenfold, and continuation of the historic 4 or 5% annual increase in demand and attendant increase in price for phosphate.

In the current economic situation in both the uranium and phosphate industries such increases cannot be projected with any degree of confidence. The resources are demonstrably in the earth's crust and unforeseen economic developments could transform them into future ore reserves.

REFERENCES

Adler, H. H., 1981, Personal Communication.

Altschuler, Z. S., 1952, "The Petrography and Uranium Contents of Phosphates from Morocco," Trace Elements Memorandum Report 238, US Geological Survey, 37 pp.

Altschuler, Z. S., 1973, "The Weathering of Phosphate Deposits—Geochemical and Environmental Aspects," *Environmental Phosphorus Handbook, John Wiley & Sons, New York, pp. 33–96.*

Altschuler, Z. S., 1980, "The Bearing of Geochemistry on the Recovery of Uranium and Rare Earths in Phosphorites," *Proceedings*, 2nd International Congress on Phosphorus Compounds, Boston, Mass., pp. 605–626.

Arsenyev, A. A., ed., 1980, "Lithology and Genesis of Phosphate-bearing Sediments." *Nauka*, Moscow, reviewed in International Geological Correlation Programme (IGCP) Project 156 Phosphorite, March 1982.

Baturin, G. M., 1978, "Fosfority no due okeane," *Nauka*, Moscow.

Baturin, G. M., Merknlova, I. K., and Chalov, P. O., "Absolute Dating of Oceanic Phosphorites by Disequilibrium Uranium," *Geochemistry International*, pp. 568–574.

Bremner, J. M., 1978, ''Trace Element Concentrations in Authigenic Minerals from S. W. Africa,'' *Abstract, Proceedings*, 10th International Congress on Sediments, Jerusalem, Israel, p. 87.

Bushinskii, 1966, ''Drevnie fosfority Azii i ikh genezis,'' *Nauka*, Moscow.

Bushinskii, 1969, ''Formatsia fosforiya,'' *Nauka*, Moscow.

Cathcart, J. B., 1956, ''Distribution and Occurrence of Uranium in the Calcium Phosphate Zone of the Land-pebble District of Florida,'' Professional Paper 300, US Geological Survey, pp. 489–494.

Cathcart, J. B., 1978, ''Uranium in Phosphate Rock,'' Professional Paper 988-A, US Geological Survey, 6 pp.

Cook, P. J., 1972, ''Petrology and Geochemistry of the Phosphate Deposits of Northwest Queensland, Australia,'' *Economic Geology*, No. 67, pp. 1198–1213.

DeVoto, R. H., and Stevens, D. M. eds., 1979, ''Uraniferous Phosphate Resources and Technology and Economics of Uranium Recovery from Phosphate Resources, United States and the Free World,'' US Department of Energy, GJBX—110 (79), Earth Sciences, Inc., Vol. 1, 744 pp.; Vol. 1A, 28 maps; Vol. 2, 672 pp.

Flicoteaux, R., Nahon, D., and Paquet, H., 1977, ''Genese des Phosphates Alumineux a Parter des Sediments Argilo—Phosphates du Tertiaire de Lam Larn (Senegal),'' *Scientific Geological Bulletin*, Vol. 30, pp. 153–174.

Flicoteaux, R., 1980, ''Genese des Phosphates Alumineux du Senegal Occidental Etapes et Guides de l'Alteration,'' D. of Sciences Thesis, 229 pp.

Frolov, V. A., and Rudyls, Y. M., 1979, ''Radioactive Elements in Phosphate—and Apatite-Bearing Rocks from Siberia and the Far East Region of the USSR,'' *Investigation of Calcium Phosphates by Physical Methods*, U. G. Gilinskays, ed., Ize. Nauka, Novosibirsk.

Gavshin, V. M., Bobrov, V. A., and Zorkina, L. S., 1974, ''Quantitative Relations Between Uranium and Phosphorous and Phosphatic Sedimentary Rocks,'' *Lit. and Min. Deposits*, No. 6, pp. 118–126 (English translation pp. 740–746).

Gilinskays, L. G., ed., 1979, *Investigation of Calcium Phosphates by Physical Methods* (in Russian), Ize. Nauka, Novosibirsk, 104 pp.

Gimmelfarb, 1965, ''Zakonomernosti razmesheniya,'' *Nedra*, Moscow.

Gulbransen, R. A., 1966, ''Chemical Composition of Phosphorites in the Phosphoria Formation,'' *Geochemica et Cosmochimica Acta*, Vol. 30, No. 8, pp. 760–778.

Kazarrinov, V. P., and Krasilnikov, N. H., 1980, ''Phosphorite and Apatite of Siberia,'' *Nauka*, Novosibirsk (in Russian), 230 pp.

Love, J. D., 1964, ''Uraniferous Phosphatic Lake Beds of Eocene Age in Intermontaine Basin of Wyoming and Utah,'' Professional Paper 474-E, US Geological Survey, 66 pp.

Manheim, F. T., 1982, Personal Communication.

Manheim, F. T., and Gulbransen, R. A., *Marine Minerals*, Chap. 6, R. G. Burns, ed., *Reviews in Mineralogy*, Vol. 6, Mineralogical Society of America, pp. 151–173.

McGinley, F. E., 1980, ''Potential By-product Uranium Production in the U.S.A.,'' International Atomic Energy Agency Report IAEA-SM-239/15, pp. 293–307.

McKelvey, V. E., and Carswell, L. D., 1956, ''Uranium in the Phosphoria Formation,'' Professional Paper 300, US Geological Survey, pp. 483–488.

Riggs, S. R., 1974, Personal Communication.

Sheldon, R. S., 1959, ''Geochemistry of Uranium in Phosphorites and Black Shales of the Phosphoria Formation,'' Bulletin 1084-D, US Geological Survey, pp. 83–115.

Strutt, R. J. 1906, ''On the Distribution of Radium in the Earth's Crust,'' *Proceedings*, Royal Society of London, Ser. A, Vol. 78, pp. 150–153.

White, W. C., and Warin, O. N., 1964, ''A Survey of Phosphate Deposits in the Southwest Pacific and Australian Waters,'' Bulletin 69, Bureau of Mineral Resources of Australia.

Zanin, Y. M., 1967, ''Classification of Phosphate Deposits Related to Weathering,'' Dokl Akad. Nauk SSSR, 176, pp. 689–692 (English translation).

Zanin, Y. M., 1969, ''Geology of the Phosphate Crusts of Weathering and Associated Phosphate Deposits,'' *Tr. Inst. Geol. Geofiz.*, Akad. Naul SSSR, Sibirsk, Otd., Vyp, 85, 158 pp.

Zverev, A. S., and Mikhailov, A. S., 1980, ''Geology of Phosphorite Deposits and Methods of Prospecting and Prognosis,'' *Nedra*, Moscow (in Russian), 245 pp.

7

Minor Elements in Mississippi Valley-Type Ore Deposits

Richard D. Hagni

Department of Geology and Geophysics,
University of Missouri-Rolla, Rolla, MO

ABSTRACT

Mississippi Valley-type ore deposits contain significant quantities of minor elements, which are present partly in the form of separate mineral phases and partly in solid solution in the major sulfide phases for which these deposits are exploited. These elements, as separate phases, are: copper, mainly in chalcopyrite and enargite; cobalt and nickel, chiefly in thiospinels and bravoite; rare earth elements associated with certain fluorites; and native gold. Many economically important minor elements occur in solid solution in sphalerite, including silver, cadmium, iron, germanium, gallium, indium, thallium, cobalt, nickel, and mercury. Significant amounts of silver and antimony, and perhaps very small amounts of gold, are present in solid solution in galena. Chalcopyrite may contain small amounts of silver, molybdenum, bismuth, cadmium, cobalt, and nickel. Marcasite and pyrite may contain silver, thallium, cobalt, and nickel.

Although some of the minor elements contained in the Mississippi Valley-type deposits are currently recovered at the smelters and refineries, many are lost in tailings, sludges, and slags; consequently, important quantities of these elements are potentially recoverable as nonconventional resources.

INTRODUCTION

Ore deposits belonging to the Mississippi Valley-type contain a wide variety of minor elements in addition to the major constituents (lead, zinc, barium, and fluorine) that are usually exploited. Some of these minor elements are routinely recovered at smelters and refineries from most of the lead and zinc concentrates. Silver, for example, is extracted from lead concentrates, and cadmium from zinc concentrates. Other elements, such as germanium and gallium, are not recovered from some zinc concentrates. As a result, significant reserves of these unrecovered elements are present but lost in recovery methods from some ore districts, and a few are important because of their rare or strategic value. Cobalt is an example of a minor element that is not recovered at present in any Mississippi Valley-type district, but which is present in significant quantities in some districts and has been recovered in the Fredericktown subdistrict of southeast Missouri since the 1860s. It is strategically important, and represents a major nonconventional mineral resource.

In this chapter, elements such as copper, gold, rare earths, cobalt, and nickel, which occur principally as separate phases, are treated first. Discussions of elements that are found in solid solutions in sphalerite, galena, chalcopyrite, marcasite, and pyrite follow. Elements that are of current or potential economic value are emphasized. The appelation "Mississippi Valley-type" is used in its broadest sense to include "Appalachian-type" (Hoagland, 1971), "Kootenay Arc-type" (Muraro, 1966), "Alpine-type" (Maucher and Schneider, 1967), and "Laisvall-type" deposits (Bjorlykke and Sangster, 1981).

MINOR ELEMENTS WHICH OCCUR AS SEPARATE MINERAL PHASES

Copper

Copper in chalcopyrite and rare enargite and luzonite crystals is present in small amounts in most Mississippi Valley-type ore deposits. The amounts are usually too small to recover except in the Southeast Missouri Lead district where copper concentrates, in addition to lead and zinc concentrates, are produced at most of the mills. In 1980 and 1981 these copper concentrates contained about 15,000 tons of copper, and placed Missouri sixth and fifth, respectively, in the United States in copper production (*Mineral Industry Surveys*, 1982). Copper contributed about 60% of the value of the zinc recovered during those years. Copper concentrates were also made, and paid for by a smelter, from the Grayville mines of the Ivey Mining Co. of Mineral Point, WI, in the Wisconsin-Illinois district in the 1960s and early 1970s (Allen V. Heyl, 1982, personal communication).

The principal copper mineral in the Southeast Missouri Lead district is chalcopyrite, which occurs as colloform masses and well-formed crystals. Copper is also found in minor quantities in bornite, chalcocite, digenite, anilite (Robert W. Potter, 1975, personal communication), djurleite, covellite, blaubleibender covellite, enargite, tennantite, fletcherite, and nickelean carrollite. The principal copper minerals recovered in Wisconsin were chalcopyrite, bornite, covellite, and enargite.

Gold

Very small amounts of native gold have been reported from two Mississippi Valley-type ore districts, one in Wisconsin-Illinois (Heyl et al., 1959), and the other in Southeast Missouri (Ohl et al., 1982). Heyl (1982, personal communication) states that in the former, gold occurs in the late chalcopyrite-barite stage of mineralization. In the Southeast Missouri Lead district, 32 oz of gold were reported to have been recovered at one of the zinc smelters treating southeast Missouri ores (Ohl et al., 1982). Trace amounts of gold have been recovered with silver at the lead smelters serving the Southeast Missouri district. It is not known whether this gold occurs as a separate phase or in solid solution.

Rare-Earth Elements

Rare-earth elements occur as separate phases in association with some fluorite in Hicks Dome in the Southern Illinois-Kentucky Fluorspar district. At the surface, rare-earth elements are associated with thorium, beryllium, columbium, and zirconium in breccias (Trace, 1960). At depth, such as in the Hamp drill hole 488 to nearly 914 m (1600 to nearly 3000 ft) below the surface, rare-earth elements are associated with dark purple fluorite in a cryptovolcanic breccia (Brown et al., 1954). Analyses of the fluorite indicate that it contains more than 1% yttrium and thorium; 1000 ppm or more dysprosium, erbium, gadolinium, ytterbium, and terbium; lesser amounts of the rare-earth elements cerium, thulium, europium, neodymium, lutetium, lanthanum, and samarium; and the elements, antimony, columbium, and scandium (Hall and Heyl, 1968). The rare-earth elements occur in separate phases, including: monazite, florencite, and fluorapatite, which are intergrown with fluorite (Trace, 1974). Exploration for rare-earth elements is continuing at Hicks Dome in an effort to outline a commercial resource of these elements.

Cobalt and Nickel in the Southeast Missouri Lead District

Small amounts of cobalt and nickel are present in most Mississippi Valley-type ore deposits, but the most important occurrence is in the Southeast Missouri Lead district. Measured resources in the district are 31.8 Mg (120 million lb) of cobalt and 54.4 Mg (170 million lb) of nickel. These cobalt resources, which are the second largest in the United States, are exceeded only by those in the Blackbird district of Idaho (Peterson et al., 1981).

Cobalt and nickel in Southeast Missouri copper-lead-zinc ores constitute a nonconventional resource that remains largely unrecovered. In the Fredericktown subdistrict, south of the Old Lead Belt, small amounts of cobalt and nickel were recovered periodically from the Madison mine from the 1860s until 1961. Ores from the Madison mine contained more than an order of magnitude higher cobalt and nickel than elsewhere in the district. Anschutz Mining Corp. is currently dewatering the Madison mine in an effort to reopen it as a cobalt mine (Brooke, 1981); however, instability in the price of cobalt has hampered progress. Although research is being continued on beneficiation techniques to recover cobalt and nickel from the ores of the Viburnum Trend in Southeast Missouri (Clifford and Higley, 1978; F. H. Sharp, 1981, personal communication), no commercial recovery has been effected except for very minor amounts in coatings on the walls of the dross furnaces formerly used in the district.

Mineralogy: Siegenite [$(Ni, Co)_3S_4$] is the principal cobalt-nickel mineral in the Southeast Missouri Lead district, although a number of other cobalt-bearing minerals are present in smaller amounts. Siegenite occurs as coatings on, and partial replacements of, chalcopyrite (Figs. 1A and B) that are especially thick in the Fredericktown subdistrict. It occurs less commonly as a replacement of sphalerite (Fig. 1C). Other cobalt-bearing minerals, which are locally present, include fletcherite [$Cu(Ni, Co)_2S_4$] (Craig and Carpenter, 1977) and nickelean carrollite

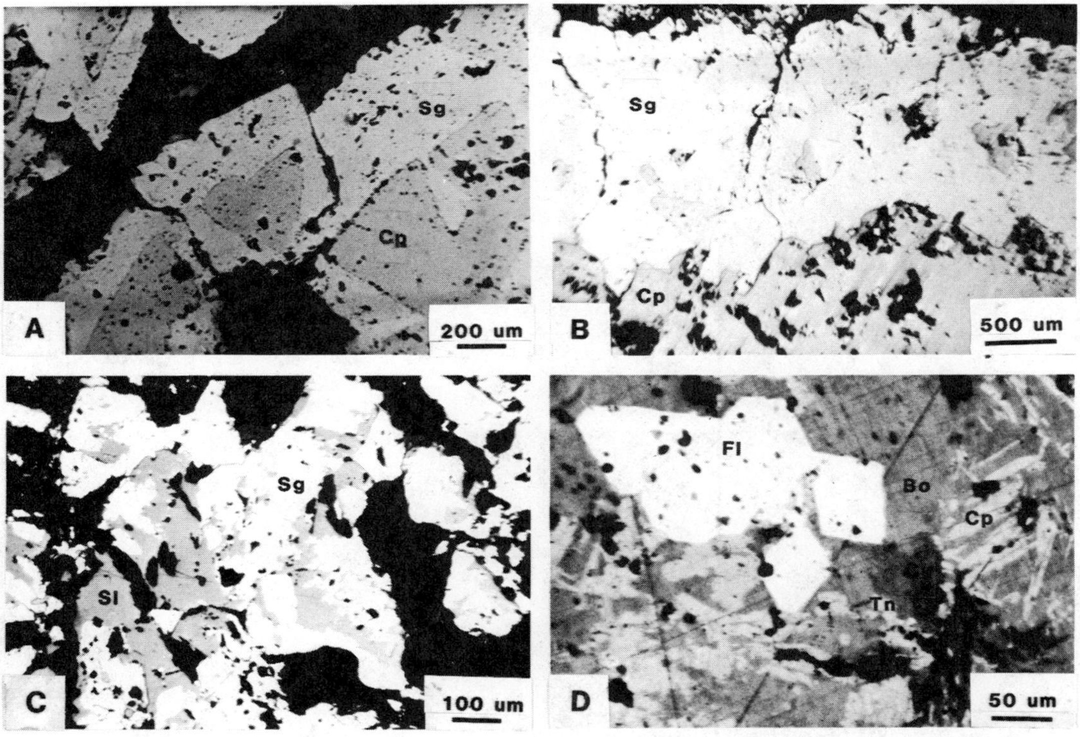

FIG. 1. Micrographs showing the character of the cobalt-nickel minerals, siegenite and fletcherite, in the Southeast Missouri Lead district. (A) Octahedrons of siegenite (Sg) have coated and partially replaced crystals of chalcopyrite (Cp). Buick mine, Viburnum trend. (B) Thick coating of siegenite (Sg) deposited upon colloform chalcopyrite (Cp). Disequilibrium between siegenite and chalcopyrite is indicated by the presence of irregularly shaped unreplaced remnants of chalcopyrite within the siegenite. Madison mine, Fredericktown subdistrict. (C) Sphalerite (Sl) that has been intensely replaced by siegenite (Sg). Abundant irregularly shaped unreplaced remnants of sphalerite occur locally in the siegenite. Magmont mine, Viburnum trend. (D) Octahedrons of fletcherite (Fl) surrounded by bornite (Bo) and chalcopyrite (Cp). The fletcherite crystals are metacrysts as shown by the presence of tiny irregularly shaped unreplaced remnants of bornite, chalcopyrite, and tennantite within the octahedrons. Chalcopyrite exhibits an exsolution intergrowth with bornite. Tennantite (Tn), and subsequently covellite (Cv), have replaced bornite and chalcopyrite. Fletcher mine, Viburnum trend.

[(Co, Ni, Cu)$_3$S$_4$] (Jessey, 1981). Fletcherite is restricted to rare bornite pods, where it occurs as euhedral replacement crystals in bornite and chalcopyrite (Fig. 1D). A few microprobe analyses show that some of the siegenitelike thiospinel associated with the copper ores at the Madison and Milliken mines is nickelean carrollite. Detailed ore microscopic studies and electron probe analyses of the Madison ores are currently in progress (Pignolet and Hagni, 1982).

Bravoite [(Fe, Ni, Co)S$_2$] occurs as tiny zoned (Fig. 2A) and unzoned crystals (Fig. 2B) and as thin layers in colloform masses of fine-grained pyrite and marcasite (Hagni and Trancynger, 1977). Some of the bravoite crystals have been shown by electron probe analyses to contain substantial amounts of cobalt (Moh and Ottemann, 1964; Taylor and Radtke, 1969).

Millerite [NiS], which is widely distributed in trace amounts in the Paleozoic rocks of Missouri, is present in the Southeast Missouri Lead district in the form of clusters of acicular crystals deposited upon octahedral galena and enclosed by the subsequently deposited cubic galena. Millerite was not in equilibrium with some of the later ore fluids. It is commonly partly or completely altered to pseudomorphs of polydymite with coatings of tiny vaesite [NiS$_2$] cubes (Fig. 2C), or rarely is replaced by pseudomorphs of siegenite (Fig. 2D).

Minor amounts of gersdorffite [NiAsS] have been observed microscopically at three mines:

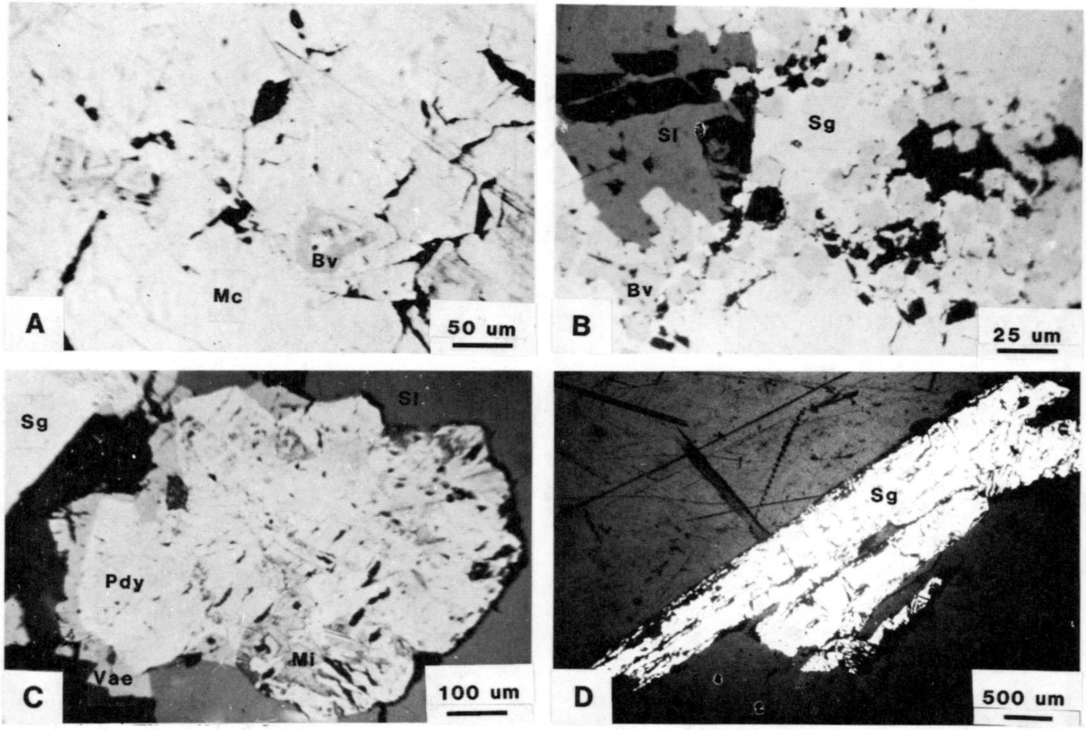

FIG. 2. Micrographs showing the character of the nickel minerals bravoite, polydymite, millerite, and vaesite, in the Southeast Missouri Lead district. (A) Four-zoned bravoite (Bv) crystals included in marcasite (Mc). Milliken mine, Viburnum trend. (B) Unzoned bravoite (Bv) crystals included within siegenite (Sg) and sphalerite (Sl). Buick mine, Viburnum trend. (C) Basal section of polydymite pseudomorph after a millerite needle-shaped crystal. Irregularly shaped replacement remnants of millerite (Mi) remain locally in the polydymite (Pdy). Subsequently deposited vaesite (Vae) forms coatings of cubic crystals deposited upon the polydymite needle. Subsequently deposited sphalerite (Sl) and siegenite (Sg) enclose the earlier formed nickel minerals. Buick mine, Viburnum trend. (D) Prismatic section of siegenite (Sg) pseudomorph after millerite needle-shaped crystals. Buick mine, Viburnum trend.

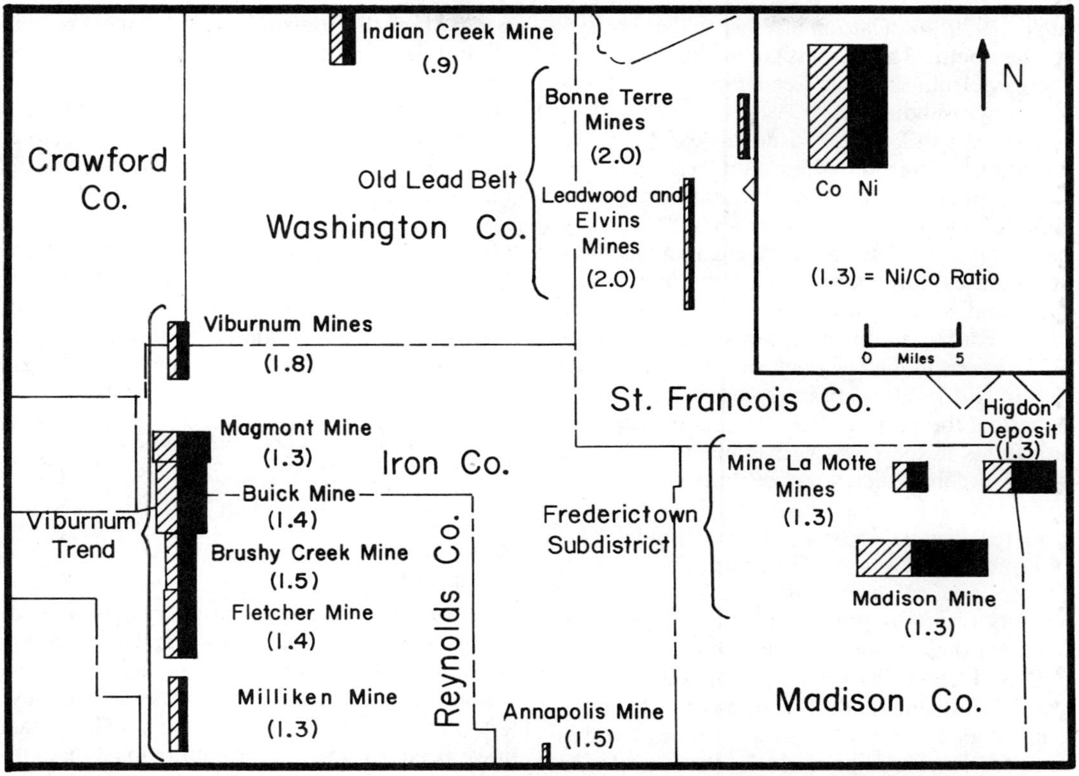

FIG. 3. Relative nickel and cobalt concentrations and Ni/Co ratios in the Southeast Missouri mining district. The widths of the bars represent relative amounts of cobalt and nickel. The heights of the bars indicate the areal extent, in a north-south direction, represented by each mine. The information presented in this figure resulted from a grant to the writer by the US Bureau of Mines and is published in the form of ratios and relative amounts in agreement with the mining companies involved.

Fletcher (Kastler, 1972), Milliken, and Viburnum 28 (Horrall, 1981). Where it is present in bornite and chalcopyrite, it occurs as ragged replacement bodies, which may exhibit thin rims of tennantite, an intermediate product in the replacement process.

Distribution Within the District: Amounts of cobalt and nickel vary from place to place in the Southeast Missouri Lead district. The greatest concentrations are in the Fredericktown subdistrict, particularly at the Madison mine, where cobalt and nickel averaged 0.23 and 0.29% respectively in the ore produced from 1944 to 1961 (Brooke, 1981).

Although the average amounts of cobalt and nickel are more than an order of magnitude smaller in the Viburnum Trend, there are significant differences in the average contents between mines (Fig. 3).* The Magmont and Buick mines in the north-central part of the trend have average concentrations that are two to three times those elsewhere in the trend. Also, a slight increase in values toward the south end of the trend, and high values for samples from some drill holes south of the Milliken mine (Erickson

*The average amounts of cobalt and nickel shown in Fig. 3 are based largely on mine production data and the variations may be partly a result of differing mining practice within the Viburnum Trend. During a specific time period cobalt-nickel-rich areas may have been avoided at one mine whereas similar areas were included in the mining at another mine. This factor may be partly responsible for the low cobalt-nickel values shown for the Bonne Terre, Leadwood, and Elvins mines and may affect the data for other mines to an unknown extent.

et al., 1979), indicate an increase in these metals to the south. This is similar to the southward increase from the Old Lead Belt to the Fredericktown subdistrict.

The ratios of nickel to cobalt exhibit small variations within the Viburnum Trend that are very similar to changes in the total concentrations of the two metals (Fig. 3). The nickel-cobalt ratio is lowest at the Magmont mine (1.3). It increases northward to the Viburnum mines (1.8) and southward to the Brushy Creek mine (1.5). Farther south, it decreases to 1.3.

Variations in total amounts of cobalt and nickel and in Ni/Co ratios are interpreted by the writer to reflect the principal movements of ore-bearing fluids in the district. It is believed that the greatest cobalt-nickel concentrations and the lowest nickel-cobalt ratios are closest to the principal source areas of the ore fluids. This interpretation suggests that one of the principal source areas of ore fluids was located in the vicinity of the Magmont and Buick mines, and another south of the Milliken mine.

Distribution Within Individual Ore Bodies: Significant spatial variations in the cobalt and nickel content of the ores are present within specific deposits. For example, the largest concentrations at the Madison mine occur in pyrite and marcasite-rich peripheral zones around the lead ore deposits. At the Buick mine, cobalt and nickel are concentrated in two north-south trending zones. The greatest concentration in the Buick mine lies along the eastern edge of the western ore body, and a smaller amount is in the central part of the eastern ore body (Rogers and Davis, 1977). The total Ni-Co values progressively decline, and the Ni-Co ratios increase, to the south in the western ore body (Rogers and Davis, 1977; Jessey, 1981).

Stratigraphic Variation: The largest amounts of cobalt and nickel are present in those ores in the Southeast Missouri Lead district that have a low stratigraphic position. At the Madison mine, the principal concentrations are associated with the copper and iron sulfide-rich ores in the sandy transition zone at the base of the Bonneterre formation, the main host for Pb-Zn ores in the district. Recent drilling has shown that some cobalt and nickel mineralization occurs as high as the top of the Bonneterre formation (Brooke, 1981). At the Mine La Motte group of mines in the Federicktown subdistrict, cobalt and nickel are more scattered, occurring mainly in the upper part of the transition zone, which is only slightly higher stratigraphically than the lowest ore at the Madison mine. At Indian Creek and Goose Creek mines, located between the Old Lead Belt and the Viburnum Trend, larger amounts of cobalt and nickel are present in the disseminated lead-zinc ores in the upper part of the underlying La Motte sandstone. These ores contain more cobalt and nickel than ores in the overlying Bonneterre formation. Within the Viburnum Trend, local pods of massive to colloform copper-iron sulfide that typically occur at the base of, or beneath, lead-zinc ores, contain significant amounts of cobalt and nickel.

MINOR ELEMENTS WHICH PRESUMABLY OCCUR IN SOLID SOLUTION

Sphalerite

The most abundant minor elements in Mississippi Valley-type sphalerite crystals are cadmium, iron, silver, germanium, gallium, indium, thallium, cobalt, nickel, and mercury. Average values for these elements in sphalerite from several deposits are given in Table 1. These data have been summarized from previously unpublished data for the Tri-State and Southeast Missouri Lead districts, and from published data for other districts. Because the data have been derived by a variety of analytical techniques and by different analysts, the writer has excluded those data which appear to him to be incorrect as a result of instrumental error or contamination. The wide variation in the number and quality of analyses available for each district greatly influences the averages. Despite these problems, the data do provide a base for general comparison.

Cadmium: Cadmium is one of the most abundant minor elements present in Mississippi Valley-type sphalerite, and is recovered from all of the zinc concentrates produced. Although cadmium may occur as greenockite (CdS), this mineral is not a primary sulfide in Mississippi Valley-type deposits; instead it forms as supergene coatings on cadmium-bearing sphalerite. Cadmium present in the primary Pb-Zn ores presumably occurs in solid solution in sphalerite, and to a much smaller extent, in other primary sulfide minerals.

Amounts of cadmium contained in most Mississippi Valley-type sphalerite average about 4000

TABLE 1. Trace Element Content (ppm) of Sphalerite from Mississippi Valley-Type Deposits

District	Cd	Fe	Ag	Ge	Ga	Ge:Ga	Co	Ni	Source
Tri-State (Missouri-Kansas-Oklahoma)	4,700 (187)	2,660 (187)	5.1 (187)	134 (187)	255 (187)	0.5	0.2 (92)	18 (92)	A
Northern Arkansas	5,433 (3)	1,933 (3)							B
Southeast Missouri	8,571 (154)	8,305 (52)	434 (28)	114 (28)	155 (2)	0.7	490 (19)	391 (19)	C
Wisconsin-Illinois	1,739 (32)	8,036 (36)	20 (32)	96 (32)	95 (6)	1.0	15 (6)	42 (6)	D
S. Illinois-Kentucky	3,743 (84)	9,712 (84)	2 (43)	286 (5)	122 (5)	2.3	11 (5)	2 (5)	E
Cent. Tennessee (Main Ores)	4,000		8	400	230	1.7	100	100	
Cent. Kentucky-Tennessee Veins	13,344 (43)	2,813 (8)	1 (8)	269 (43)	98 (43)	2.8	0 (8)	2 (8)	F
East Tennessee	4,185 (13)	3,644 (9)	36 (2)	60 (4)	220 (9)	0.3			G
Austinville, Virginia	1,619 (27)	1,025 (4)							H
Friedensville, Pennsylvania	1,250 (2)								I
Pine Point, NWT	527 (73)		1.5						J
Other W. Canada	2,000 (16)	21,700 (16)							K
Silesia, Poland	4,616 (138)	18,548 (138)	163 (138)		15 (138)			6 (138)	L
Alpine, Europe	1,400 (10)	3,500 (10)	25 (10)	110 (10)					M
S. Pennine, England	3,000	15,000	200						N
Laisvall, Sweden	3,000								O

Sources:
A. Hagni, (this chapter)
B. McKnight, 1935
C. Rogers and Davis, 1977; Hall and Heyl, 1968; Rickman, 1981; Horrall, 1982; Hagni, this chapter; Sverjensky, 1981; Krantz, 1972; Bhatia and Hagni, 1980
D. Bradbury, 1961; Hall and Heyl, 1968; McLimans et al., 1980
E. Hall and Heyl, 1968; Maniocha, 1974; Hughes, 1967; Hatch, 1976
F. Jolly and Heyl, 1968
G. Lenker, 1962; Maher & Fagan, 1970; Jolly & Heyl, 1968
H. Warren and Thompson, 1945; Callahan, 1968; Maher & Fagan, 1970
I. Evans et al., 1968
J. Kyle, 1981; Evans et al., 1968
K. Kaiser et al., 1954
L. Haranczyk, 1979; Schroll, 1953
M. Fruth and Maucher, 1966
N. El Shazley et al., 1957
O. Rickard, et al., 1980
( ) Number of analyses.

to 5000 ppm, and are greater than those in sphalerite from many other types of deposits. Sphalerite from volcanogenic massive sulfide deposits (Mercer, 1976), for example, contain about 2023 ppm cadmium. Sphalerite from the Wisconsin-Illinois district contains substantially less cadmium (1739 ppm; Hall and Heyl, 1968). Sphalerite crystals from Austinville, Friedensville, Pine Point, and Alpine deposits also have a low cadmium content, but the numbers of available analyses for most of those areas are small. The average amount of cadmium in sphalerite from the Southeast Missouri Lead district is much higher (8571 ppm), and sets the ore of this district apart from other Mississippi Valley-type ores. High cadmium content (13,344 ppm) is characteristic of sphalerite in the near-surface veins in the central Kentucky-Tennessee area (Jolly and Heyl, 1968), in contrast to the lower (4000 ppm) cadmium content in the associated stratabound ore at depth. The mineralogical character of the ores of the Northern Arkansas Zinc district are very similar to that of the Tri-State district, and the sphalerite in the two districts has a similar cadmium content. Tri-State sphalerite has an average cadmium content of 4700 ppm and was the principal source of United States cadmium throughout most of its history. Northern Arkansas sphalerite contains about 5433 ppm cadmium. This feature suggests that the two districts may be genetically related.

TABLE 2. Areal Distribution of Cadmium in Sphalerite Within the Southeast Missouri Lead District

	Cd, ppm	No. of Analyses
Subdistricts:		
Viburnum Trend, Missouri	7,587	53
Old Lead Belt, Missouri	9,088	101
Old Lead Belt (Davis, 1960):		
Very near major faults	12,805	24
Near major faults	11,356	5
Further from major faults	9,635	45
Far from major faults	8,087	26
Viburnum Trend:		
Viburnum 28 Mine (Horrall, 1982)	6,086	7
Buick Mine (Rogers & Davis, 1977; Sverjensky, 1981)	7,649	25
Milliken (Ozark Lead) Mine (Rickman, 1981)	8,014	21

Significant spatial variations in the cadmium content of sphalerite have been determined for the ores in some of the Mississippi Valley-type districts. For example, in the Southeast Missouri Lead district, a large number of analyses (154) indicate that sphalerite from the Viburnum Trend contains a smaller average amount (7587 ppm) of cadmium than does sphalerite in the Old Lead Belt where the average is 9088 ppm (Table 2). Furthermore, Davis (1960) has shown that the cadmium content of sphalerite deposited close to faults, which are believed to have been ore-fluid plumbing systems, is higher than it is in sphalerite deposited more distant from the faults (Table 2). In the Viburnum Trend, information currently available from three mines (Table 2) indicates that the cadmium content varies significantly from one mine to another. This may reflect the general movement of ore fluid in the Viburnum Trend, with the highest values found closer to the source of the fluids.

Sphalerite crystals of different paragenetic position within a given district also exhibit significant variations in cadmium content. Sphalerite was repetitively deposited in most Mississippi Valley-type ore deposits, and each generation of sphalerite is characterized by its color and textural relationship with other minerals. Although the causes for the colors may be a function of several factors (Nassau, 1980), the minor element content, especially that of iron, is a major determining factor.

In the Tri-State district, three of four main generations of sphalerite have a similar cadmium content that averages about 5000 ppm, whereas late red sphalerite is distinguished by a substantially lower average of about 3660 ppm (Table 3). Even less cadmium (2667 ppm) is present in the sphalerite that occurs in iron sulfide and siderite concretions in Pennsylvanian shale in and near the district, and which occurs stratigraphically above the principal zinc deposits in the Mississippian host rocks. Stalactitic sphalerite from the Joplin field of the Tri-State district has the lowest cadmium content (1580 ppm) of all the Tri-State sphalerites.

The average cadmium content of the five principal generations of sphalerite deposited in the Viburnum Trend of the Southeast Missouri Lead district exhibit even greater variation (Table 4) than do Tri-State sphalerites; however, the number of analyses is smaller. Some of the early black sphalerite (black A) and later brown sphal-

TABLE 3. Trace Element Content (ppm) of Sphalerite, Tri-State District

	Paragenetic Position	Cd	Fe	Ag	Ge	Ga	In	Mn	No. of Analyses
Yellow	1A*	5040	2,580	2.5	62	104	2	300	25
Orange	1B*	4760	2,475	3.5	128	374	5	125	16
Yellow-Brown	2	9300	2,300	0	34	95	0	0	2
Brown	3*	5088	2,403	4.1	82	213	2	319	94
Brown	4	4930	3,667	7.7	171	370	0	666	3
Red	5*	3660	2,883	4.5	298	400	13	262	42
Red-Brown	6	5050	5,100	4.3	164	200	0	0	4
Brown	7	2900	2,500	2.0	17	68	0	0	1
Stalactitic (Joplin Area)		1580	2,580	54	926	65	0	18000	5
Concretions		2667	14,000	0.7	<20	90	0	0	3

*Main stages of sphalerite deposition.

TABLE 4. Cadmium Variation with Sphalerite Color, Viburnum Trend, Southeast Missouri

	Cd, ppm	No. of Analyses
Black A (oldest)	13044	9
Black B	3774	7
Early Yellow	3461	15
Brown	9784	16
Late Yellow (Youngest)	2775	4

erite contain the largest amounts of cadmium; late yellow sphalerite has the lowest cadmium content. Such variations are believed by the writer to be due principally to changes in the chemical composition of the ore fluid through time.

A comparison of the average cadmium content of sphalerite belonging to several generations of deposition from three Mississippi Valley-type deposits is given in Table 5. Early light-colored (yellow) sphalerite is distinctly different in its average cadmium content in the three districts. Subsequently deposited darker (brown) sphalerite is also markedly different in average cadmium content. Only the late yellow sphalerite deposited in the Southeast Missouri and Wisconsin-Illinois districts have similar average cadmium contents. The dissimilarity of cadmium contents of the earlier sphalerites in the three districts indicates that the minerals were deposited by ore fluids containing different amounts of cadmium, and that the sphalerite in these three districts probably was deposited from ore fluids arising from different sources.

Iron: Iron content is one of the principal factors determining the color of Mississippi Valley-type sphalerite. For zinc concentrates with excessive iron content, iron is an important deleterious constituent, requiring zinc refineries to blend sphalerite from different deposits in order to keep the iron content below 2%.

The average iron content of sphalerite shows significant variations between districts. Sphalerite from the Tri-State, Northern Arkansas, East Tennessee, Central Kentucky-Tennessee veins, and Alpine districts has a low average iron content of between 1933 to 3644 ppm (Table 1). Sphalerite from the Southeast Missouri, Wisconsin-Illinois, and Southern Illinois-Kentucky districts has a higher average iron content of between 8036 to 9712 ppm. The highest average content for Mississippi Valley-type sphal-

TABLE 5. Cadmium and Iron Variations with Sphalerite Color, Three Mississippi Valley-Type Districts (Ppm)

	SE Missouri		Tri-State		Wisconsin-Illinois	
	Cd	Fe	Cd	Fe	Cd	Fe
Black A	13044	8122 (9)				
Black B	3774	8117 (9)				
Early Yellow	3461	7857 (15)	4760	2475 (22)	2188	3475 (8)
Brown	9784	10562 (16)	5088	2403 (94)	1299	12433 (12)
Late Yellow and Red	2775	2835 (4)	3660	2883 (42)	2550	3158 (12)

( ) = Number of analyses.

erite occurs in the Silesian and Pine Point deposits with about 2 to 3%.

The iron content of sphalerite crystals of different generations of deposition within a district also varies significantly (Table 5). The dark brown crystals in the Southeast Missouri and Wisconsin-Illinois districts generally have a high average content, whereas similarly colored sphalerite in the Tri-State district are relatively depleted in iron. The dark color of these low-iron sphalerites may be the result of the presence of several other trace elements as well as iron. The black color of some of the sphalerite in the Southeast Missouri Lead district is due to inclusions of organic matter (Sverjensky, 1981) as well as the presence of iron in solid solution. Early dark sphalerite bands in the Wisconsin-Illinois and Pine Point districts contain as much as 10% iron (McLimans et al., 1980; Kyle, 1981).

Silver: The silver content of sphalerite in most Mississippi Valley-type deposits is low, typically only a few parts per million (Table 1); however, in some of the deposits, in Southeast Missouri, Silesia (Poland), and the South Pennines (England), sphalerite contains substantial amounts of silver. Average silver values in these deposits are several times greater than the average (68 ppm) in massive sulfide deposits (Mercer, 1976). The largest average silver content (434 ppm) in sphalerite from Mississippi Valley-type deposits is from the Southeast Missouri Lead district. An unusually silver-rich sphalerite specimen from the Old Lead Belt contained as much as 5000 ppm (Taylor and Radke, 1969).

Because of the large tonnages [52 kt (575,000 st) in 1980; Ohl et al., 1982; *Mineral Industry Surveys*, 1981] of lead ore mined in Southeast Missouri, the district produces large amounts of associated metals in addition to being the world's largest lead producer. For example, zinc recovered in the district in 1980 accounted for 19.5% of United States production of that metal and placed Missouri second among the zinc producing states. With production of large tonnages of ore, recovery of a minor element, such as silver, can be a significant contribution to domestic resources. Southeast Missouri produced 2.36 million troy oz of silver from both zinc and lead concentrates in 1980. The silver recovered in 1980 amounted to 7.5% of United States silver production, and placed Missouri fifth among the silver-producing states. The value of recovered silver was about $49 million, almost as much as the value ($54 million) of zinc recovered.

The average silver content of sphalerite crystals from the four principal depositional stages in the Tri-State district exhibits small systematic variations (Table 3). Early yellow sphalerite (stage 1A) contains 2.5 ppm, orange sphalerite (stage 1B) 3.5 ppm, subsequent brown sphalerite (stage 3) 4.1 ppm, and late red sphalerite (stage 5) 4.5 ppm.

In contrast to the typical averages of Tri-State sphalerites (2.5 to 4.5 ppm, Table 3), the silver content of stalactitic sphalerite from the area is exceptionally high (54 ppm), whereas sphalerite crystals in concretions have an exceptionally low silver content (0.7 ppm, Table 3).

Germanium: Sphalerite crystals in Mississippi Valley-type deposits contain significant amounts of germanium (Table 1). Crystals in the Southern Illinois-Kentucky, Central Kentucky, and Central Tennessee districts contain the largest average amounts (200 to 400 ppm), and those in the Tri-State, Southeast Missouri, and Wisconsin-Illinois districts have the smallest (96 to 134 ppm). Germanium has been recovered and paid for from zinc concentrates in the Tri-State and Southern Illinois-Kentucky districts, and as recently as 1978–1979 it has been extracted from stockpiled zinc residues in the Tri-State district and at that time served as the only domestic source (Lucas, 1980). In other districts, such as Pine Point, significant amounts of germanium are present, but are not recovered.

Sphalerite deposited during different generations of ore deposition within a given district may have large differences in germanium content. In the Tri-State district, for example, early yellow sphalerite of paragenetic stage 1A, and later brown sphalerite of stage 3, have low average germanium contents of 62 and 82 ppm respectively (Table 3). Early orange sphalerite has a higher average content (128 ppm). Late red sphalerite (stage 5) contains two to five times as much germanium (298 ppm) as the earlier three generations. In addition to these four major periods of sphalerite deposition, a limited number of analyses for yellow-brown (stage 2), brown (stages 4 and 7), and red-brown (stage 6) indicate that the sphalerite in these stages has average germanium contents that range from 17 to 171 ppm. These distinct differences in the germanium content of sphalerite crystals deposited during different periods are interpreted to be the result of introduction of ore fluids in several pulses, each with differing fluid compositions.

Gallium: The gallium content of sphalerite in many Mississippi Valley-type deposits is distinctively high (Table 1). Sphalerite from the Tri-State, Central Tennessee, East Tennessee, and Southeast Missouri districts have especially high average contents (150 to 255 ppm). Gallium has been recovered from zinc concentrates in the Tri-State and Southeast Missouri Lead districts. Sphalerite from the Southern Illinois-Kentucky, Wisconsin-Illinois, and Central Kentucky-Tennessee districts has a lower average content (95 to 122 ppm), and the content in Silesia is very low (15 ppm). The gallium content of sphalerite has a general tendency to follow that of germanium. Thus, the ratios for germanium to gallium do not depart greatly from one district to another. The ratios in those districts for which both germanium and gallium data are available show a range from 0.3 for the East Tennessee district to a maximum of 2.8 for the Central Kentucky-Tennessee veins.

Sphalerite crystals deposited at different times within a given district may contain different gallium contents. For the four principal sphalerite generations in the Tri-State district, gallium variations with time are the same as for germanium, but the magnitudes are different (Table 3). Early yellow sphalerite (stage 1A) has the lowest average content of gallium (104 ppm). Brown sphalerite (stage 3) contains about twice as much (213 ppm) as the earlier yellow sphalerite, and orange (stage 1B) and red (stage 5) sphalerite contain the most, 374 and 400 ppm, respectively.

The gallium content of sphalerite in the iron sulfide concretions, coal, and Pennsylvanian shales at Pleasanton, KS, is very low (31 to 58 ppm, Table 6), but the amount in the sphalerite contained within the siderite concretions near the Tri-State district is relatively high (180 ppm).

Stalactitic sphalerite in the Joplin field has an exceptionally high germanium content of about 926 ppm (Table 3). However, because this form of sphalerite is rare in the Tri-State district, it is not considered to be an important source of germanium.

Although the Tri-State ores are largely confined to strata of Mississippian age, minor amounts are present in overlying Pennsylvanian shales. Epigenetic mineralization in the overlying shales, such as that at the Jack Bin mine in the Picher field, contains quantities of germanium (60 ppm) and cadmium (5800 ppm) in the brown sphalerite that are very similar to those in the yellow and brown sphalerite in underlying Mississippian rocks (Table 6). In contrast, the brown sphalerite crystals deposited in shrinkage cracks of iron sulfide and siderite concretions in Pennsylvanian shales near the district have much lower contents of germanium (< 20 ppm) and cadmium (3500 ppm) than the brown sphalerite in the Mississippian rocks. Brown sphalerite that occurs in fractures in Pennsylvanian coal at Springfield, MO, near the easternmost extent of the Tri-State district, also has a low germanium content (< 20 ppm), and an even relatively lower cadmium content (900 ppm). These data indicate that sphalerite crystals formed in place during diagenesis of the Pennsylvanian shales can be distinguished by their trace element concentrations from subsequently introduced sphalerite crystals. An exception is that of the lead-zinc ores introduced into Pennsylvanian shales at Pleasanton, KS,

TABLE 6. Trace Element Content of Brown Sphalerite in Pennsylvanian Host Rocks, Tri-State District (ppm)

	Ge	Ga	Cd	Fe	Ag
Mississippian Host Rock:					
Yellow Sphalerite	62	104	5040	2,580	2.5
Brown Sphalerite	82	213	5088	2,403	4.1
Pennsylvanian Host Rock:					
Jack Bin Mine (Picher Field)	60	87	5800	5,200	0
Pleasanton, Kansas	<20	56	5900	4,500	0
Springfield, Missouri (in coal)	<20	31	900	2,000	0
Chautauqua Co., Kansas (FeS_2 concretion)	<20	58	3600	15,000	0
Lyon Co., Kansas (Siderite concretion)	<20	180	3500	25,000	2

100 km (65 miles) north of deposits in Mississippian rocks of the Tri-State district. Although the brown sphalerite at the Pleasanton locality contains an amount (5900 ppm) of cadmium similar to that of the brown sphalerite in the Tri-State district, its germanium content (< 20 ppm) is very low.

Indium

Indium is present in small amounts in the Tri-State (4.7 ppm, Table 3) and central Tennessee (about 10 ppm) districts. Undetermined small amounts are reported in concentrates from the Southeast Missouri, Illinois-Kentucky, and Wisconsin-Illinois districts (Parker, 1978), and it has been found in the zinc concentrates in the Tri-State and Southeast Missouri districts.

Variation of the average indium content in sphalerite crystals belonging to the four principal depositional stages in the Tri-State district follows the general pattern shown by germanium and gallium (Table 3). Early yellow sphalerite (stage 1A) and brown sphalerite (stage 3) have a low indium content (2 ppm), whereas orange (stage 1B) and red (stage 5) sphalerite have higher indium contents, 5 and 13 ppm, respectively.

Although the stalactitic sphalerite in the Joplin area contains exceptionally high amounts of germanium and silver, it does not have detectable amounts of indium (Table 3). Similarly, no indium has been detected in the sphalerite contained in concretions in Pennsylvanian shale.

Thallium: Although thallium has not been detected in sphalerite in most Mississippi Valley-type deposits, it is a significant minor element in sphalerite in the Silesia deposits, where results of 138 analyses show that sphalerite contains an average of 340 ppm thallium (Haranczyk, 1979; Schroll, 1953). Thallium has been detected in flue dust in the Tri-State district (Waring, 1917), and in lead concentrates from Southeast Missouri (Parker, 1978).

Cobalt and Nickel: Small amounts of cobalt and nickel, usually only a few parts per million, are present in sphalerite in many Mississippi Valley-type districts (Table 1). A noteworthy exception is sphalerite from the Southeast Missouri Lead district, where the average contents are 490 ppm cobalt and 391 ppm nickel. These high concentrations create an important beneficiation problem for the zinc refineries (Hagni, 1981), because they cause poor current efficiency. It is necessary to keep total cobalt and nickel in the zinc concentrates below 0.25%. Cobalt and nickel in solid solution in the sphalerite, and occurring as crystals of siegenite included within sphalerite, exceed this limit periodically.

The Southeast Missouri Lead district, which has the highest concentrations of cobalt and nickel in sphalerite, also has the largest amounts of cobalt and nickel deposited as siegenite. This indicates that ore fluids in this district were characterized by high cobalt and nickel contents during those periods when sphalerite as well as siegenite was being deposited. Sphalerite from veins in Central Kentucky and Tennessee contain no cobalt and rarely contain nickel (Jolly and Heyl, 1968).

Mercury: Jolly and Heyl (1968) detected small amounts of mercury in sphalerite in several Mississippi Valley-type districts and large amounts in veins in the Central Kentucky and Central Tennessee districts. Central Kentucky sphalerites with mercury contents up to 1000 ppm are among the largest known in the world (Table 7). Although the number of analyses are less for other districts, sphalerite in the Southern Illinois-Kentucky district has an average mercury content that is an order of magnitude greater than the mercury content of sphalerite in the Tri-State and Wisconsin-Illinois districts (Jolly and Heyl, 1968).

Mercury has not been recovered from sphalerite in any of the ores from Mississippi Valley-type districts. It has been recovered from sphalerite from deposits at Balmat-Edwards, NY, and in Ireland, which were formerly thought by some individuals to belong to the Mississippi Valley-

TABLE 7. Mercury Content of Sphalerite

District	Hg, ppm	No. of Analyses	Type of Analysis
Tri-State	0.6 to 1.0	2	AA
Wisconsin-Illinois	0.1 to 0.7	6	AA
Southern Illinois-Kentucky	0.1 to 20.1	2	AA
East Tennessee	ND to 5	11	Spec.
Cent. Tennessee Veins	0.3 to 300	20	Wet chemical & spec.
Cent. Kentucky Veins	0.2 to 1000	23	Wet chemical & spec.

Data from Jolly & Heyl (1968).

type of deposits (Paul Gerdemann, 1982, personal communication).

Variations of Minor Elements Within Single Crystals: Sphalerite crystals belonging to a single generation of protracted ore deposition may exhibit significant variations in their minor element content from their cores outward. Minor element variations within two brown sphalerite crystals belonging to stage 3 in the Grace B mine in the Tri-State district are shown in Fig. 4. The two crystals have similar patterns. The amount of cadmium generally decreases distally from the cores of the crystals. This type of pattern is similar to that of compatible elements whose concentrations in a liquid decrease with time as a result of their consumption by a growing crystal. Cadmium readily substitutes for zinc in the sphalerite lattice, and in the crystals' early stage of development, the substitution of cadmium for zinc may have consumed a large part of the cadmium from the ore fluid. Like cadmium, iron content also decreases outward through much of the sphalerite crystals; however, its concentration increases markedly in the outer regions of both crystals. This increase in iron content during the later deposition of brown sphalerite may reflect the introduction of another pulse of ore fluids containing additional iron. The occurrence of tiny pyrite and marcasite crystals commonly coating the outer brown sphalerite portions of the crystals supports this interpretation.

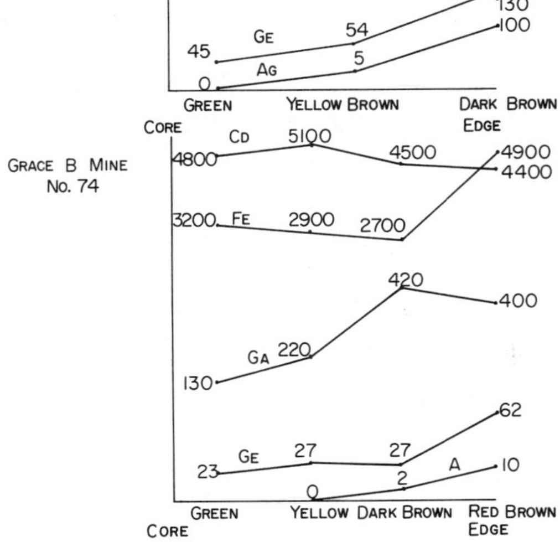

FIG. 4. Trace element variations within single sphalerite crystals, Tri-State district.

The amount of germanium and silver increases outward toward the surfaces of the crystals, and the rate of increase also increases. This pattern is typical of incompatible elements that substitute for zinc with some degree of difficulty, and the concentrations of which had increased in the ore fluids from which the sphalerite crystals had been precipitated.

Gallium content, which also increases distally, exhibits a pattern similar to that of germanium and silver, suggesting that its concentration had been increased in the ore fluid as a result of its relative incompatibility in the sphalerite structure. Gallium content decreases in the outermost brown sphalerite regions because of the influx of cadmium-poor fluids.

Galena

Silver: The low silver content of galena in Mississippi Valley-type deposits is characteristic. For example, the average silver content of galena from the four principal districts in the Mississippi Valley is 65 ppm (based on data compiled here). This is more than an order of magnitude less than that recorded for volcanogenic deposits (850 ppm) [Mercer, 1976].

Although the silver content of Mississippi Valley-type galena is low, there are significant differences in the average silver content between districts (Table 8). The lowest (9 to 15 ppm) occurs in galena from the Northern Arkansas, Tri-State, Austinville, and Wisconsin-Illinois districts. The close similarity of silver content in Northern Arkansas galena to that in Tri-State is additional support for the concept that those two districts are extensions of the same deposit.

The average silver content in galena from the Southeast Missouri Lead district is much higher (85 ppm) than that of the previously mentioned districts. In addition to the relatively high silver content of its galena, the Southeast Missouri Lead district is set apart from other Mississippi Valley-type districts by the high cadmium, cobalt, silver, and nickel contents of its sphalerite,

TABLE 8. Silver and Antimony Content of Galena and Sphalerite (ppm)

District	Galena			Sphalerite	Ratio of Ag in Gn:					
	Ag	Sb	Sb/Ag	Ag	Ag in Sl	As	Bi	Cu	Cd	Source
Tri-State (Missouri-Kansas-Oklahoma)	11.2(63)*	49(16)*	4.3	5.1(187)*	2.2		52(41)*	9.6(58)*		A
N. Arkansas	9 (2)									B
Se. Missouri	85 (318)	171(195)	2.0	434 (28)	0.2	340(86)	50(86)	59 (11)*	27 (96)	C
Wisconsin-Illinois	15.7(283)	283(42)	18.0				1(9)	12 (36)	7.5(2)	D
S. Illinois-Kentucky	149 (91)	813(87)	5.5	1.5(43)	99.3	234(59)	139(59)	105 (12)	15 (16)	E
Austinville, Virginia	15 (12)									F
Silesia, Poland	173 (159)	213(159)	1.2	163 (159)	1.1	609(159)	10(44)	37 (159)	20 (44)	G
S. Pennine, England	<50			200	<0.3					H
Laisvall, Sweden	192									I

*( ) Number of analyses; Gn, galena; Sl, sphalerite.

Sources:
A. Hagni, (this chapter); Blackburn and Schwendeman, 1977; Marshall and Joensuu, 1961
B. McKnight, 1935
C. Mantei, 1977; Bhatia and Hagni, 1980; Krantz, 1972; Horrall, 1982; Hall & Heyl, 1968
D. Hall and Heyl, 1968; Bradbury, 1961; Marshall and Joensuu, 1961
E. Hall and Heyl, 1968; Hughes, 1967; Maniocha, 1972
F. Blackburn and Schwendeman, 1977
G. Harańczyk, 1979; Ridge and Smolarska, 1972
H. El Shazey et al., 1957
I. Rickard et al., 1974

and by the presence of certain minerals, such as siegenite. Galena from the Southern Illinois-Kentucky, Silesia, and Laisvall areas contains a still greater (149 to 192 ppm) content of the silver. Silver has been recovered as a byproduct in all of these districts.

Comparison of the average silver contents of galena with sphalerite reveals a wide range of ratios (Table 8). In some districts, such as Tri-State, galena contains more than twice as much silver as sphalerite. In other districts, such as Southern Illinois-Kentucky, galena contains an order of magnitude more silver than sphalerite. In the Silesia district, the two minerals have nearly equal average silver contents. However, this relationship is reversed in other districts. Sphalerite from the Wisconsin-Illinois district has a slightly greater content of silver than galena. Sphalerite from the Southeast Missouri Lead district contains five or more times the amount of silver in galena. The total amount of silver recovered from Southeast Missouri galena is not much less than the quantities recovered from sphalerite, because of the much larger amounts of galena in the deposits.

Antimony: The antimony content of galena in Mississippi Valley-type deposits is low as compared with that from higher temperature deposits. The average for the five districts for which antimony data are available (Table 8) is 306 ppm. Galena from the Southern Illinois-Kentucky district has the largest (813 ppm) average. Galena from the Tri-State district has the lowest (49 ppm). The antimony to silver ratio is much higher for the Wisconsin-Illinois district than for the other districts for which data are available (Table 8). These features of the minor element content may reflect the influence of some igneous activity in the Southern Illinois-Kentucky district, and the lack of such influence in the Tri-State district.

Significant lateral variations in the silver and antimony content in galena occur in three Mississippi Valley districts. The analyses of galena in the Southern Illinois-Kentucky district show that the highest concentrations of both minor elements is greatest in the north central part of the district near the Hicks Dome alkaline igneous complex (Hall and Heyl, 1968). Galena in the Wisconsin-Illinois district shows a small but regular increase to the east and northeast (Hall and Heyl, 1968). A large number of recent unpublished analyses of galenas from that district have substantiated the trend (Allen V. Heyl, 1982, personal communication).

A total of 182 analyses of galena crystals from the Viburnum Trend in southeast Missouri show that the greatest average concentrations of silver and antimony occur at the Magmont mine (Mantei, 1970). These two minor elements gradually decrease both north and south from the Magmont mine; however, there is an increase in silver in galena cubes at the south end of the trend (Table 9). Changes in the pattern of minor element concentrations in galena are similar to that of cobalt and nickel in the ores. The patterns reflect control by the plumbing system, and indicate that one of the major sources of the ore fluids was in the vicinity of the Magmont mine, and that a second source was located south of the Milliken mine.

Chalcopyrite

Analyses for minor elements in chalcopyrite from the Mississippi Valley-type districts are less common than for sphalerite and galena. The average abundances of several minor elements in chalcopyrite from the Southeast Missouri Lead district are given in Table 10. Significant amounts of molybdenum, bismuth, and cadmium are present. The average amount of silver (38 ppm) in the chalcopyrite is nearly one-half that present in the average galena crystal (85 ppm) from the lead district. These data indicate that there is a small but significant silver resource in the copper concentrates and mattes from the Southeast Missouri Lead district.

Southeast Missouri chalcopyrite also contains small amounts of nickel in solid solution. The copper concentrates contain this nickel together with larger amounts of nickel in siegenite, which

TABLE 9. Variations of the Silver and Antimony Content of Cubic Galena in the Viburnum Trend, Southeast Missouri

Mine	Ag, ppm	Sb, ppm	No. of Analyses
Indian Creek (North)	44	59	30
Viburnum 27, 28, 29	78	106	28
Magmont	164	309	21
Buick	100	273	29
Brushy Creek	83	291	26
Fletcher	132	139	10
Milliken	106	48	38

Data from Mantei (1970).

TABLE 10. Trace Element Content of Chalcopyrite, Southeast Missouri Lead District (ppm)

Ag	Mo	Bi	Ni	Cd	Source
38	234	257	33	272	Bhatia & Hagni, 1980

is present partly as inclusions within chalcopyrite, and partly as free particles (Hagni, 1981) whose flotation properties are similar to those of chalcopyrite. Nickel is recovered from some copper mattes produced from southeast Missouri copper concentrates (Frank H. Sharp, 1982, personal communication).

Marcasite and Pyrite

Investigations of the minor elements present in the iron sulfide minerals in Mississippi Valley-type deposits are few. This is partly due to the fact that iron sulfides are not recovered in these districts and, therefore, have not been of economic interest. However, because the iron sulfides are locally abundant, and because some contain minor elements that may be of economic interest, the most abundant of these, marcasite and pyrite, merit further study.

The abundances of some minor elements in marcasite are summarized in Table 11. The average nickel and copper contents for marcasite from deposits in the Mississippi Valley are about 100 to 300 ppm. Small amounts (19–77 ppm) of cobalt occur in marcasite in at least two of the districts. Some marcasite in the Southern Pennines has an extremely high (3%) nickel content (Ixer and Townley, 1979). The thallium content of Silesian marcasite is especially high (559 ppm) and exceeds the content in both sphalerite (341 ppm) and galena (34 ppm) from the district.

The average silver content of marcasite varies widely from one district to another. Marcasite from Wisconsin-Illinois contains less than 1 ppm silver (Bradbury, 1961). Tri-State marcasite has 9.4 to 15.6 ppm silver (Barrett, 1940). Preliminary analyses of marcasite from southeast Missouri indicate that the average silver content is about 50 ppm, and that the maximum amounts are nearly an order of magnitude greater. Further studies could define a nonconventional resource of silver in the southeast Missouri marcasite.

SUMMARY AND CONCLUSIONS

Ore deposits of Mississippi Valley-type contain a variety of minor elements. Available data have been summarized for those minor elements of greatest current or potential economic interest. Average values determined for these minor elements show significant variations between the various districts, within portions of some districts, between multiple generations of deposition of minerals, and within individual sulfide crystals.

Some of these minor elements have been routinely recovered from the ores and concentrates produced in the districts. Other minor elements have been recovered from only some of the districts, and yet other minor elements have not been recovered from any Mississippi Valley-type district.

Those minor elements which have not been recovered represent a nonconventional source of those metals. For many of these minor elements, their presence is very important because of their rare or strategic value.

ACKNOWLEDGMENTS

Thanks are due to J.S. Bryson for his spectrographic analyses of the author's Tri-State sphalerite specimens. The author gratefully acknowledges the help of J.R. Craig, D.R. Rothbard and S.H. Stow, who provided electron probe and spectrographic analyses of Southeast Missouri sphalerite and galena specimens taken from the author's collections and from those of his

TABLE 11. Trace Element Content of Marcasite (ppm)

	Ag	Co	Ni	Mo	Cu	Mn	Tl	No. of Analyses	Source
Tri-State	12.5		195	10	100				Barrett, 1940; McKnight & Fischer, 1970; Hagni (this chapter)
Se. Missouri	50	19	294	2	149			65	Bhatia & Hagni, 1980
Wisconsin-Illinois	<1	77	99	20	288			12	Bradbury, 1961
Silesia, Poland			23	3.4	29	71	559	72	Haranczyk, 1979
S. Pennine, England			32,750					1	Ixer & Townley, 1979

students, D.L. Rickman, D.R. Jessey, K.B. Horrall, and S. Pignolet. D.R. Jessey served as the graduate research assistant on a US Bureau of Mines grant to the writer and drafted an earlier version of Fig. 3.

BIBLIOGRAPHY

Barrett, F. B., 1940, "Impurities in Minerals of Picher District," *Engineering & Mining Journal*, Vol. 141, No. 1, p. 47.

Bhatia, D. M. S., and Hagni, R. D., 1980, "Laser Probe Determinations of Trace Element Concentrations in Sulfide Minerals from the Magmont Mine, Viburnum Trend, Southeast Missouri," *AIME-SME Trans.*, Vol. 268, pp. 1847–1855.

Bjorlykke, A., and Sangster, D. F., 1981, "An Overview of Sandstone Lead Deposits and Their Relationship to Red-Bed Copper and Carbonate-Hosted Lead Zinc Deposits," *Economic Geology Seventy-Fifth Anniversary Volume 1905–1980*, B. J. Skinner, ed., Economic Geology Publ. Co., Lancaster, PA, pp. 179–213.

Blackburn, W. H., and Schwendemen, J. F., 1977, "Trace-Element Substitution in Galena," *Canadian Mineralogist*, Vol. 15, pp. 365–373.

Bradbury, J. C., 1961, "Mineralogy and the Question of Zoning, Northwestern Illinois Zinc-Lead District," *Economic Geology*, Vol. 56, pp. 132–148.

Brooke, G. L., 1981, "Madison Cobalt Mine, Madison County, Missouri," AIME-SME Preprint No. 81–324, Denver, Colo., Nov. 18–20, 3 pp.

Brown, J. S., Emery, J. A., and Meyer, P. A., Jr., 1954, "Explosion Pipe in Test Well on Hicks Dome, Hardin County, Illinois," *Economic Geology*, Vol. 49, pp. 891–902.

Callahan, W. H., 1968, "Geology of the Friedensville Zinc Mine, Lehigh County, Pennsylvania," *Ore Deposits of the United States 1933–1967*, J. D. Ridge, ed., Vol. 1, AIME, New York, pp. 96–107.

Clifford, R. K., and Higley, L. W., Jr., 1978, "Cobalt and Nickel Recovery from Missouri Lead Belt Chalcopyrite Concentrates," Report of Investigations, 8321, US Bureau of Mines, 14 p.

Craig, J. R., and Carpenter, A. B., 1977, "Fletcherite, $Cu(Ni, Co)_2S_4$, a New Thiospinel from the Viburnum Trend (New Lead Belt), Missouri," *Economic Geology*, Vol. 72, pp. 480–486.

Davis, J. H., 1960, "Mineralization in the Southeast Missouri Lead District," Ph.D. dissertation, University of Wisconsin, 68 pp.

El Shazley, E. M., Webb, J. S., and Williams, D., 1957, "Trace Elements in Sphalerite, Galena and Associated Minerals from British Isles," *Transactions*, Institution of Mining & Metallurgy, Vol. 66, pp. 241–72.

Erickson, R. L., Mosier, E. L., and Viets, J. G., 1978, "Generalized Geologic and Summary Geochemical Maps of the Rolla 1° × 2° Quadrangle," Missouri, Miscellaneous Field Studies Map MF-1004, US Geological Survey.

Evans, T. L., Campbell, F. A., and Krouse, H. R., 1968, "A Reconnaissance Study of Some Western Canadian Lead-Zinc Deposits," *Economic Geology*, Vol. 63, pp. 349–359.

Fruth, I., and Maucher, A., 1966, "Spurenelemente und Schwefel-Isotope in Zinkblenden der Blei-Zinc-Lagerstatte von Gorno," *Miner. Deposita*, Vol. 1, pp. 238–250.

Hagni, R. D., 1981, "Applications of the Ore Microscope to Ore Dressing Problems in the Viburnum Trend, Southeast Missouri, U.S.A.," *Proceedings*, International Conference on Applied Mineralogy in the Mineral Industry, Johannesburg, South Africa, 18 pp.

Hagni, R. D., and Trancynger, T. C., 1977, "Sequence of Deposition of the Ore Minerals at the Magmont Mine," *Economic Geology*, Vol. 72, pp. 451–464.

Hall, W. E., and Heyl, A. V., 1968, "Distribution of Minor Elements in Ore and Host Rock, Illinois-Kentucky Fluorite District and Upper Mississippi Valley Zinc-Lead District," *Economic Geology*, Vol. 63, pp. 655–670.

Harañcyzk, C., 1979, "Metallogenic Evolution of the Silesia-Cracow Region," *Research on the Genesis of Zinc-Lead Deposits of Upper Silesia, Poland*, Prace Instytutu Geologicznego, Tom SCV, pp. 109–132.

Hatch, J. R., Gluskoter, H. J., and Lindahl, P. C., 1976, "Sphalerite in Coals from the Illinois Basin," *Economic Geology*, Vol. 71, pp. 613–624.

Heyl, A. V., et al., 1959, "The Geology of the Upper Mississippi Valley Zinc-Lead District (Ill.-Iowa-Wis.)," Professional Paper 309, US Geological Survey, 310 pp.

Hoagland, A. D., 1971, "Appalachian Strata-Bound Deposits: Their Essential Features, Genesis and the Exploration Problem," *Economic Geology*, Vol. 66, pp. 805–810.

Horrall, K. B., 1982, "Mineralogical, Textural, and Paragenetic Studies of Selected Ore Deposits of the Southeast Missouri Lead-Zinc-Copper District and Their Genetic Implications," Ph.D. dissertation, University of Missouri-Rolla, 650 pp.

Hughes, T. H., 1967, "Minor Elements in Sulfide Minerals of the Hill Mine, Cave-in-Rock District, Illinois," Ph.D. dissertation, Colorado School of Mines, 79 pp.

Ixer, R. A., and Townley, R., 1979, "The Sulphide Mineralogy and Paragenesis of the South Pennine Orefield, England," *Mercian Geology*, Vol. 7, pp. 51–63.

Jessey, D. R., 1981, "Mineralogical and Compositional Variations of the Nickel-Cobalt Mineralization in the Southeast Missouri Mining District," *Process Mineralogy: Extractive Metallurgy, Mineral Exploration, Energy Resources*, D. M. Hausen and W. C. Park, eds., AIME, New York, pp. 159–177.

Jolly, J. L., and Heyl, A. V., 1968, "Mercury and Other Trace Elements in Sphalerite and Wall Rocks from Central Kentucky, Tennessee, and Appalachian Zinc District," Bulletin 1252-F, US Geological Survey, pp. F1–F29.

Kaiser, E. P., Herring, B. F., and Rabbit, J. C., 1954, "Minor Elements in Some Rocks, Ores, and Mill and Smelter Products," TEI-415, US Geological Survey, 119 pp.

Kastler, E. J., 1972, "Mineralization of the Fletcher Lead-Zinc Mine in Southeast Missouri," M.S. Thesis, University of Missouri, 118 pp.

Kline, H. D., Calhoun, W. A., and Reynolds, B. M., 1961, "Mining and Milling Methods and Costs, Madison Mine, National Lead Company, St. Louis Smelting and Refining Division, Madison County, Missouri," Information Circular 8028, US Bureau of Mines, 26 pp.

Krantz, D. V., 1972, "Selected Trace Element Distribution in Galena and Sphalerite, Magmont Mine, Iron County, Missouri," M.S. Thesis, Southern Illinois University, 84 pp.

Kyle, J. R., 1981, "Geology of the Pine Point Lead-Zinc District," *Handbook of Stratabound and Stratiform Ore Deposits*, Vol. 9, K. H. Wolf, ed., Elsevier, Amsterdam, pp. 643–741.

Lenker, E. S., 1962, "A Trace Element Study of Selected Sulfide Minerals from the Eastern United States," Ph.D. dissertation, Pennsylvania State University, 151 pp.

Lucas, J. M., 1980, "Germanium," *Metals and Minerals*, Vol. 1, *Minerals Yearbook 1978–1979*, US Bureau of Mines, pp. 1038–1040.

Maher, S. W., and Fagan, J. M., 1970, "Trace Element Content of Some Ore Deposits in the Southeastern States," Information Circular 16, Tennessee Division of Geology.

Maniocha, M. L., 1972, "Trace Element Distribution in Sphalerite, Galena, Calcite and Barite, Barnett Mine, Pope County, Illinois," M.S. Thesis, Southern Illinois University.

Mantei, E. J., 1977, "Antimony and Silver Content Variation of the Galena Cubes of the 'Viburnum Trend,' and Indian Creek Mines, Missouri," Geoscience Series No. 4, Southwest Missouri State University, 13 pp.

Marshall, R. R., and Joensuu, O., 1961, "Crystal Habit and Trace Element Content of Some Galenas," *Economic Geology*, Vol. 56, pp. 758–771.

Maucher, A., and Schneider, H. J., 1967, "The Alpine Lead-Zinc Ores," *Genesis of Stratiform Lead-Zinc-Barite-Fluorite Deposits in Carbonate Rocks*, J. S. Brown, ed., Monograph 3, Economic Geology Publishing Co., Lancaster, PA, pp. 71–89.

McKnight, E. T., 1935, "Zinc and Lead Deposits of Northern Arkansas," Bulletin 853, US Geological Survey, 311 pp.

McKnight, E. T., and Fischer, R. P., 1970, "Geology and Ore Deposits of the Picher Field, Oklahoma and Kansas," Professional Paper 588, US Geological Survey, 165 pp.

McLimans, R. K., Barnes, H. L., and Ohmoto, H., 1980, "Sphalerite Stratigraphy of the Upper Mississippi Valley Zinc-Lead District, Southwest Wisconsin," *Economic Geology*, Vol. 75, pp. 351–361.

Mercer, W., 1976, "Minor Elements in Metal Deposits in Sedimentary Rocks: A Review of the Recent Literature," *Handbook of Stratabound and Stratiform Ore Deposits*, Vol. 2, K. H. Wolf, ed., Elsevier, Amsterdam, pp. 1–28.

Mineral Industry Surveys, 1981, "The Mineral Industry of Missouri in 1980," US Bureau of Mines and Missouri Dept. of Natural Resources, Div. of Geology & Land Survey, 3 pp.

Mineral Industry Surveys, 1982, "The Mineral Industry of Missouri in 1981," US Bureau of Mines and Missouri Dept. of Natural Resources, Div. of Geology & Land Survey, 3 pp.

Moh, G. H., and Ottemann, J., 1964, "X-Ray Fluorescence and Electron-Probe Analysis of Some Pyrite-Type Minerals," *Year Book, 1963–64*, Carnegie Inst. of Washington, pp. 214–216.

Muraro, 1966, "Metamorphism of Zinc-Lead Deposits in Southeastern British Columbia," *A Symposium on the Tectonic History and Mineral Deposits of the Western Cordillera in British Columbia and Neighboring Parts of the United States*, Spec. Vol. No. 8, Canadian Institute of Mining & Metallurgy, pp. 239–247.

Nassau, K., 1980, "The Causes of Color," *Scientific American*, Vol. 243, No. 4, pp. 124–154.

Ohl, J. P., Wharton, H. M., and Rueff, A. W., 1982, "The Mineral Industry of Missouri," Preprint, *Area Reports: Domestic*, Vol. 2, *Minerals Yearbook 1980*, US Bureau of Mines, 18 pp.

Parker, J. G., 1978, "The Occurrence and Recovery of Certain Minor Metals in the Processing of Lead and Zinc," Open File Report, US Bureau of Mines, 111 pp.

Peterson, G. R., Bleiwas, D. I., and Thomas, P. R., 1981, "Cobalt Availability-Domestic: A Minerals Availability System Appraisal," Mines Report of Investigations 8848, US Bureau of Mines, 31 pp.

Pignolet, S., and Hagni, R. D., 1982, "Characterization of the Cobalt-Nickel Mineralization at the Madison Mine, Fredericktown, Missouri," Missouri Academy Science annual meetings, Point Lookout, MO.

Rickard, D. T., et al., 1979, Studies on the Genesis of the Laisvall Sandstone Lead-Zinc Deposit, Sweden, *Economic Geology*, Vol. 74, pp. 1255–1285.

Rickman, D. L., 1981, "A Thermochemical Study of the Ore Deposits of the Milliken Mine, New Lead Belt, Missouri," Ph.D. dissertation, University of Missouri-Rolla, 310 pp.

Ridge, J. D., and Smolarska, I., 1972, "Factors Bearing on the Genesis of the Silesian-Cracovian Lead-Zinc Deposits in Southern Poland," *Proceedings*, 24th International Geological Congress, Montreal, Section 6, pp. 216–229.

Rogers, R. K., and Davis, J. H., 1977, "Geology of the Buick Mine, Viburnum Trend, Southeast Missouri," *Economic Geology*, Vol. 72, pp. 372–380.

Schroll, E., 1953, "Über Unterschiede im Spurgengehalt bei Wurtziten, Schalenbalendes und Zinkblenden," Akad. Wissenschaft. Mathem.-Naturw. Kl. Abt. I Bd 162, H. 5.

Sverjensky, D. A., 1981, "The Origin of a Mississippi Valley-Type Deposit in the Viburnum Trend, Southeast Missouri," *Economic Geology*, Vol. 76, pp. 1848–1872.

Taylor, C. M., and Radtke, A. S., 1969, "Micromineralogy of Silver-Bearing Sphalerite from Flat River, Missouri," *Economic Geology*, Vol. 64, pp. 306–318.

Trace, R. D., 1960, "Significance of Unusual Mineral Occurrence at Hicks Dome, Hardin County, Illinois," Professional Paper 400-B, US Geological Survey, pp. B63–B64.

Trace, R. D., 1974, "Illinois-Kentucky Fluorspar District," *A Symposium on the Geology of Fluorspar*, D. W. Hutcheson, ed., Special Publ. 22, Kentucky Geological Survey, pp. 58–76.

Waring, W. G., 1913, "The Zinc Ores of the Joplin District: Their Composition, Character, and Variation," *AIME Trans.*, Vol. 57, pp. 657–670.

Warren, H. V., and Thompson, R. M., 1945, "Sphalerites from Western Canada," *Economic Geology*, Vol. 40, pp. 309–335.

8

Geochemical Processes That Control Minor and Trace Element Composition of United States Coals

Joseph R. Hatch

US Geological Survey, Denver, CO

ABSTRACT

When compared to average shale composition, average coal is enriched in sulfur and selenium, has similar amounts of arsenic, beryllium, lead, antimony and molybdenum, and is depleted in at least 26 other elements. Within a coalfield, or region, some coals are enriched (10X average shale) in silver, arsenic, cadmium, cobalt, germanium, molybdenum, nickel, lead, selenium, uranium, and zinc. Important factors controlling element distributions in coal include geochemical conditions (primarily pH) in the peat swamp, composition and depositional environment of coal roof rocks, thermal maturity (rank), nature and intensity of any epigenetic mineralization, composition of ground waters that come in contact with the coal, and degree of weathering.

INTRODUCTION

The minor and trace elements present in coals are of interest as possible environmental contaminants, in their effects on coal utilization, and as possible unconventional sources for metals. To aid intelligent decision making and economic evaluations with respect to coal element composition, the US Geological Survey (USGS), at least 25 state geological surveys, and a number of colleges and universities have cooperated in producing a data base of detailed elemental analyses for about 9000 well-documented samples of United States coals. Examples of publications listing and summarizing coal analytical data include Glass (1975); Swanson et al. (1976); Gluskoter et al. (1977); Hatch and Swanson (1977); Moore (1977); Botoman and Stith (1978); Henderson, Oman, and Coleman (1979); Hatch, et al. (1979); Wedge and Hatch (1980); Khalsa and Ladwig (1981); and Trent et al. (1982). These 9000 analyses serve as a basis for evaluating elemental abundances in coals from the various coal provinces, regions, and fields. By combining existing data with additional geological and geochemical data, insight may be gained into when elements were introduced into coals and which physical and geochemical parameters are most important in determining coal composition. Knowledge of the origin of elements is essential to the prediction of where minimum or maximum element contents would be found.

The following sections briefly discuss average United States coal composition and the factors controlling the element composition of coal.

AVERAGE UNITED STATES COAL COMPOSITION

When the average composition of coal in the United States (799 samples summarized in

TABLE 1. Comparison of the Average Content of 38 Elements in Shale and Coals

Element	Average shale*, %	Average coal⁺, %
S	0.24	2.0
Si	27.3	2.6
Al	8.0	1.4
Ca	2.21	0.54
Mg	1.5	0.12
Na	0.96	0.06
K	2.66	0.18
Fe	4.72	1.6
Ti	0.46	0.08
	Parts per million	
As	13	15
B	100	50
Ba	580	150
Be	3	2
Co	19	7
Cr	90	15
Cu	45	19
F	740	74
Ga	19	7
Hg	0.4	0.18
Li	66	20
Mo	2.6	3
Nb	11	3
Ni	68	15
Pb	20	16
Sb	1.5	1.1
Sc	13	3
Se	0.6	4.1
Sr	300	100
Th	12	4.7
U	3.7	1.8
V	130	20
Y	26	10
Yb	2.6	1
Zn	95	39
Zr	160	30

*Turekian and Wedepohl (1961).
⁺Swanson et al. (1976).

Swanson et al., 1976) is compared (Table 1) with the average compositon of shale (Turekian and Wedepohl, 1961) coals are enriched in sulfur and selenium; have similar amounts of antimony, arsenic, beryllium, lead, and molybdenum; and are depleted in silicon, aluminum, calcium, magnesium, potassium, iron, titanium, boron, barium, cobalt, chromium, copper, fluorine, gallium, mercury, lithium, columbium, nickel, scandium, strontium, thorium, uranium, vanadium, yttrium, zinc, and zirconium. Depletion of most elements in coal is due to dilution by organic matter composed primarily of carbon, hydrogen, oxygen, nitrogen, and sulfur.

Locally, within a field or region, individual coal samples have anomalously high contents (arbitrarily set at 10X average shale composition) of one or more elements (Table 2). Elements that are relatively enriched in coals are either organically associated [germanium, molybdenum?, selenium, or uranium] (Zubovic et al., 1960, 1961) or form highly insoluble sulfides (antimony, arsenic, cadmium, cobalt, copper, molybdenum, nickel, lead, silver, and zinc). Most high element contents listed in Table 2 are in coal samples collected from the Middle Pennsylvanian age Cherokee group, from the Western Interior coal region (Iowa, Nebraska, Missouri, Kansas, Oklahoma, and Arkansas).

FACTORS CONTROLLING ELEMENTAL COMPOSITION OF COAL

Factors that are important in controlling the elemental composition of coals include the following: (1) geochemical conditions in the peat swamp, (2) composition and depositional environment of coal roof rocks, (3) thermal maturity [rank], (4) nature and intensity of any

TABLE 2. Maximum Content of 13 Elements in Coals in the United States [Data are Reported on a Whole Coal Basis]

Element	Maximum content, ppm	Sample locality and reference source
Ag	3	Wappelo, County, Iowa¶
As	360	Walker County, Alabama*
Cd	170	Howard County, Missouri⁺
Co	300	Wappelo County, Iowa¶
Cu	910	Tazwell County, Virginia*
Ge	100	Randolph County, Missouri⁺
Mo	150	Wappelo County, Iowa¶
Ni	700	Wappelo County, Iowa¶
Pb	470	Marion County, Iowa¶
Sb	37	Wappelo County, Iowa¶
Se	75	Wappelo County, Iowa*
U	6,200	Sandoval County, N. Mex.§
Zn	51,000	Crawford County, Kansas≠

* Swanson et al. (1976).
⁺ Wedge and Hatch (1980).
≠ Hatch et al. (1976).
§ Bachman et al. (1959).
¶ Hatch et al. (in press).

epigenetic mineralization, (5) composition of ground waters in contact with the coals, and (6) degree of weathering.

In theory it is easy to list the processes and times of trace-element fixation in coals. In practice, however, it is quite difficult to demonstrate the "how much" and "when" for a given element, as any given element can be distributed among several mineral or organic phases. Coal should be considered an open system with respect to elemental composition, with the potential for change existing continually from the initial stages of peat formation until the coal is mined or weathers away on outcrop.

Syngenesis and Early Diagenesis

pH: Although the various geochemical and microbial factors affecting introduction and fixation of elements during, and immediately following, the time of peat deposition are interrelated, the pH of waters in the swamp and immediate post-swamp environments appears to be the most critical (Cecil et al., 1982). Under low pH conditions (3–4.5), leaching of most metal ions is favored and the activity of sulfate-reducing bacteria is minimal; where pH conditions are near neutral (6–8), the activity of sulfate-reducing bacteria reaches a maximum (Baas Becking et al., 1960).

A strong direct relationship exists between sulfur content in coals and $CaCO_3$ content of associated rocks (Cecil et al., 1982). Low $CaCO_3$ contents would result in minimal carbonate buffering of depositional and early diagenetic sedimentary waters and would indicate relatively low pH conditions; high $CaCO_3$ contents would indicate relatively high pH conditions. Lower Pennsylvanian coals of the Pocahontas basin of West Virginia are associated with noncalcareous rocks (Cecil et al., 1982), and have low mineral matter, sulfur, and minor and trace element contents. The same is true of the vast coal deposits of Paleocene age from the Rocky Mountain and Northern Great Plains coal provinces. In contrast, coals occurring in sections with carbonates (e.g., midcontinent, Middle Pennsylvanian age coals) have relatively high mineral matter, sulfur, and minor and trace element contents.

In Table 3, the elemental compositions of coals in carbonate-bearing sections (Vermillion Creek coal and Lexington coal) are compared with the compositions of coals associated with noncalcareous rocks (unnamed Williams Fork formation coals and Sudduth coal). These four data sets represent coals of about the same rank (subbituminous A and high-volatile C bituminous coal). The Vermillion Creek and Lexington sample sets have relatively high contents of sulfur and significantly higher contents of calcium, potassium, iron, arsenic, beryllium, cadmium, cobalt, chromium, copper, manganese, molybdenum, nickel, antimony, scandium, selenium, uranium, vanadium, yttrium, ytterbium, and zinc. Some elements in this list (iron, arsenic, cadmium, cobalt, copper, nickel, antimony, and zinc) form insoluble sulfides. Others might be present as carbonates (calcium and manganese), phosphates (scandium, yttrium, and ytterbium), or illite or smectite clays (potassium, calcium), mineral phases that are more stable at near neutral pH's. A third group of elements (beryllium, chromium, molybdenum, selenium, uranium, and vanadium) are an element assemblage found in uranium roll-type deposits (Harshman, 1974) and probably are precipitated from depositional or early diagenetic waters by reduction with H_2S with subsequent formation of insoluble compounds.

Roof Rock Lithology: The geochemical and physical conditions of environments that immediately follow the peat swamp environment may significantly affect final coal composition. In the midcontinent region, for example, coals that are immediately overlain by black phosphatic shales are significantly enriched in uranium, molybdenum, selenium, vanadium, antimony, and germanium, and occasionally chromium and mercury. These black shales represent deposition under anoxic marine conditions (Heckel, 1977). Uranium and the other metals apparently were more readily available in the black-shale depositional environment when compared with other post-coal depositional environments.

In Illinois, relatively thick coals overlain by these black shales have higher contents of uranium, molybdenum, selenium, vanadium, antimony, and germanium near the top of the bed. Figs. 1A and 1B (Gluskoter et al., 1977) show the distribution of uranium and germanium in a set of bench samples of the Herrin No. 6 coal from southern Illinois. These six elements also may be enriched to a greater or lesser degree in the lowermost benches of a coalbed. This indicates that solutions moving upward through the peats are the source of these elements; so-

TABLE 3. Geometric Mean Contents and Geometric Deviations of Ash and 36 Elements for Four Sets of Rocky Mountain and Midcontinent Subbituminous A and High-Volatile C Bituminous Coals¶

	Vermillion Creek coal, Wasatch Formation, Sweetwater County, Wyoming,* 6 samples		Lexington coal, Marmaton Group, North-Central Missouri,⁺ 23 samples		Unnamed coalbeds, Williams Fork Fm, Northwestern Colorado,≠ 63 samples		Sudduth coal, Coalmont Formation, Jackson County, Colorado,§ 21 samples	
Element	**GM**	**GD**	**GM**	**GD**	**GM**	**GD**	**GM**	**GD**
				%				
Ash	16.3	1.7	11.2	1.2	8.4	1.7	6.8	1.8
S	5.5	1.2	3.6	1.3	0.6	1.5	0.3	1.4
Si	3.0	2.7	1.7	1.3	2.2	2.3	1.4	2.2
Al	0.77	2.7	0.76	1.3	1.2	1.8	0.85	1.9
Ca	1.7	2.8	0.60	2.5	0.37	1.5	0.45	1.4
Mg	0.097	2.0	0.050	1.4	0.066	1.8	0.047	1.7
Na	0.073	1.5	0.015	1.5	0.035	2.6	0.016	4.2
K	0.071	6.6	0.11	1.3	0.061	2.9	0.017	3.3
Fe	1.8	2.8	1.5	1.6	0.25	2.1	0.23	1.4
Ti	0.023	3.5	0.039	1.3	0.052	1.6	0.047	2.0
				Parts per million				
As	66	1.6	8.7	2.0	0.5	3.3	1.7	1.5
B	70	1.4	150	1.2	70	1.7	50	2.2
Ba	150	1.4	50	1.4	150	1.9	150	1.3
Be	1	1.8	2	1.2	0.5	2.9	0.15	4.6
Cd	0.23	1.3	1.6	17	0.06	3.3	<0.10	---
Co	3.7	2.0	3	1.8	1	2.1	0.7	2.2
Cr	12	2.1	20	1.5	3	2.4	1.5	2.3
Cu	14	1.7	12	1.3	5.6	1.7	6.8	1.8
F	130	2.3	51	2.0	95	2.1	44	1.7
Hg	0.05	1.3	0.13	1.7	0.04	2.3	0.04	2.3
Li	7.7	2.7	4.9	1.6	8.4	1.9	4.8	2.6
Mn	320	2.2	66	2.0	17	2.8	16	2.8
Mo	20	1.3	10	2.0	0.3	2.9	0.5	2.5
Ni	15	1.7	15	1.4	1.5	2.0	1	1.8
P	380	3.3	---	---	120	5.6	200	2.5
Pb	<5.0	---	41	2.0	4.4	2.2	3.3	2.0
Sb	3.6	1.5	3.5	2.1	0.2	1.9	0.2	1.8
Sc	3	1.8	3	1.5	1.5	1.8	1	1.8
Se	4.8	1.6	2.4	3.4	0.9	1.6	0.7	2.3
Sr	150	1.5	20	1.7	100	2.0	100	1.9
Th	1.9	2.8	---	---	1.5	3.6	1.9	2.0
U	16	1.8	2.7	2.2	1.0	1.8	0.4	2.9
V	70	2.0	30	1.6	7	1.9	7	1.9
Y	10	1.7	7	1.3	5	2.0	3	2.1
Yb	1	2.0	1	1.2	0.5	2.0	0.2	2.1
Zn	27	1.5	240	8.6	7.3	2.5	6.5	2.1
Zr	15	1.7	15	1.4	20	1.9	10	3.1

*From Hatch (in press, Tables 5 and 6)
⁺From Wedge and Hatch (1980, Tables 10A and 10C)
≠From Hildebrand et al. (1981, Tables 7A and 9A)
§From Hatch et al. (1979a, Tables 11 and 13)
<Means less than the value shown; ---, not determined; GM = geometric mean, GD = geometric deviation.

lutions derived from dewatering of sediments below the peats.

For contrast, also in Illinois, coals that are overlain by thick, grayshale sequences (> 6 m of shale) have significantly lower sulfur and lower metal contents than do beds overlain by thin gray shale sequences or other lithologies (Fig. 2) (Gluskoter and Simon, 1968; Gluskoter and

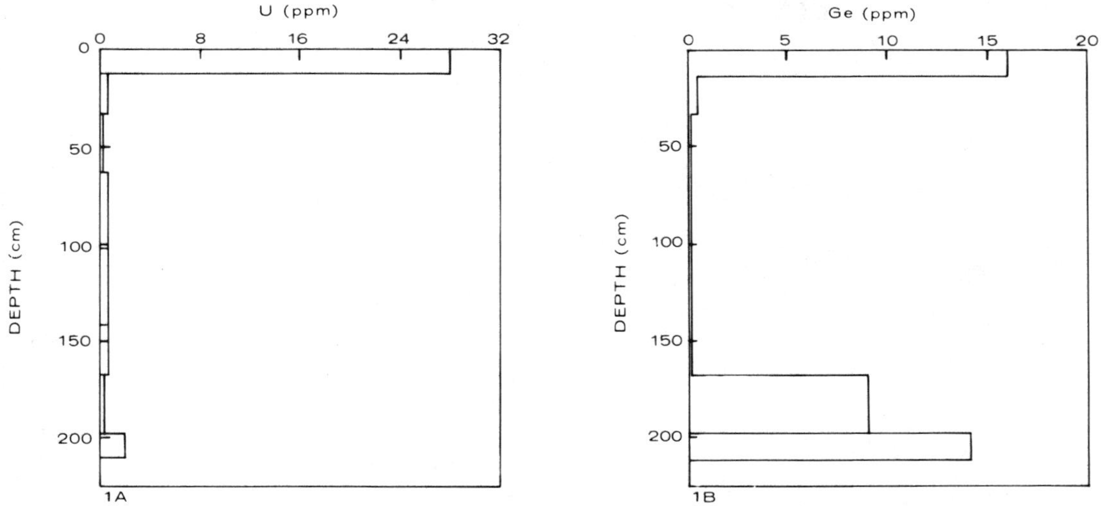

FIG. 1. Distribution of uranium (A), and germanium (B), in a set of bench samples from the Herrin (No. 6) coal, Illinois. Data from Gluskoter et al. (1977, Tables 13 and 14, bench set 1).

Hopkins, 1970). Apparently, the thick fluvial derived sediments that produced the gray shales limited interaction of peats with marine waters, the probable source of much sulfur and many minor and trace elements in Illinois Basin coals.

Late Diagenesis

With increasing depth of burial, the peat undergoes continuous chemical changes to adjust to higher temperatures and pressures. Organic geochemical changes that occur include condensation, polymerization, aromatization, and loss of functional groups containing oxygen, sulfur, and nitrogen (Tissot and Welte, 1978). The loss of polar functional groups, primarily OH^- and $COOH^-$ (Whitehurst, 1978) that act as cation-exchange sites on the organic matter is shown by decreases in contents of calcium,

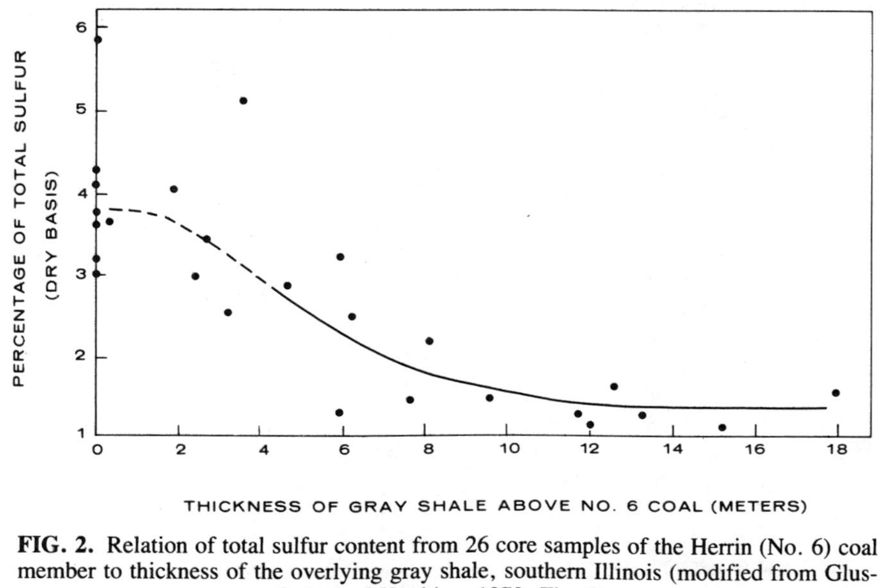

FIG. 2. Relation of total sulfur content from 26 core samples of the Herrin (No. 6) coal member to thickness of the overlying gray shale, southern Illinois (modified from Gluskoter and Hopkins, 1970, Fig. 2).

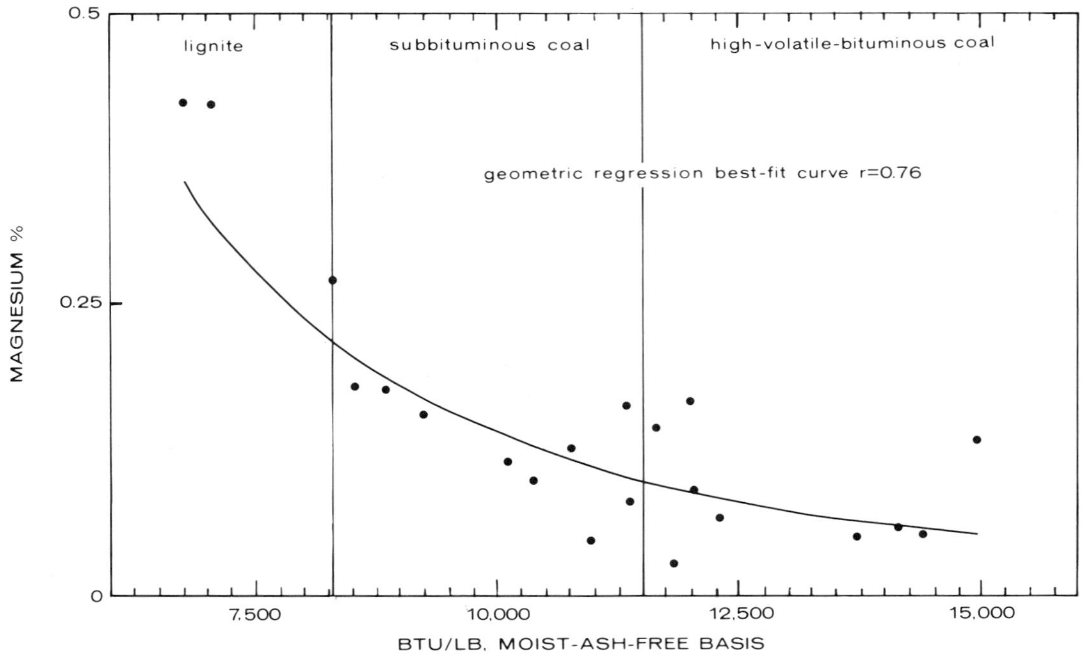

FIG. 3. Relation of magnesium content in coal to Btu per pound (moist, mineral-matter-free basis) and rank of coal for 21 sets (770 samples total) of Rocky Mountain and Northern Great Plains coal samples (data from Hildebrand and Hatch, 1977).

magnesium, barium, strontium, barium, and sodium (and possibly manganese and boron) in Rocky Mountain and Northern Great Plains coals as coal rank increases from lignite B to high-volatile C bituminous coal (Hildebrand and Hatch, 1977; Swanson et al., 1976, Table 4C). As an example, Fig. 3 is a plot of the geometric mean magnesium content (whole-coal basis) vs. heat of combustion (British thermal unit (Btu) per pound on a moist, mineral-matter-free basis) and rank for 21 groups of Rocky Mountain and Northern Great Plains coal samples.

Epigenesis

Coals are subject to epigenetic enrichment of elements by precipitation from ground waters or hydrothermal solutions moving through the coals. Coals are good potential host rocks, since they have good fracture permeability and porosity and can be a source of reduced sulfur.

Sphalerite in Midcontinent Coals: The best examples of epigenetic element enrichments in coal are in the Eastern and Western regions of the Interior coal province. Samples of Pennsylvanian age coals from northwest, northeast, and southeast Illinois (Eastern region); and southeast Iowa, northeast, north-central, and southwestern Missouri, southeast Kansas, northeast Oklahoma, and southeast Nebraska (Western region) contain as much as 51,000 ppm zinc, 170 ppm cadmium, and sometimes have much higher than average contents of lead, nickel, cobalt, mercury, and molybdenum (Hatch, 1976a; Hatch et al., 1976; Hatch, 1979). Areas in these two regions where coals have high zinc contents are illustrated in Figs. 4A and 4B. Elemental analyses of the Lexington bed listed in Table 3 are typical of coals from these areas (note the higher zinc, cadmium, lead, and mercury contents for this set of samples).

At least six different lines of evidence show that zinc, cadmium, lead, nickel, cobalt, and molybdenum in Interior province coals are primarily of epigenetic origin.

1) As shown in Figs. 4A and 4B, areas of high zinc content in coal are geographically related to known mining districts and (or) to major structural features of the midcontinent area (Hatch 1976a; Hatch, 1979). In the Western region, coals in areas adjacent to the Oklahoma-Missouri-Kansas Tri-State district, central Missouri

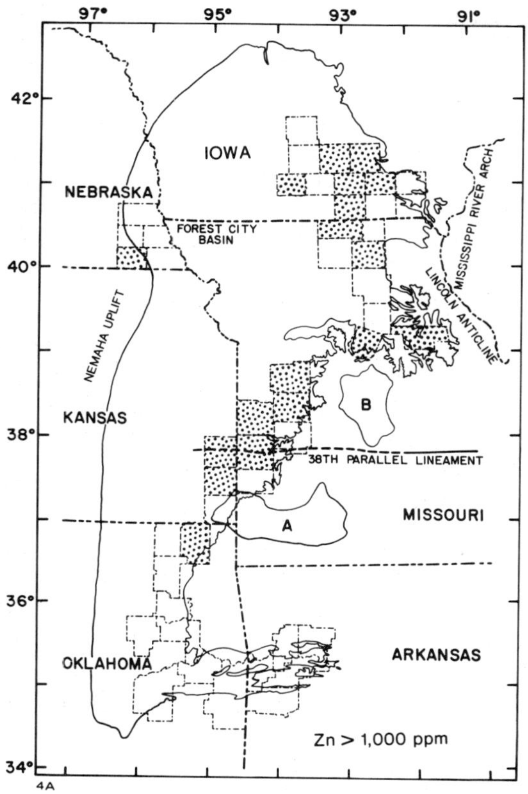

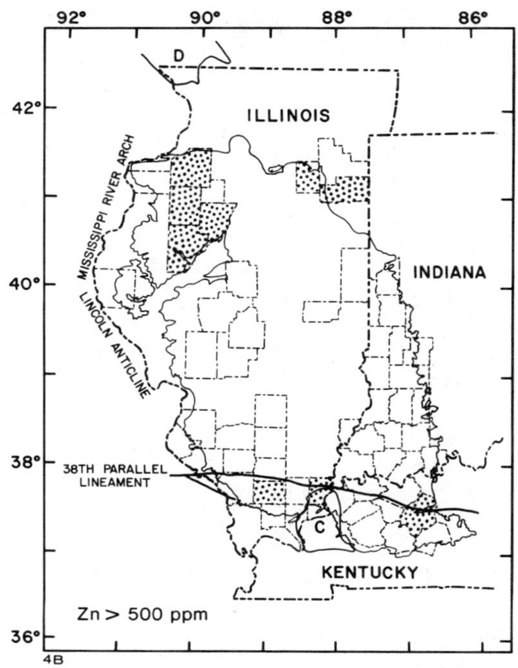

FIG. 4. Zinc distribution in coals from the western region (A), and eastern region (B), of the Interior coal province. Sampled counties are outlined; stipled pattern represents counties with high zinc content (> 1000 ppm Zn; western region, > 500 ppm Zn, eastern region) in a coal sample. A = Tri-state district; B = Central Missouri galena-barite district; C = Illinois-Kentucky fluorspar district; D = Upper Mississippi Valley lead-zinc district.

galena-barite district, 38th parallel lineament, Nemaha uplift, and Lincoln anticline all have high zinc contents. In the Eastern region, analyses of coal samples from areas adjacent to the Illinois-Kentucky fluorspar district, 38th parallel lineament, and Kankakee arch show high zinc contents (Gluskoter et al., 1977; Hatch, 1979).

2) Zinc is present in the coal as sphalerite (ZnS) which is found along cleats and fractures associated with calcite, pyrite, kaolinite, and barite, a mineral assemblage common to Mississippi Valley lead-zinc deposits (Gluskoter and Lindahl, 1973; Hatch et al., 1976a; Hatch, et al., 1976; Cobb, 1981). Cadmium substitutes for zinc in the sphalerite structure, so cadmium distribution closely parallels that of zinc.

3) Studies of coal inclusions in the sphalerite and associated minerals show that the coal was at a slightly lower rank (subbituminous B coal) when the minerals were deposited (Cobb, 1981). Present day coal ranks are high-volatile C, B, and A bituminous coal.

4) Within the high zinc areas, zinc, cadmium, lead, nickel, cobalt, and molybdenum contents in coal are irregularly distributed and have content differences as much as 1000X between samples from coalbeds that are stratigraphically closely related, or even between benches within the same bed (Hatch, 1979).

5) An apparent vertical zoning of zinc/cadmium molecular ratios is evident in five core holes from southeastern Iowa; cadmium is relatively enriched in the stratigraphically higher coals (Hatch, et al., 1976).

6) Homogenization temperatures for the fluid inclusion in sphalerite from both the Western and Eastern regions range from 75° to 113°C (Cobb, 1981). Salinities of inclusion fluids in sphalerite, pyrite, and calcite range from 11.1 to 36.1 wt % dissolved solids (Cobb, 1981). Both homogenization temperatures and fluid-

inclusion salinities are similar to those in Mississippi Valley lead-zinc deposits.

Supergene Enrichment

Uranium in Western Coals: Although the great majority of western coals, in particular the thicker minable deposits, average 1 ppm uranium or less (Hatch and Swanson, 1977), the uranium content of some coals of Tertiary age, at a number of locations in the western United States, range from about 50 to 200 ppm. Within these deposits, smaller irregular masses of coal containing 1000 ppm uranium or more have been found (Denson, 1959).

A partial listing of studies on these deposits would include Denson et al., (1959); Gill et al., (1959); and Zeller and Schopf (1959) who described and discussed occurrences in Harding and Perkins Counties, South Dakota, and in Bowman County, North Dakota; Moore et al., (1959), Golden Valley, Billings, and Slope Counties, North Dakota; Gill (1959), Carter County, Montana; Pipiringos (1961), Masursky (1962), and Roehler (1979), Sweetwater County, Wyoming; Mapel and Hail (1959) Cassia County, Idaho; and Bachman et al., (1959), Sandoval County, New Mexico.

The following characteristics are shared by these deposits: (1) Uranium content is higher (a) in beds with high ash contents, (b) at the top of thick beds (> 1.0 m), and (c) in the highest bed in a stratigraphic section. (2) Uraniferous coals are associated with uranium-bearing tuffaceous sandstone units; ground waters associated with these sandstones units have from 10 to 100 times more dissolved uranium than does normal ground water (Denson et al., 1956; Fix, 1955; Gill and Moore, 1955). (3) Uranium content of coals is directly related to permeability of the enclosing rocks, i.e., coals bounded by sandstones, fractured or faulted coals, or coals in highly jointed areas have higher uranium contents. (4) Uranium distribution commonly is related to local structures. (5) Higher molybdenum, selenium, and sometimes germanium, vanadium, nickel, cobalt, lead, and arsenic contents are associated with the higher uranium contents.

The relationships listed suggest to most investigators (best summarized by Denson (1959); Denson et al., (1959, Fig. 9); and Denson and Gill (1965)) that the uranium and associated elements are of supergene origin. Ground waters leached the elements from the overlying tuffaceous units and carried them to the coals where they accumulated either through reduction and precipitation or by adsorption on the organic matter.

Analyses listed in Table 3 for the Vermillion Creek coalbed show uranium, vanadium, molybdenum, selenium, and arsenic contents at significantly higher levels than those in the Lexington bed. One explanation is that additional amounts of these elements were introduced by ground waters (Hatch, in press).

Local changes in cation composition (sodium, magnesium, and calcium) of coals of similar rank from the Powder River and Fort Union coal regions, Montana and North Dakota, have been shown by R. T. Hildebrand (written communication, 1981) to reflect differences in present-day ground-water compositions. Gluskoter and Rees (1964) found that chlorine content of the Herrin (No. 6) coal from Illinois is closely related to the chlorine content of the ground water.

Surface Weathering

To compare the geochemistry of weathered and unweathered coals, Hatch and Affolter (1981), collected stratigraphically equivalent sets of Paleocene age subbituminous C coal samples from outcrops and cores in the Red Rim area, Sweetwater and Carbon Counties, Wyoming. The samples were collected on extremely steep slopes and at depths of from 0.8 to 2.0 m. Analytical results showed that in the semiarid, relatively cool climate of south-central Wyoming, contents of 16 of 42 elements changed significantly. In the weathered coals, zinc, cadmium, nitrogen, and oxygen contents increased, whereas sulfur, hydrogen, silicon, potassium, sodium, boron, chromium, gallium, lanthanum, lithium, molybdenum, and vanadium contents decreased. In this area, chemical changes due to weathering were noticeable at depths up to 10 m. Hatch and Affolter (1981) suggested that for coals of lower rank and/or weathered under more severe conditions (more moisture, warmer temperatures, and less steep slopes) additional chemical changes would be expected.

SUMMARY

1) When compared to average shale composition, average coal has higher sulfur and selenium contents, similar amounts of arsenic, beryllium, lead, antimony, and molybdenum, and is depleted in at least 26 other elements.

Depletion is due to dilution by organic matter composed primarily of carbon, hydrogen, oxygen, nitrogen, and sulfur.

2) Changes in element composition of coal can occur just about any time during the geologic history of a deposit, from initial stages of peat formation through the time the coal is mined or weathers away on outcrop.

3) Within a coal field or region, some coals are enriched (10X shale average) in silver, arsenic, cadmium, cobalt, copper, germanium, molybdenum, nickel, lead, selenium, uranium, and zinc. Element enrichments result from particularly favorable geochemical and geological conditions. Examples include: (a) coalbeds overlaid by marine black shale lithologies (for high contents of uranium, molybdenum, selenium, vanadium, chromium, silver, and germanium); (b) coalbeds from areas near known metaliferous mining districts (for high contents of zinc, cadmium, lead, nickel, and cobalt); and (c) the uppermost coalbed in a coal-bearing section overlain by tuffaceous sandstone units (for high contents of uranium, molybdenum, and selenium).

REFERENCES CITED

Baas Becking, L. G. M., Kaplan, I. R., and Moore, D., 1960, "Limits of the Natural Environment in Terms of pH and Oxidation-Reduction Potentials," *Journal of Geology*, Vol. 68, pp. 243–284.

Bachman, G. O., et al., 1959, "Uranium-Bearing Coal and Carbonaceous Shale in the La Ventana Mesa Area, Sandoval County, New Mexico," Bulletin 1055-J, US Geological Survey, pp. 295–308.

Botoman, G., and Stith, D. A., 1978, "Analyses of Ohio Coals," Information Circular 47, Ohio Dept. of Natural Resources, Division of Geological Survey, 148 pp.

Cecil, C. B., et al., 1982, "Geologic Factors that Control Mineral Matter in Coal," *Proceedings*, ANS/ACS Atomic and Nuclear Methods in Fossil Energy Research Symposium, Plenum Press, pp. 323–336.

Cobb, J. C., 1981, "Geology and Geochemistry of Sphalerite in Coal," Ph.D. Thesis, University of Illinois, 204 pp.

Denson, N. M., 1959, "Introduction," *Uranium in Coal in the Western United States*, Bulletin 1055, US Geological Survey, pp. 1–10.

Denson, N. M., Bachman, G. O., and Zeller, H. D., 1959, "Uranium-Bearing Lignite in Northwestern South Dakota and Adjacent States," Bulletin 1055-B, US Geological Survey, pp. 11–58.

Denson, N. M., and Gill, J. R., 1965, "Uranium-Bearing Lignite and Carbonaceous Shale in the Southwestern Part of the Williston Basin—A Regional Study," Professional Paper 463, US Geological Survey, 75 pp.

Denson, N. M., Zeller, H. D., and Stephens, J. G., 1956, "Water Sampling as a Guide in the Search for Uranium Deposits and Its Use in Evaluating Widespread Volcanic Units as Potential Source Beds for Uranium," Professional Paper 300, US Geological Survey, pp. 673–681.

Fix, P. P., 1955, "Hydrogeochemical Exploration for Uranium," Professional Paper 300, US Geological Survey pp. 667–671.

Gill, J. R., 1959, "Reconnaissance for Uranium in the Ekalaka Lignite Field, Carter County, Montana," Bulletin 1055-F, US Geological Survey, pp. 167–180.

Gill, J. R., and Moore, G. W., 1955, "Carnotite-Bearing Sandstone in Cedar Canyon, Slim Buttes, Harding County, South Dakota," Bulletin 1009-I, US Geological Survey, pp. 249–264.

Gill, J. R., Zeller, H. D., and Schopf, J. M., 1959, "Core Drilling for Uranium-Bearing Lignite, Mendenhall Area, Harding County, South Dakota," Bulletin 1055-D, US Geological Survey, pp. 97–146.

Glass, G. B., 1975, "Analyses and Measured Sections of 54 Wyoming Coal Samples," Report of Investigations No. 11, Wyoming Geological Survey, 219 pp.

Gluskoter, H. J., and Hopkins, M. E., 1970, "Distribution of Sulfur in Illinois Coals," Guidebook Series No. 8, Illinois State Geological Survey, pp. 89–95.

Gluskoter, H. J., and Lindahl, P. C., 1973, "Cadmium—Mode of Occurrence in Illinois Coals," *Science*, Vol. 188, pp. 264–266.

Gluskoter, H. J., and Rees, O. W., 1964, "Chlorine in Illinois Coal," Circular 372, Illinois State Geological Survey, 23 pp.

Gluskoter, H. J., 1977, "Trace Elements in Coals: Occurrence and Distribution," Circular 499, Illinois State Geological Survey, 154 pp.

Gluskoter, H. J., and Simon, J. A., 1968, "Sulfur in Illinois Coals," Circular 432, Illinois State Geological Survey, 28 pp.

Harshman, E. N., 1974, "Distribution of Elements in Some Roll-Type Uranium Deposits: Formation of Uranium Ore Deposits," *Proceedings*, International Atomic Energy Agency Symposium, Athens, Greece, pp. 169–183.

Hatch, J. R., 1976, "Sphalerite in Coals from Southeastern Iowa, Missouri, and Southeastern Kansas," Open-File Report 76-796, US Geological Survey, 26 pp.

Hatch, J. R., 1979, "Distribution of Zinc, and Cadmium in Coals from the Eastern and Western Regions, Interior Coal Province," *Abstracts of Papers*, 9th International Congress of Carboniferous Stratigraphy and Geology, p. 86.

Hatch, J. R., in press, "Element Geochemistry," *Geological Investigations of the Vermillion Creek Coal Bed in the Eocene (Niland Tongue) of the Wasatch Formation, Sweetwater County, Wyoming*, Professional Paper, US Geological Survey.

Hatch, J. R., and Affolter, R. H., 1981, "Weathering Induced Chemical Changes in Surface Coals from the Paleocene Fort Union Formation, Red Rim Area, Sweetwater and Carbon Counties, Wyoming," *Bulletin*, American Association of Petroleum Geologists, Vol. 65, No. 5, p. 936.

Hatch, J. R., Affolter, R. H., and Davis, F. D., 1979, "Chemical Analyses of Coal from the Blackhawk Formation, Wasatch Plateau Coal Field, Carbon, Emery, and Sevier Counties, Utah," Special Studies 49, Utah Geological and Mineral Survey, pp. 69–102.

Hatch, J. R., Avcin, M. J., and Van Dorpe, P. E., (in press) "Element Geochemistry of Cherokee Group Coals

(Middle Pennsylvanian) from South-Central and Southeastern Iowa," Technical Paper 5, Iowa Geological Survey.

Hatch, J. R., Gluskoter, H. J., and Lindahl, P. C., 1976a, "Sphalerite in Coals from the Illinois Basin," *Economic Geology*, Vol. 71, pp. 613–624.

Hatch, J. R., Madden, D. H., and Affolter, R. H., 1979a, "Chemical Analyses of Coal and Coal-Associated Rock Samples from the Coalmont Formation, McCallum and Coalmont Areas, North Park, Jackson County, Colorado," Open-File Report 79-1099, US Geological Survey, 42 pp.

Hatch, J. R., and Swanson, V. E., 1977, "Trace Elements in Rocky Mountain Coals," *Geology of Rocky Mountain Coal—A Symposium*, D. K. Murray, ed., Research Series No. 1, Colorado Geological Survey, pp. 143–165.

Heckel, P. H., 1977, "Origin of Phosphatic Black Shale Facies in Pennsylvanian Cyclothems of Mid-continent North America," *Bulletin*, American Association of Petroleum Geologists, Vol. 61, No. 7, pp. 1045–1068.

Henderson, J. A., Oman, C. L., and Coleman, S. L., 1979, "Analyses of Coal Samples Collected 1976–1977," Research Publication 33, Virginia Division of Mines, 135 pp.

Hildebrand, R. T., et al., 1981, "Geology and Chemical Analyses of Coal and Coal-Associated Rock Samples, Williams Fork Formation (Upper Cretaceous), Northwestern Colorado," Open-File Report 81-1348, US Geological Survey, 94 pp.

Hildebrand, R. T., and Hatch, J. R., 1977, "The Distribution of Sodium and Alkaline-Earth Elements in Coal of the Rocky Mountain and Northern Great Plains Provinces," *Abstracts with Programs*, Geological Society of America, Vol. 9, No. 7, pp. 1015–1016.

Khalsa, N. S., and Ladwig, L. R., 1981, "Colorado Coal Analyses, 1976–1979," Information Series 10, Colorado Geological Survey, 364 pp.

Mapel, W. J., and Hail, W. J., Jr., 1959, "Tertiary Geology of the Goose Creek District, Cassia County, Idaho, Box Elder County, Utah, and Elko County, Nevada," Bulletin 1055-H, US Geological Survey, pp. 217–254.

Masursky, H., 1962, "Uranium-Bearing Coal in the Eastern Part of the Red Desert Area, Wyoming," Bulletin 1099-B, US Geological Survey, 149 pp.

Moore, G. W., Melin, R. E., and Kepferle, R. C., 1959, "Uranium-Bearing Lignite in Southwestern North Dakota," Bulletin 1055-E, US Geological Survey, pp. 147–166.

Moore, R. T., 1977, "Chemical Analyses of Coal Samples from the Black Mesa Field, Arizona," Circular 18, The Arizona Bureau of Mines, 14 pp.

Pipiringos, G. N., 1961, "Uranium-Bearing Coal in the Central Part of the Great Divide Basin," Bulletin 1099-A, US Geological Survey, 104 pp.

Roehler, H. W., 1979, "Vermillion Creek Coal Bed, High-Sulfur, Radioactive Coal of Paludal-Lacustrine Origin in Wasatch Formation of Vermillion Creek Basin, Wyoming and Colorado," *Bulletin*, American Association of Petroleum Geologists, Vol. 63, No. 5, p. 839.

Swanson, V. E., et al., 1976, "Collection, Chemical Analysis, and Evaluation of Coal Samples in 1975," Open-File Report 76-468, US Geological Survey, 503 pp.

Tissot, B. P., and Welte, D. A., 1978, *Petroleum Formation and Occurrence*, Springer-Verlag, New York, 538 pp.

Trent, V. A., et al., 1982, "Chemical Analyses and Physical Properties of 12 Coal Samples from the Pocahontas Field, Tazwell County, Virginia, and McDowell County, West Virginia," Bulletin 1528, US Geological Survey, 37 pp.

Turekian, K. K., and Wedepohl, K. H., 1961, "Distribution of the Elements in Some Major Units of the Earth's Crust," *Bulletin*, Geological Society of America, Vol. 72, No. 2, pp. 175–192.

Wedge, W. K., and Hatch, J. R., 1980, "Chemical Composition of Missouri Coals," Geological Survey Report of Investigations 63, Missouri Department of Natural Resources, 102 pp.

Whitehurst, D. D., 1978, "A Primer on the Chemistry and Constitution of Coal," *Organic Chemistry of Coal*, J. W. Larsen, ed., Symposium Series 71, American Chemical Society, pp. 1–35.

Zeller, H. D., and Schopf, J. M., 1959, "Core Drilling for Uranium-Bearing Lignite in Hardin and Perkins Counties, South Dakota, and Bowman County, North Dakota," Bulletin 1055-C, US Geological Survey, pp. 59–96.

Zubovic, P. Stadnichenko, T. and Sheffery, N. B., 1960, "The Association of Some Minor Elements with Organic and Inorganic Phases of Coal," Professional Paper 400-B, US Geological Survey, pp. B84–B87.

Zubovic, P., Stadnichenko, T., and Sheffery, N. B., 1961, "Chemical Basis of Minor-Element Associations in Coal and Other Carbonaceous Sediments," Professional Paper 424-D, US Geological Survey, pp. D345–D348.

9

Metal Concentrations in Some Marine Black Shales of the United States

George A. Desborough and Forrest G. Poole

US Geological Survey, Denver, CO

ABSTRACT

Marine black-shale samples of Paleozoic age in the western United States that exceed 10 wt % organic carbon contain as much as 5000 ppm vanadium, 5000 ppm molybdenum, 1300 ppm nickel, 600 ppm cobalt, 16 ppm silver, 11,700 ppm chromium, 18,800 ppm zinc, 280 ppm cadmium, and 350 ppm selenium. Depositional settings include: (1) continental-rise and marginal-ocean basins, (2) continental-shelf and foreland basins, and (3) cratonic-platform embayments. Most of these metalliferous organic-rich shales have not been examined for resources because rapid weathering and poor exposures preclude easy sampling of unoxidized rock necessary to determine actual concentration and residence of indigenous metals. Generally, when interval thickness is considered, the marine black shales of Paleozoic age in the western United States are richer in metals than marine upper Paleozoic black shales in the eastern United States.

INTRODUCTION

For more than half a century, certain black organic-rich shales have been known to contain anomalously high concentrations of several metals. Metal concentrations in black shales have been studied in many parts of the world as indicated by the large annotated bibliography compiled by Tourtelot (1970). In the United States, direct production of metals from organic-rich shales has not been accomplished, although some localities have been prospected on the surface or explored by drilling during the past 30 years and samples from a few deposits have been evaluated for extraction of metals in commercial and government laboratories.

Our studies have emphasized the potential for metal resources in organic-rich Paleozoic marine shales of the western United States. We are concerned specifically with shales that (1) contain more than 10 wt % of organic carbon, (2) have certain valuable metals which exhibit relatively high solubilities in seawater, such as vanadium, molybdenum, nickel, and zinc, and (3) were not derived from nearby hydrothermal or volcanogenic sources. Holland (1979) recently discussed the relative solubilities of these and other elements in seawater, and pointed out that their relative abundance in marine shales is generally proportional to their solubility in seawater.

The Upper Devonian and Mississippian Chattanooga shale and equivalent strata which underlie large areas in eastern and southeastern United States "has been regarded as a classic example of a syngenetic metal-rich black shale" (Vine and Tourtelot, 1970, p. 266) for 20 years. A large number of samples from the Chattanooga shale have been studied recently for their metal content (Leventhal, 1978, 1979; Shaffer et al., 1981). These data, as well as those for metalliferous Pennsylvanian black shales in Illinois and Indiana (Coveney, 1981), are compared with the metalliferous shales in the western United States.

Organic-rich metalliferous marine strata are found in three general depositional settings in the western United States: (1) continental-rise and marginal-ocean basins, (2) continental-shelf and foreland basins, and (3) cratonic-platform embayments. Conditions necessary for accumulation of significant metal concentrations include: (1) marine basin setting, (2) high content of organic carbon, (3) low content of calcite and dolomite, (4) high content of sapropel (mostly marine algae) and low content of terrestrial plant material, (5) slow rate of deposition of mineral detritus and (or) chemical precipitates, and (6) anoxic environments of deposition and diagenesis (Poole and Desborough, 1981b).

This report is a summary of several aspects of the three depositional types of metalliferous organic-rich shales in the western United States represented by the following units:

I. Continental-rise and marginal-ocean basin	Gibellini and Cole Creek facies of the Woodruff formation, southern Fish Creek central and north-central Nevada
II. Continental shelf and foreland basin	Meade Peak member of the Phosphoria formation, southeastern Idaho and adjacent states.
III. Cratonic-platform embayment	Heath formation, Little Snowy Mountains, Fergus County, central Montana.

Subsequent order of discussion of these shale units is based on the chronological discovery of their anomalously high content of metal. Fig. 1 shows the general location of these three occurrences. We discuss general mineralogy, concentration and occurrence of metals, alteration by ground water and weathering, types of associated strata, and thermal maturity.

Consideration of the usual parameters critical for resource development, such as grade, thickness, homogeneity, and tonnage, is made more difficult because these rocks are so fine grained that physical liberation or separation and preconcentration of metalliferous phases has not so far been feasible. Possible extraction method(s) must include treating the rock in bulk by hydrometallurgical methods. The host mineral or material for each metal of potential value must be known in order to design for extraction. Both optical microscopy and electron microprobe techniques have been used here to study the mineralogical and chemical associations of the metals in black shales.

MEADE PEAK MEMBER OF THE PHOSPHORIA FORMATION, SOUTHEASTERN IDAHO AND WESTERN WYOMING

The most extensively studied metalliferous and organic-rich unit in the western United States is the Meade Peak member of the Permian Phosphoria formation, which occupies large areas of southeastern Idaho, southwestern Montana, northern Utah, and western Wyoming. For more than 30 years, the member has been known to contain, in addition to phosphate, relatively high concentrations of vanadium (Ravitz et al., 1947). Although several metallurgical extraction methods for vanadium have been evaluated for these beds since the 1940s, recovery of vanadium has been feasible only as a byproduct of phosphate production. Impediments to direct mining and processing of the vanadium-rich beds associated with phosphate include their very thin (e.g., <1 m) nature, and the expensive roasting processes for recovery of vanadium.

These metalliferous organic-rich beds are associated with a variety of strata including pelletal phosphorite, dolomitic and calcareous mudstone and siltstone, dolomite, and limestone. Beds range in thickness from a few centimeters to several meters, and the thicker beds have been correlated throughout the region (McKelvey, 1946, plate 7). The most metal- and organic-rich beds of the Meade Peak member occur in a north-trending belt about 170 km long and 85 km wide along the Idaho-Wyoming state line where the member has a total thickness of 20 to 80 m. Regional studies of the metal and organic-carbon content by Maughan (1976) showed that the highest organic carbon content is essentially coincident with the highest concentrations of several metals, including silver, vanadium, zinc, chromium, nickel, molybdenum, and copper.

Exposures of the Meade Peak member in the eastern Idaho-western Wyoming region occur in north- and northwest-trending fault blocks composed of strata that generally dip east or northeast. These Permian strata are allochthonous and

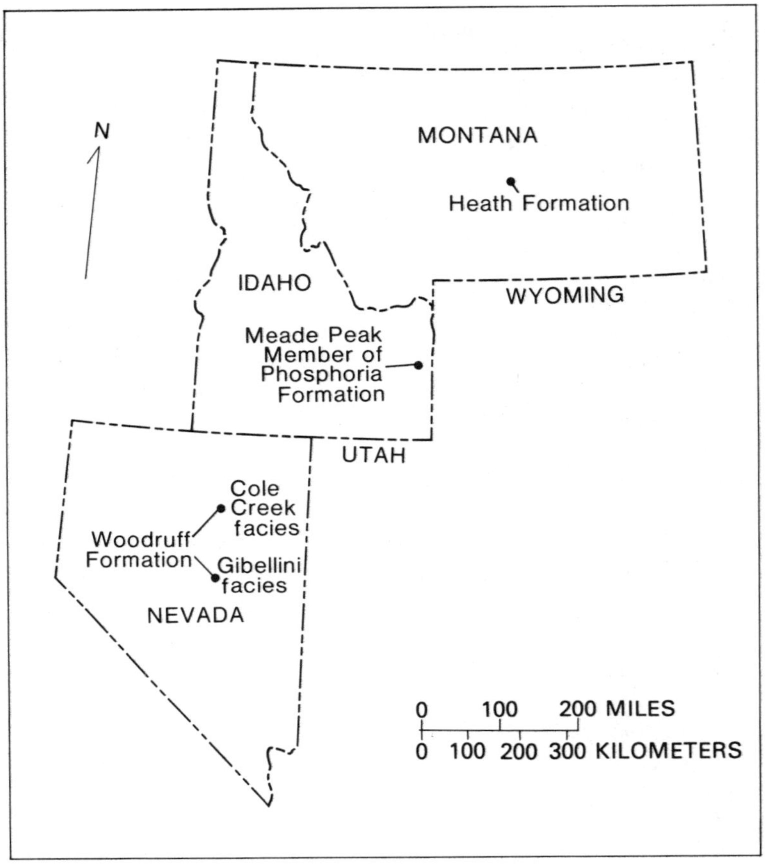

FIG. 1. Map showing locations of metalliferous organic-rich shales of the Woodruff formation in Nevada, Meade Peak member of the Phosphoria formation in Idaho, and the Heath formation in Montana.

were moved eastward many tens of kilometers during the Cretaceous Sevier orogeny. Sevier folding and thrusting are important factors because the inclined attitude of the beds, as well as the associated fractures and faults, have permitted significant ground-water movement and mobilization of metals, as well as dissolution of calcite in many areas. In addition, the relatively incompetent organic-rich beds were plastically deformed during tectonism which resulted in significant lateral discontinuities in unit thickness. The thermal effect of burial—possibly to 6100 m—caused the original organic matter or kerogen to be converted to finely crystalline graphite (Powell and others, 1975, fig. 3).

Gulbrandsen (1966, p. 774) recognized that "weathering effects are probably very important" in metal distribution, but he (1975, 1979) and others (Maughan, 1976; Desborough, 1977) did not determine the specific enrichment or depletion effects that ground water and (or) weathering can have on metal concentration and redistribution. This omission was due in part to the paucity of fresh samples. Our most recent chemical, mineralogical, and microscopic studies show that unaltered samples are coherent, have no porosity on a microscopic scale, have high physical strength and contain some iron sulfides, and minor but detectable amounts of dolomite and (or) calcite (i.e., greater than 0.2 wt % mineral carbon). A good example of metal enrichment of an organic-rich bed by ground water is the "carbon seam," which is a thin (≥ 1 m) bed recognized throughout the region (shale bed of McKelvey, 1946, plate 7). A channel sample collected across this 1-m-thick bed at the Mabie Canyon mine contained 117 ppm (117 g/t) silver, 0.84% vanadium, 1.6% chromium, and 32% organic carbon (Desborough and Maughan, 1977). Comparable values for silver have never been reported from beds in the Meade Peak member; the values of both vanadium and chromium also are as high as, or higher than, any others reported. This sample was noncoherent (poorly consolidated) and contained no sulfides or mineral carbon (Desborough, 1977). We believe that these high metal values are due in part to enrichment by ground water.

The most recent mineralogical and quantitative chemical studies were made on 31 individual lithologic units aggregating 20.8 m in thickness collected by E. K. Maughan in 1976 from an underground crosscut tunnel near Pritchard Creek, Caribou Range, Idaho (cen. W½ of Sec. 6, T. 1 N., R. 43 E., Bonneville County). All samples from each lithologic unit were free of effects of ground-water alteration and (or) weathering. The highest metal concentrations generally occur in organic-rich beds less than 10 cm thick, and are as follows, in parts per million: silver 22, vanadium 5200, zinc 8000, cadmium 870, selenium 280, chromium 11,700, nickel 1400, uranium 125, and molybdenum 870. A summary of the concentrations of metals, phosphate, sulfur, and organic carbon is given in Table 1 for all beds, as well as those beds which are rich in organic carbon and phosphate. The concentration of silver is not consistently related to the organic carbon concentration, although the highest values of 21 and 22 ppm occur in two beds with the highest concentrations of organic carbon (21.6 and 22.8 wt %). Uranium is associated with both phosphate-rich and organic-rich beds.

X-ray diffraction studies indicate that the dominant minerals in the organic-rich beds are quartz, apatite, dolomite, or calcite; illite is always present. Potassium feldspar, pyrite, albite, fluorite, and sphalerite are generally present in detectable quantities.

Microscopic and electron microprobe studies of polished slabs and polished thin sections of these unaltered samples suggest that the principal residences of several elements are as follows:

Element	*Principal Residence*
Vanadium	Organic material, unidentified iron sulfide, unidentified titanium-chromium-iron-vanadium oxides
Chromium	Organic material, 10Å mica, iron sulfide and oxide
Nickel	Unidentified iron sulfide
Zinc, cadmium	Sphalerite
Selenium	Pyrite
Uranium	Apatite, organic material
Molybdenum	Organic material

We consider present knowledge of metal concentration, occurrence, and distribution in strata of the Meade Peak member to be only in a preliminary state.

WOODRUFF FORMATION, CENTRAL NEVADA

The Devonian Woodruff formation in central Nevada (Fig. 1) has been prospected and drilled for evaluation of potential vanadium resources since 1960 (Roberts et al., 1967, p. 86). The eugeosynclinal Upper Devonian Gibellini facies of the Woodruff formation is a vanadiferous deposit that caps ridge 7129 in the southern Fish Creek Range in the southern part of Eureka County (Fig. 1). On the basis of about 50 holes drilled by Noranda Exploration, Inc. in the mid-1970s, the dimensions of the deposit are about

Table 1. Summary of Concentration of Elements in Lithologic Units from the Meade Peak Member of the Phosphoria Formation*

Ag	V	Zn	Cd	Se	Cr	Ni	U	Mo	P_2O_5	S	Organic carbon
Weighted mean concentration for all 31 lithologic units; aggregate thickness = 20.8 m											
8.3	540	1670	75	77	1550	223	43	102	8.6	1.8	6.2
Weighted mean concentration for 6 lithologic units with greater than 10 wt % organic carbon; aggregate thickness = 2.46 m											
11	1400	2850	188	144	2880	567	80	244	10.2	3.0	12.8
Weighted mean concentration for 9 lithologic units with greater than 15 wt % P_2O_5; aggregate thickness = 5.32 m											
10	348	1780	67	54	870	175	79	38	26.1	1.65	5.28

*Collected from an underground crosscut at Pritchard Creek near the Snake River in the Caribou Range, southeastern Idaho. Values are in ppm, except for P_2O_5, S, and organic carbon, which are in weight percent.

700 m long, 300 m wide, and 100 m thick. The deposit is part of an allochthonous mass in which the Gibellini facies is separated by a low-angle fault from the underlying allochthonous Bisoni facies of the Lower Mississippian Webb formation and locally from the autochthonous Mississippian Antelope Range formation (Fig. 2). Both allochthonous masses are exotic slide blocks emplaced during Mississippian tectonic activity.

Detailed studies of surface exposures and of drill cuttings from the vanadiferous deposit were made recently by Desborough et al. (1979) and Böhlke et al. (1981). The deposit consists of interbedded organic-rich siliceous mudstone, siltstone, and chert as much as 100 m thick that has been oxidized and (or) leached by surficial weathering to depths of 45 m (Fig. 2). The fresh rock is a low-grade oil shale yielding an average of 25 L/t. The oxidized zone is enriched in vanadium and selenium, but depleted in iron, nickel, zinc, arsenic, and molybdenum. The underlying eugeosynclinal-like Bisoni facies consists chiefly of dolomitic and (or) argillaceous siltstone, siliceous mudstone, chert, and minor limestone, sandstone, and grit that have a much lower content of metals and organic carbon.

The bedded siliceous rocks of the Gibellini facies, where unaltered, are irregularly laminated and very fine grained olive-black, olive-gray, brownish-black, and brownish-gray rocks, whereas the oxidized strata are much lighter shades of gray and brown. The most intensely oxidized rocks are very light colored with pale-brown, red, or yellow tints due to the presence of vanadium, selenium, and iron oxides. On the basis of these color differences, the oxidized zone is easily recognized in surface exposures and in drill-hole cuttings which were taken as 1.5 m composite samples during the drilling program conducted by Noranda.

Most fresh rock of the deposit is fractured and veined with bluish-white chalcedonic quartz and sparse black solid bitumen. Due to the high content of dispersed kerogen and the fine-grained character of the rock, other minerals are identifiable only by microscopy or X-ray diffraction methods. Polished thin sections show that sparse pyrite, marcasite, sphalerite, round and flattened algal cysts, and chalcedony-filled veinlets are the most conspicuous microscopic features. Subhedral and euhedral pyrite and minor marcasite occur in some of the opaline silica-filled fractures, which are apparently of diagenetic origin. X-ray diffraction studies of 20 composite samples from two drill holes show, in addition to quartz, minor amounts of apatite, potassium feldspar, illite, iron-rich chlorite, sphalerite, and pyrite. The sapropelic organic matter is dominantly very fine grained kerogen consisting of flaky and stringy fragments of organisms smaller than 15 mμ and oriented parallel to the bedding.

A representative fresh sample contained 11.5 wt % of organic carbon (0.7% chloroform-extractable bitumen and 10.8% kerogen). Fischer assay oil yield for this sample was 51.7 L/t, and total sulfur was 2.2 wt %. Conodonts in the sample are pale yellow (CAI = 1), which suggests less than 60% fixed carbon and a maximum rock temperature of less than 60°C since deposition. The vitrinite reflectance mean value of 0.43% for this sample is compatible with the conodont color alteration and other organic geochemical data and indicates that the organic matter in the rock is immature and has not reached the postdiagenetic stage of catagenesis. Metal concentrations in parts per million in this fresh sample are as follows: vanadium 3140, zinc 2570, molybdenum 252, nickel 315, and selenium 68. Microscopic, electron microprobe, and X-ray diffraction studies show that zinc occurs in fine grains of sphalerite; nickel occurs in the marcasite and pyrite; selenium occurs chiefly in organic matter, but also was detected in pyrite; vanadium and molybdenum occur in the organic material; and a single 5-mμ-long grain of molybdenite was identified in the sample.

The mean concentrations of several metals in samples from 15 m of the oxidized zone and 15 m of the unoxidized zone in a representative drill hole are given in Table 2. Vanadium and selenium are enriched in the oxidized zone, but iron, nickel, zinc, arsenic, and molybdenum are depleted. The relatively low uranium concentrations show no significant differences in the two zones. The range in metal values in parts per million for 20 3.0-m composite samples from the unoxidized strata in two drill holes are as follows: vanadium 1500–3700, zinc 3300–18,800, molybdenum 60–700, nickel 210–1330, selenium 20–85, and arsenic 35–105.

We estimate that the contained vanadium resource in both the oxidized and unoxidized vanadiferous strata at this central Nevada locality is between 90 and 154 million kg of V_2O_5. The major uncertainty in these numbers reflects variation in volume estimates owing to the undu-

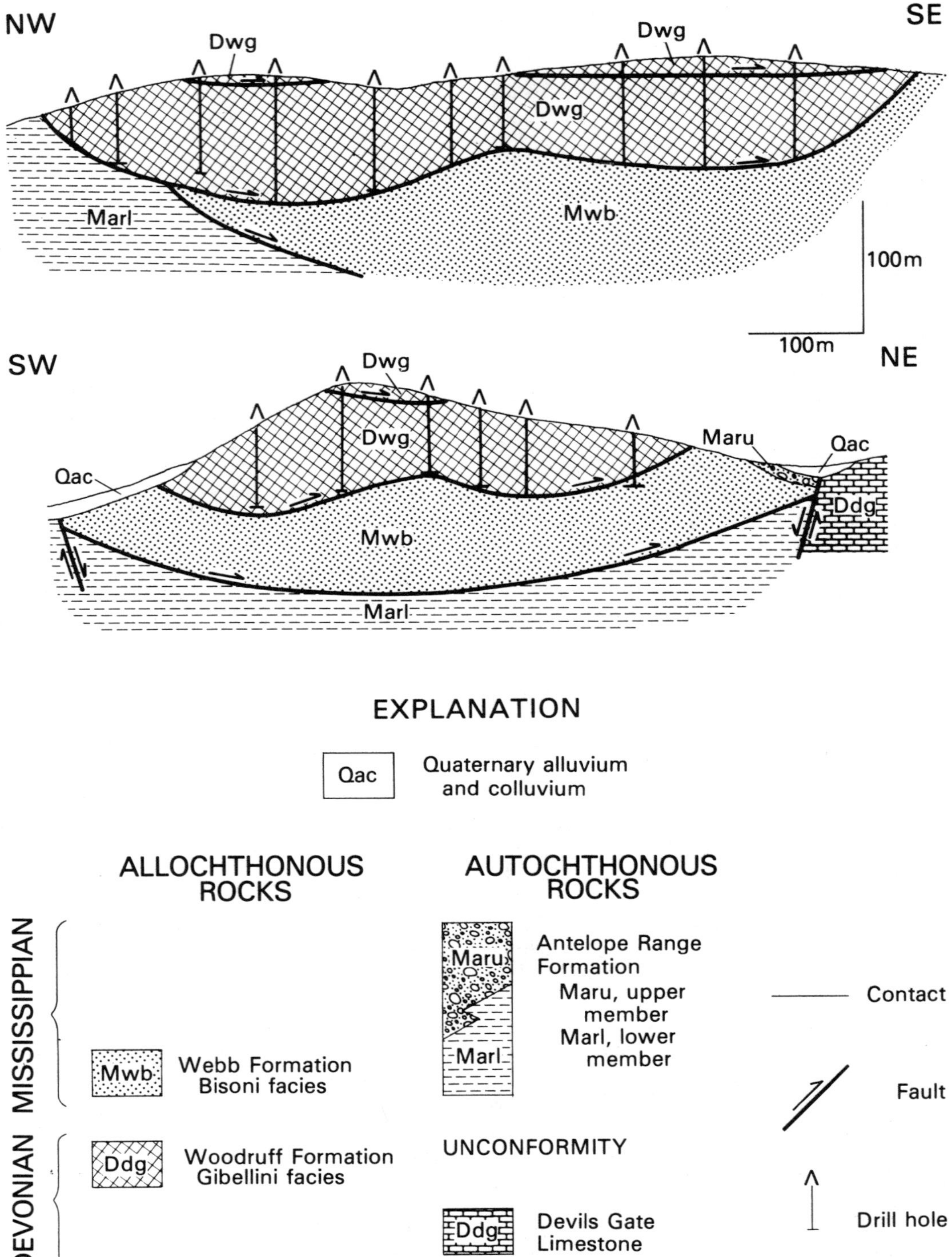

FIG. 2. Schematic axial and medial cross sections of the metalliferous Gibellini facies of the Woodruff formation, and Bisoni facies of the Webb formation at the Gibellini and Bisoni properties, southern Fish Creek Range, Eureka County, Nevada.

Table 2. Metal Distribution in Oxidized and Unoxidized Strata of the Woodruff Formation in Twenty Samples from Drill Hole NG-12, Gibellini Claims, Southern Fish Creek Range, Eureka County, NV*

Depth m	V	Fe	Ni	Zn	As	Se	Mo	U
			Mean concentration in oxidized strata					
30.5–45.7	3890	5310	<100	370	<30	340	45	50
			Mean concentration in unoxidized strata					
45.7–61	2260	12,800	765	10,900	70	90	330	40

*Sampling interval 1.5 m, composited to 3.0 m. Quantitative X-ray fluorescence analysis by G.N. Green. Values are in ppm.

latory nature of the fault contact with underlying Mississippian strata (Fig. 2). Militating against near-future development of the resource is that (1) it is contained in both oxidized and unoxidized zones, (2) has an average grade of probably less than 0.5% V_2O_5, and (3) the volume of the deposit is relatively small. It is encouraging to note, however, that the occurrence of syngenetic low-grade vanadiferous strata in two widely separated (150 km) allochthonous masses of the Woodruff formation (Cole Creek and Gibellini facies shown on Fig. 1) suggests that similar deposits may occur in the formation elsewhere in Nevada.

Recovery of vanadium has been studied by the US Bureau of Mines (Brooks and Potter, 1974) for one occurrence of the Lower Devonian Cole Creek facies of the Woodruff formation in north-central Nevada about 12 km south of Carlin in southwestern Elko County (Fig. 1). USBM concluded that as much as 20 million tons of shale averaging 0.8% V_2O_5 is amenable to open-pit mining. The potential resource of vanadium is substantial when compared to current consumption. Conversely, complex geologic relations at the Cole Creek locality are poorly understood, and the principal occurrence is a mixture of oxidized and unoxidized strata. Also, some rock units are dolomitic, which makes the sulfuric acid consumption high for conventional extraction processes.

At most localities the fresh Woodruff formation is dark brown or black, owing to its high content of organic carbon, and is chiefly interbedded siliceous mudstone, siltstone, and chert that generally have been called "black shales." Some of the beds appear to be diagenetically silicified fine-grained turbidites. It was noted only recently that these strata are low-grade "oil shales" (Desborough et al., 1979, 1981; Poole and Desborough, 1980, 1981a). Several geologic factors have deterred development of these vanadiferous occurrences in the Woodruff formation. All of these occurrences are allochthonous, with attendant complex folding and faulting both on a large and small scale; these complications generally have limited both the lateral and vertical extent of beds due to tectonic dismemberment and subsequent erosion. Large- and small-scale fracturing both across and along bedding planes permitted relatively deep oxidation at most localities, resulting in two types of "mineralized" zones: a postdiagenetic oxidized zone and an original unoxidized part which contains a high content of organic carbon. Recovery of metals from each of these two types of occurrences would require different pretreatment prior to metal recovery. Little attention has been given to potentially useful energy and metal byproducts that might be considered in conjunction with the recovery of vanadium. For the organic-rich unoxidized strata, no viable technology has been demonstrated for economic extraction of vanadium and other metals.

HEATH FORMATION, CENTRAL MONTANA

The Upper Mississippian Heath formation occurs in an east-trending belt about 320 km long and 30 km wide in central Montana. The formation consists chiefly of limestone, calcareous shales and mudstones, claystones, siltstones, and minor gypsum and metalliferous organic-rich shales. Strata of the Heath formation were deposited on the cratonic platform in an east-trending marine embayment known as the Big Snowy trough. The Heath formation reaches a thickness of about 120 m along the trough axis, but thins abruptly toward the northern and southern trough margins (Smith and Gilmour, 1979, p. X21). After deposition of the Heath formation, extensive erosion resulted in as much as 45 m of relief on the unconformable surface separating the Heath formation from the overlying Tyler formation of Mississipian and Pennsylvanian age. Some black shales in the Heath formation are physically similar to the black shales in the lower part of the Tyler formation; however, the Tyler shales contain relatively low amounts of metals. In addition, along the northern margin of the Big Snowy trough, the Ellis group of Jurassic age unconformably truncates the Heath formation (Smith and Gilmour, 1979, p. X21). Our preliminary studies of the thin, metalliferous organic-rich shales in the Heath formation have been limited to an outcrop belt along the northern flank of the Little Snowy Mountains in Fergus County about 30 to 40 km southeast of Lewistown, MT. We selected this region for further study because oil shales were reported by Miller (1953) and Cox (1973). Moreover, faunal studies had been conducted in the area by Easton (1962), stratigraphic studies were done by Maughan and Roberts (1967) and Harris (1972), and geologic mapping in the area was done by Gardner (1950) and Lindsey (1980).

The first report mentioning metal content of the Heath black shales was that of Vine and Tourtelot (1969) who used the semiquantitative emission spectrographic method in analyzing 28 outcrop samples from both the Heath and Tyler formations in Montana. Miller (1953) and Cox (1979) reported that many black shales in the Heath formation in Fergus County, Montana, were oil shales, based on distillate collected from shale heated in a closed retort and condenser apparatus. They obtained oil yield values up to 108 L/t, but most values were less than about 45 L/t for intervals less than 2 m thick. A recent correlation chart by Cox and Cole (1981) shows that all of the organic-rich shales they studied lie just below, or just above a zone of prominent gypsum beds in the middle of the Heath formation. Our recent preliminary report (Desborough et al., 1981, Tables 1 and 2) on metalliferous Heath oil shales presented values for Fischer assay oil yield and assay of several metals for two outcrop samples: one sample is from a unit in the middle part of the formation, and the other sample is from a unit in the lower part of the formation. The highest metal values and oil yields were from a 1-m-thick zone in the lower Heath as follows (in parts per million): vanadium 4500, nickel 1100, zinc 6900, selenium 120, molybdenum 900, and oil yield of 50 L/t. Subsequent analyses of the same sample from the lower Heath revealed 600 ppm cobalt, 690 ppm chromium, 15 ppm silver, 23.4% organic carbon, and 1.9% sulfur. Reassay of the sample using the material balance Fischer assay method gave an oil yield of 65 L/t. The vitrinite reflectance mean value of 0.38% for the sample from the lower part of the Heath formation indicates that the organic matter in the rock is thermally immature and that the rock has a low thermal history.

Preliminary study of a 9-m interval, exposed partly by bulldozing and backhoe digging, reveals complex lithologic variation as shown on Fig. 3. These strata are in the lower part of the Heath formation and occur in a stream bank in the SW¼NE¼ of Sec. 14, T. 13 N., R. 21 E., on the N Bar Ranch, Fergus County, Montana. Twenty-two individual lithologic units were identified and include limestones, mudstones, gray and black shales, and coal (Fig. 3). Seven of these units are metalliferous black shale beds ranging in thickness from 0.08 to 1.3 m. Channel samples of 5 and 7 kg were taken from a 0.25-m-thick bed between 2 and 3 m, and a 1.3-m-thick bed at 4 m; all of the others were intensely weathered. Fischer assay oil yield from the 0.25-m-thick bed was 120 L/t and the 1.3-m-thick bed gave 50 L/t. These oil-yield values are comparable to the range of 45 to 108 L/t reported by Cox (1973) for beds in the middle part of the Heath formation.

Vertical fracture surfaces and bedding surfaces that showed evidence of leaching and (or) weathering were chipped from the larger rock fragments in each of the two samples. This altered-rock sample aggregated less than 10% by weight of the original 5- and 7-kg samples. Fresh and altered fractions were analyzed by atomic absorption photometry after acid digestion, and the results are shown in Table 3. These data indicate a very significant effect of near-surface processes on the metal content of the black shales of the Heath formation: vanadium concentrations were reduced by 25%, and molybdenum concentrations were reduced by about 75%. These data demonstrate that samples obtained at or near the surface can give only a minimum estimate of the resource potential of these strata. In regard to vanadium, the reaction of the metalliferous Mississippian Heath shales to groundwater leaching and (or) weathering is opposite to that of the organic-rich shales of the phos-

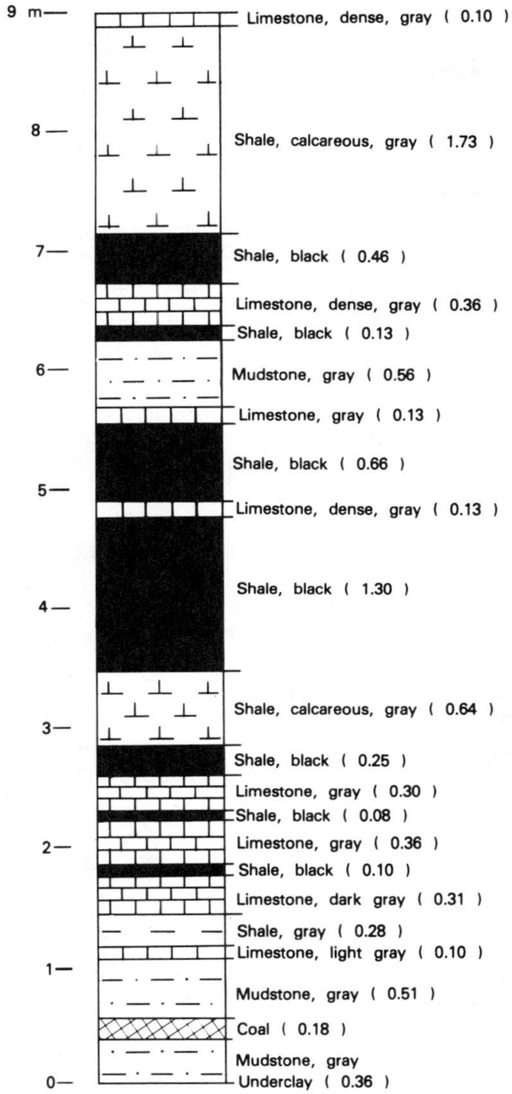

FIG. 3. Metalliferous black shales and associated strata in the lower part of the Heath formation exposed in SW¼ NE¼ Sec. 14, T. 13 N., R. 21 E., N Bar Ranch, Fergus County, Montana. Individual unit thickness in meters.

phatic Meade Peak member of the Permian Phosphoria formation, and also opposite to that of the Gibellini facies of the Devonian Woodruff formation, both of which become enriched in vanadium when subjected to oxidation. The causes of variation in metal enrichment and depletion are not fully understood.

Mineralogical data obtained by X-ray diffraction analysis show that the most metal- and organic-rich shales reported by Desborough et al. (1981) consist chiefly of illite, mixed-layer illite-montmorillonite, quartz, calcite, albite, siderite, pyrite, and gypsum. The host phases of metals of potential economic interest have not yet been directly established by microscopic or electron microprobe analysis. However, the mobilization indicated for both vanadium and molybdenum in samples affected by oxidation (Table 3) suggests that in fresh rock these metals are in either sulfide minerals or organic material, just as in the Meade Peak member of the Phosphoria formation, rather than within the lattice of silicate minerals.

COMPARISON OF METAL CONTENT OF PALEOZOIC SHALES IN THE EASTERN AND WESTERN UNITED STATES

The Upper Devonian and Lower Mississippian Chattanooga shale and equivalent strata in the central, eastern, and southeastern United States have been regarded as "metal-rich" (Vine and Tourtelot, 1970). It seems appropriate to make a comparison of metal values in some Paleozoic strata of the eastern United States with metal values in some Paleozoic strata of the western United States emphasized in this report.

Very thin (≤ 1 m) beds of metalliferous black shale of Pennsylvanian age in Illinois, Indiana, Iowa, Kansas, and Missouri (Coveney, 1981), as well as the much thicker Chattanooga shale and equivalent strata of Late Devonian and Early Mississippian age in several eastern states (Leventhal, 1978, 1979; Shaffer et al., 1981) have recently been studied in detail. The highest metal concentrations reported for these thin Paleozoic beds in the eastern United States are compared with metal values obtained for beds of similar thickness in the western United States in Table 4. The highest contents of molybdenum, nickel, vanadium, and zinc are essentially in the same range, except that one of the western shales is significantly higher in molybdenum (5000 ppm) than any of the others (Table 4). Concentrations of uranium are higher in the eastern shales than those in the nonphosphatic Woodruff and Heath formations, and in the Phosphoria as well. Silver, chromium, and selenium are higher in the

Table 3. Concentration of Vanadium and Molybdenum in Bulk Channel Samples and the Ground Water Leached and (or) Weathered Chips Extracted from the Bulk Samples of Heath Formation Metalliferous Shales

	0.25-m-thick bed		1.30-m-thick bed	
Element	Bulk sample	Altered by ground water or weathering	Bulk sample	Altered by ground water or weathering
Vanadium, ppm	5000	3700	2500	1800
Molybdenum, ppm	5000	1300	1000	270

black shales of the western United States (Table 4). It seems highly unlikely that any economic production will be realized in the near future from such thin beds in either western or eastern shales, unless it is done in conjunction with the production of another commodity, such as coal, oil shale, or phosphate.

Table 5 gives the concentrations of organic carbon and several metals for "average black shale," "metal-rich black shale," and for the high values (95th percentile), and median percentile for the Chattanooga shale (Vine and Tourtelot, 1970). The first four columns of concentration data in Table 5 are from Vine and Tourtelot (1970, Tables 2–4). Their data for the Chattanooga shale were obtained from 51 core samples from five localities in Tennessee and Alabama. The last two columns of data in Table 5 list quantitative analytical results on fresh samples of the Meade Peak member of the Phosphoria formation and the Gibellini facies of the Woodruff formation. Our data in Table 5 are for the weighted mean average in each of the entire 20.8- and 15.7-m-thick intervals, respectively.

The Meade Peak member is significantly more enriched in silver, chromium, molybdenum, nickel, vanadium, and zinc than the Chattanooga shale (Table 5). The Woodruff formation contains significantly higher concentrations of molybdenum, nickel, vanadium, and zinc than the Chattanooga shale (Table 5). If content of relatively high-value metals is the determinant factor in resource development, the organic-rich middle and upper Palezoic marine shales of the western United States are more attractive than

Table 4. Highest Concentrations of Several Metals in Thin Beds of Metalliferous Middle and Upper Paleozoic Black Shales in Western and Eastern United States*

	Western United States			**Eastern United States**		
					Chattanooga shale and equivalents	
	Phosphoria formation, Idaho (this study)	**Woodruff formation, Nevada (this study)**	**Heath formation, Montana (this study)**	**Pennsylvanian shales, Illinois and Indiana (Coveney, 1981)**	**Indiana (Shaffer et al., 1981)**	**Kentucky, New York, Ohio, Tennessee, Virginia, and West Virginia (Leventhal, 1978, 1979)**
Strata thickness, m	**0.3-1**	**0.3-3**	**0.3-1+**	**≤1**	**<1**	**<1**
Element						
Ag	16	--	15	5	--	7
Cr	11,700	500	690	--	200	150
Mo	660	700	5000	1,000	670	207
Ni	780	1,330	1100	--	830	595
Se	230	190	120	--	--	11
U	80	--	--	150	--	91
V	2,000	3,140	5000	5,000	5,240	1800
Zn	7,100	18,800	6900	10,000	17,300	4360

*Values are in ppm; -- = not available; localities for western US are those given in text.

Table 5. Concentrations of Organic Carbon and Several Metals in Average Black Shale, Metal-Rich Black Shale, and Chattanooga Shale Compared with Two Sequences of Metalliferous Paleozoic Shales in the Western United States*

			Chattanooga shale, V&T		Western USA	
			Percentile		Mean for interval	
Element	Average black shale, V&T	Metal-rich black shale, V&T	95th	Median	Meade Peak member 20.8 m	Woodruff formation 15.7 m
C_{org}	3.2	--	14.5	10.4	6.2	--
Ag	<1	5	1	<1	8.3	--
Cr	100	500	100	70	1550	--
Mo	10	150	100	70	102	330
Ni	50	200	150	100	223	765
V	150	700	500	150	540	2,260
Zn	<300	1000	300	<300	1670	10,900

*V&T = Vine and Tourtelot, 1970; m = meters; values are in ppm except organic carbon which is in weight percent; -- = not reported.

the Upper Devonian and Lower Mississippian Chattanooga shale and equivalent strata in the eastern United States.

Appropriate evaluation of the resource potential of these strata will require a substantial effort in sedimentological, tectonic, metallurgical extraction, and analytical investigations, including assessment of the combustible components of these strata which might be used in sequence or concomitantly for economic utilization of metal and energy resources.

ACKNOWLEDGMENTS

This present paper is a summary of data derived from our study of several metalliferous marine shales in the western United States. Raymond Coveney, Jr., and Nelson Shaffer provided comparative data on metal concentrations of some upper Paleozoic marine shales in the mid-continent and eastern United States. Coveney, Shaffer, Harry Tourtelot, and Russell Tysdal critically reviewed the manuscript and provided many helpful suggestions that benefited the paper.

REFERENCES

Böhlke, J. K., et al., 1981, "Spectroscopic Geochemical Study of Vanadiferous Marine Sediments of the Gibellini Claims, Southern Fish Creek Range, Eureka County, Nevada," Open-File Report 81-32, US Geological Survey, 37 pp.

Brooks, P. T. and Potter, G. M. 1974, "Recovering Vanadium from Dolomitic Nevada Shale," Report of Investigations No. 7932, US Bureau of Mines, 20 pp.

Coveney, R. M., Jr., 1981, "Thin Black Pennsylvanian-Age Shales of the Mid-Continent U.S.—A Neglected Metallic Resource?," *Abstracts with Programs*, Geological Society of America, Vol. 13, No. 7, p. 432.

Cox, W. E., 1973, "Fergus County Research Project; Evaluation of Heath-Tyler Oil Shales," with *Laboratory Analyses* by J. W. Blumer, Open-File Report No. 15, Montana Bureau of Mines and Geology, 83 pp.

Cox, W. E. and Cole, G. A., 1981, "Oil Shale Potential in the Heath and Tyler Formations, Central Montana," Geologic Map 19, Montana Bureau of Mines and Geology.

Desborough, G. A., 1977, "Preliminary Report on Certain Metals of Potential Economic Interest in Thin Vanadium-Rich Zones in the Meade Peak Member of the Phosphoria Formation in Western Wyoming and Eastern Idaho," Open-File Report 77-341, US Geological Survey, 27 pp.

Desborough, G. A. and Maughan, E. K., 1977, "Concentration and Residence of Metals in the Meade Peak Member of the Phosphoria Formation, Eastern Idaho—Potential Metal Resources of the Future," *Program with Abstracts*, Geological Association of Canada-Mineral Association of Canada-Society of Economic Geologists-Canadian Geophysical Union, Vol. 2, p. 15.

Desborough, G. A., Poole, F. G., and Green, G. N., 1981, "Metalliferous Oil Shales in Central Montana and Northeastern Nevada," Open-File Report 81-121, US Geological Survey, 14 pp.

Desborough, G. A., et al., 1979, "Metals in Devonian Kerogenous Marine Strata at Gibellini and Bisoni Properties in Southern Fish Creek Range, Eureka County, Nevada," Open-File Report 79-530, US Geological Survey, 31 pp.

Easton, W. H., 1962, "Carboniferous Formations and Faunas of Central Montana," Professional Paper 348, US Geological Survey, 126 pp.

Gardner, L. S., 1950, "Geology of the Button Butte-Forest Grove Area, Fergus County, Montana," Oil and Gas Investigations Preliminary Map 106, US Geological Survey.

Gulbrandsen, R. A., 1966, "Chemical Composition of Phosphorites of the Phosphoria Formation," *Geochemica et Cosmochimica Acta*, Vol. 30 pp. 769–778.

Gulbrandsen, R. A., 1975, "Analytical Data on the Phosphoria Formation, Western United States," Open-File Report 75-554, US Geological Survey, 45 pp.

Gulbrandsen, R. A., 1979, "Preliminary Analytical Data on the Meade Peak Member of the Phosphoria Formation at Hot Springs Underground Mine, Trail Canyon Trench and Conda Underground Mine, Southeastern Idaho," Open-File Report 79-369, US Geological Survey, 35 pp.

Harris, W. L., 1972, "Upper Mississippian and Pennsylvanian Sediments of Central Montana," Unpublished Ph.D. thesis, University of Montana, Missoula, 252 pp.

Holland, H. D., 1979, "Metals in Black Shales—A Reassessment," *Economic Geology*, Vol. 74, pp. 1676–1680.

Leventhal, J. S., 1978, "Trace Elements, Carbon and Sulfur in Devonian Black Shale Cores from Perry County, Kentucky; Jackson and Lincoln Counties, West Virginia; and Cattaraugus County, New York," Open-File Report 78-504, US Geological Survey, 32 pp.

Leventhal, J. S., 1979, "Chemical Analysis and Geochemical Associations in Devonian Black Shale Core Samples from Martin County, Kentucky; Carroll and Washington Counties, Ohio; Wise County, Virginia; and Overton County, Tennessee," Open-File Report 79-1503, US Geological Survey, 52 pp.

Lindsey, D. L., 1980, "Reconnaissance Geologic Map of the Big Snowies Wilderness and Contiguous Rare II Study Areas, Montana," Map MF-1243A, US Geological Survey.

Maughan, E. K., 1976, "Organic Carbon and Selected Element Distribution in the Phosphatic Shale Members of the Permian Phosphoria Formation, Eastern Idaho and Parts of Adjacent States," Open-File Report 76-577, US Geological Survey, 92 pp.

Maughan, E. K. and Roberts, A. E., 1967, "Big Snowy and Amsden Groups and the Mississippian-Pennsylvanian Boundary in Montana," Professional Paper 554-B, US Geological Survey, pp. B1–B27.

McKelvey, V. E., 1946, "Preliminary Report on Stratigraphy of the Phosphatic Shale Member of the Phosphoria Formation in Western Wyoming, Southeastern Idaho, and Northern Utah," Unpublished Ph.D. Thesis, University of Wisconsin, Madison, 162 pp.

Miller, R. N., 1953, "Oil Shales in Montana," Unpublished B.S. Thesis, Montana College of Mineral Science and Technology, Butte, 78 pp.

Poole, F. G. and Desborough, G. A., 1980, "Oil and Metals in Ordovician and Devonian Kerogenous Marine Strata of Central Nevada [Abstract]," *Bulletin*, American Association of Petroleum Geologists, Vol. 64, No. 5, p. 767.

Poole, F. G., and Desborough, G. A., 1981a, "Metals in Organic Matter in Devonian Rocks of Central Nevada [Abstract]," Geophysical Laboratory Abstracts for *The Geochemistry of Organic Matter in Ore Deposits*, Carnegie Institution of Washington, p. 112.

Poole, F. G., and Desborough, G. A., 1981b, "Geologic Factors Controlling Occurrence of Low-Temperature Metal Deposits in Paleozoic Marine Shales in Nevada," *Abstracts with Programs*, Geological Society of America, Vol. 13, No. 7, p. 530.

Powell, T. G., Cook, P. K., and McKirdy, D. M., 1975, "Organic Geochemistry of Phosphorites; Relevance to Petroleum Genesis, *Bulletin*, American Association of Petroleum Geologists, Vol. 59, No. 4, pp. 618–632.

Ravitz, S. F., et al., 1947, "Treatment of Idaho-Wyoming Vanadiferous Shales," AIME Technical Publication, No. 2178, 14 pp.

Roberts, R. J., Montgomery, K. M., and Lehner, R. E., 1967, "Geology and Mineral Resources of Eureka County, Nevada, Bulletin 64, Nevada Bureau of Mines, 152 pp.

Shaffer, N. R., et al., 1981, "Heavy Metals in Organic-Rich New Albany Shale of Indiana," *Abstracts with Programs*, Geological Society of America, Vol. 13, No. 7, p. 551.

Smith, D. L. and Gilmour, E. H., 1979, "The Mississippian and Pennsylvanian (Carboniferous) Systems in the United States—Montana," Professional Paper 1110-X, US Geological Survey, pp. X1–X32.

Tourtelot, E. B., 1970, "Selected Annotated Bibliography of Minor-Element Content of Marine Black Shales and Related Sedimentary Rocks, 1930–65," Bulletin 1293, US Geological Survey, 118 pp.

Vine, J. D. and Tourtelot, E. B., 1969, "Geochemical Investigations of Some Black Shales and Associated Rocks," Bulletin 1314-A, US Geological Survey, 43 pp.

Vine, J. D., and Tourtelot, E. B., 1970, "Geochemistry of Black Shale Deposits—A Summary Report," *Economic Geology*, Vol. 65, pp. 253–272.

10

Nonbauxite Alumina Resources

Haydn H. Murray

Department of Geology, Indiana University, Bloomington, IN

ABSTRACT

Although alumina constitutes about 15% of the earth's crust, it is expensive to separate and purify for the production of aluminum with the exception of the alumina in bauxite. The United States has a variety of nonbauxite resources that could substitute for bauxite, including high alumina clays, alunite, dawsonite in oil shale, anorthosite, and nepheline syenite. Each of these resources is discussed from the standpoint of geology, geography, processing, and potential recovery. Potential byproducts include silica from clays, potash and sulfuric acid from alunite, soda ash from dawsonite, and soda ash and potash from anorthosite and nepheline syenite.

INTRODUCTION

At the present time nearly all of the aluminum metal produced in the United States, and for that matter in the world, is obtained by the electroreduction of alumina extracted from bauxite by the Bayer process. The Bayer process is an alkaline leach under pressure and at moderately high temperature. Bauxite, a rock consisting chiefly of aluminum hydroxide minerals, is the principal ore used by the world's aluminum industry. The aluminum industry utilizes about 90% of the bauxite mined and the remaining 10% is used in making refractories, catalysts, cements, chemicals, abrasives, and other materials. The types of bauxite are classified on the basis of the composition of the aluminum hydroxide minerals present which are (1) trihydrate consisting mainly of gibbsite, $Al_2O_3 \cdot 3H_2O$; (2) monohydrate consisting mainly of boehmite, $Al_2O_3 \cdot H_2O$; (3) mixed bauxite comprised of both gibbsite and boehmite.

Bauxite deposits were formed by the weathering of aluminous rock (Table 1) and most are residual accumulations but some others have been transported to their present location. Bauxites were formed in tropical or subtropical areas and range in age from Precambrian to Recent. Patterson (1965) estimated 5.8 billion tons of reserves and another 9.6 billion tons of potential resources of bauxite in the world. As for the reserves he estimated 45 million tons in the United States, 600 million tons in Jamaica, 390 million tons in South America, 850 million tons in Europe, 1.5 billion tons in Africa, 2 billion tons in Australia, and 280 million tons in Asia. These reserve estimates were made in 1965 and sizable additional reserves have been proven in Brazil, central and west Africa, Australia, and India since that time.

The bauxite reserves in the world are adequate to sustain the growing demand for aluminum for several hundred years. Unfortunately the United States has very limited reserves of bauxite and must rely on foreign sources for more than 90% of its bauxite ore. World demand for bauxite is increasing and this factor combined with production levies and the nationalization of many foreign bauxite mining operations has placed the

Table 1. Chemical Analyses of a Typical Bauxite and its Parent Rock (1) from Saline County, Arkansas

	Nepheline Syenite	Bauxite
Al_2O_3	21.11	59.8
SiO_2	56.51	5.3
Fe_2O_3	1.80	1.9
FeO	1.20	-
MgO	0.48	-
CaO	0.91	-
Na_2O	9.13	-
K_2O	6.34	-
TiO_2	0.50	2.0
ZRO_2	0.04	-
P_2O_5	0.06	-
SO_3	0.15	-
Cl	0.39	-
MnO	0.24	-
H_2O-	0.09	-
H_2O+	1.50	-
Loss of ignition	-	30.6
	100.45	99.9

United States in a precarious position in which the costs of imported bauxite are escalating and there are constraints on the availability. Political activities by many of the major bauxite-producing countries have forced the nationalization of the industry, imposed sharply increased taxes and royalties on the commodity, and produced strong efforts for cartel type price control. Although Patterson (1977) has indicated several reasons why bauxite will continue to be the major source of aluminum for the United States in the foreseeable future, all of the foregoing factors have combined to force the United States to thoroughly investigate the possibility of using nonbauxite resources to produce alumina.

NONBAUXITE RESOURCES FOR ALUMINA PRODUCTION

There has been renewed interest during the past several years in the recovery of aluminum from nonbauxitic ores. Although the cost of producing alumina from nonbauxitic ores may be higher than from bauxite, lower shipping costs and the stability of the sources of supply may offset any cost disadvantage and also improve our balance of payments. The United States is abundantly endowed with a variety of nonbauxitic aluminum resources. Those that are most suitable for the processing and recovery of cell-grade alumina are high-alumina clays, alunite, dawsonite in oil shales, anorthosites, and nepheline syenites. The nature of the resource determines the process required to produce a refined alumina product and many of these processes have been evaluated by Peters et al. of the US Bureau of Mines (1962, 1963, 1964). A report prepared by the committee on accessory elements of the National Academy of Sciences focused attention on the redistribution of accessory elements and compounds from mining and processing nonbauxitic alumina ores (Apps et al., 1979). Each of the most suitable nonbauxitic alumina resources will be discussed from the viewpoint of geology, geography, processing, and potential recovery of alumina.

High Alumina Clays

High-alumina clays consist predominantly of kaolinite and contain over 30% alumina. The principal US source of kaolinite is in Georgia and South Carolina, where up to 9 billion tons are estimated to be present (Husted, 1974). Arkansas and Oregon contain smaller deposits of high alumina clays totaling about 500 million tons each. Kaolin is processed for use as a filler and coating pigment in the paper, paint, plastics, rubber, ceramic, catalyst, ink, and many other industries (Murray, 1963)

The mineral kaolinite contains 38% Al_2O_3 and is therefore regarded as a potential source of aluminum. Many deposits of kaolin are relatively pure and contain only small amounts of quartz, muscovite, and minor accessory minerals. Table 2 shows a typical chemical analysis of a Georgia kaolin along with a low and high

Table 2. Chemical Analysis of Kaolin

	Typical, wt %	Low, wt %	High, wt %
SiO_2	44.0	50.2	40.6
Al_2O_3	38.5	31.5	44.4
Fe_2O_3	0.45	1.21	0.94
TiO_2	1.10	1.82	1.42
K_2O	0.18	0.24	0.24
H_2O	15.77	16.03	12.40

alumina kaolin as a range of those found in Georgia and South Carolina.

The Georgia-South Carolina kaolin belt (Fig. 1) extends southwest-northeast across Georgia and into South Carolina. The kaolin deposits occur as irregular, elongate, lenticular sedimentary bodies in thick sand units of late Cretaceous and early Tertiary age (Fig. 2). The kaolin deposits average between 4.7 to 9.1 m (15 and 30 ft) in thickness with some as much as 24.3 m (80 ft) thick. Individual deposits extend laterally for distances ranging from a few tens of feet to several miles and 80% or more of their total volume consists of the mineral kaolinite. At the present time over 5 million tons of kaolin are mined and processed yearly for the filler and coating pigment markets. All present mining is in open pits. The overburden ranges from 7.5 to 45.5 m (25 to 150 ft) in thickness and consists of red sands, white sands, fuller's earth (calcium montmorillonite), and sandy clays. The sands are aquifers and artesian conditions in some areas of the kaolin belt can present serious mining problems. The kaolin beds dip 1° to 2° southeastward toward the Atlantic Ocean from the contact of the coastal plain sedimentary rocks with the crystalline rocks of the Piedmont plateau. If kaolin ever becomes an economic source of alumina, the Georgia-South Carolina kaolin belt will be the primary mining area in the United States.

Other areas of kaolin deposits in the United States are shown in Fig. 3. All contain very much smaller potential tonnages of kaolin than the Georgia-South Carolina belt and have a lower aluminum content as summarized in Table 3 taken from data of the US Bureau of Mines (1967). Of these other kaolin deposits only the ones in Arkansas and Oregon can be considered potential sources of alumina in the foreseeable future. The kaolins in Arkansas are associated with bauxites and are both residual on the weathered nepheline syenite and sedimentary of Tertiary age (Tracey, 1944). Some of the high alumina bauxitic kaolins are now being mined and mixed with the bauxite. The Oregon kaolins

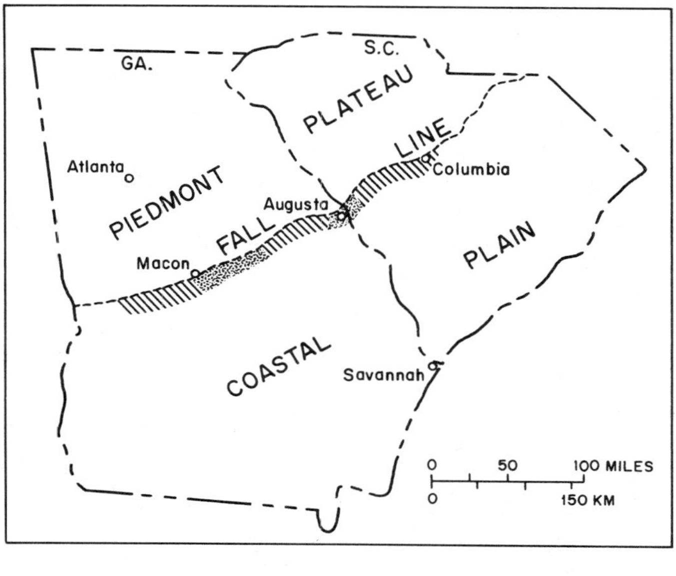

FIG. 1. Location of kaolin deposits in Georgia and South Carolina.

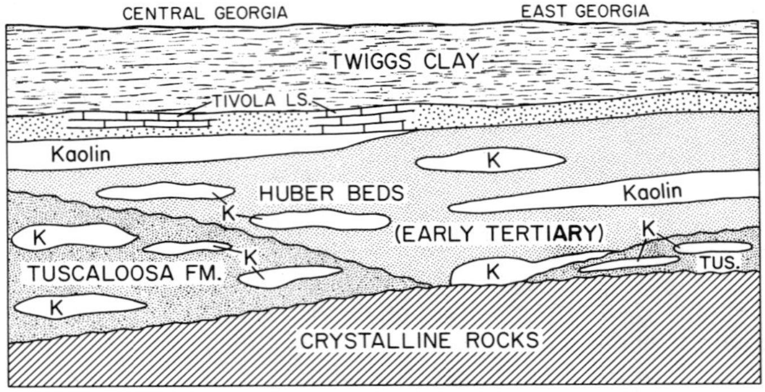

FIG. 2. Diagrammatic cross section showing kaolin beds in sands of Cretaceous and Tertiary age in Georgia and South Carolina.

are Eocene in age and are associated with laterites and bauxites and are relatively thick (Allen et al., 1951).

The kaolinitic clays that occur below the coal seams in several eastern and midwestern states, California, and Montana (Fig. 3) are another potential source of alumina. The US Bureau of Mines (1967) estimates that there are over 8 billion tons of fire clay containing an average of 27% alumina. Because these underclays could be mined along with the overlying coal they should be considered if high alumina clays are ever used as an ore of aluminum. The state of Ohio contains the largest of the US underclay resources with more than 4 billion tons. Table 4 shows chemical analyses of four clays associated with Pennsylvanian age coals.

High alumina clays have been evaluated as a source of aluminum in several foreign countries as well as in the United States. There is, however, no known commercial plant anywhere in the world. The current mining and processing practices used by the kaolin industry do not appear to mobilize any hazardous accessory materials. Acid processes are considered the most promising for alumina production from clays. Five recovery processes have been or are being evaluated and are the major probable recovery processes: (1) hydrochloric acid, (2) nitric acid, (3) sulfurous acid, (4) sulfuric acid, and (5) Pechiney "H" sulfuric acid process. Depending on the acid chosen as the solvent, the kaolin might or might not be calcined prior to extraction. Many of these processes have been evaluated on a pilot-plant scale and comparative economics produced (Nunn et al., 1979). The conclusion reached by Nunn et al. (1979) was that the hydrochloric acid process has signifi-

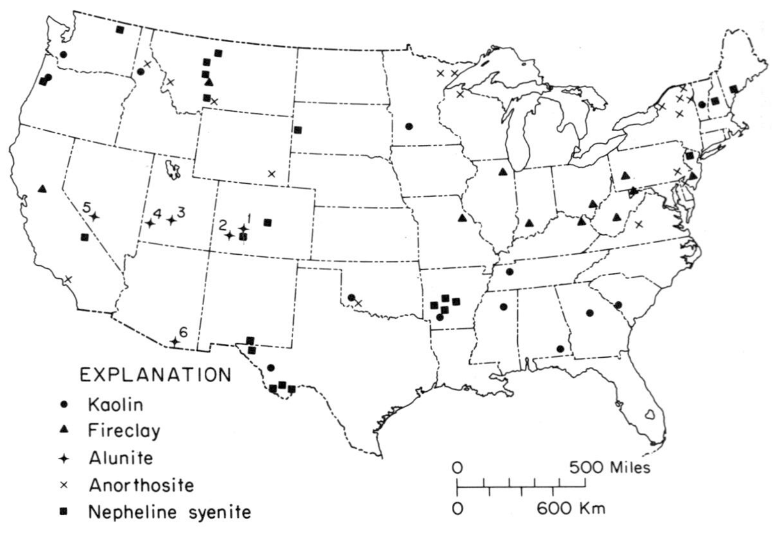

FIG. 3. Map of the US showing locations of kaolin, fire clay, alunite, amorthosite, and nepheline syenite resources.

Table 3. Estimated Kaolin Tonnage and Grade in States Other Than Georgia and South Carolina

	Millions of tons, estimated	Average Al_2O_3 content, %
Alabama	120	33
Arkansas	545	31
Idaho	120	25
Minnesota	127	33
Mississippi	30	33
Oklahoma	300	25
Oregon	518	28
Tennessee	30	32
Texas	212	33
Vermont	31	29
Washington	80	31

cantly lower capital and operating costs than the four other processes considered.

The major residue is silica with minor quantities of iron oxide and titanium dioxide. The tonnage of silica residue would exceed that of recovered alumina by 10 to 20%. The silica particles have a very high surface area because of their extremely fine grain size. Work should be done to improve the color by removing the iron and titanium because this silica byproduct would make an excellent filler for newsprint and rubber. If the silica residue could be developed into a marketable product this would improve the economics and might make the extraction of alumina from high alumina clays an economically feasible process.

Alunite

Alunite is a hydrous sulfate of aluminum and potassium $[KAl_3(SO_4)_2(OH)_6]$ which in its pure form contains about 37% alumina, 11% potassium oxide, 39% sulfate, and 13% water. Naturally occurring alunite contains some sodium substituting for potassium. When the sodium content exceeds that of potassium the mineral is called natroalunite. Alunite occurs in sedimentary, metamorphic, and igneous rocks; it occurs as veins and bedded sedimentary deposits of virtually the pure mineral and as disseminated aggregates in igneous rocks.

In the United States alunite was utilized as a source of potassium sulfate during World War I and evaluated as a source of alumina during World War II. The first description of alunite resources was Thoenen (1941). A recent study of alunite resources of the United States and the world are described by Hall (1978). Deposits of alunite disseminated in hydrothermally altered volcanic and subvolcanic igneous rocks have been identified that are estimated to contain from

Table 4. Analyses of Fire Clays from Pennsylvania and Flint Clays from Pennsylvania and Missouri

	Fire clay from Armstrong Co., PA, %	Fire clay from Clearfield Co., PA, %	Flint clay from Clearfield Co., PA, %	Flint clay from Missouri, %
SiO_2	70.30	55.15	40.70	42.68
Al_2O_3	17.50	28.70	40.57	38.49
Fe_2O_3	1.17	1.06	0.53	1.55
FeO	0.91	0.80	0.72	
CaO	0.09	0.15	0.16	
MgO	0.37	0.46	0.19	0.08
K_2O	1.27	2.20	1.35	0.49
Na_2O	0.39	0.40	0.60	0.28
TiO_2	1.50	1.67	2.57	2.90
Loss on ignition	6.45	9.36	12.59	
S	0.20	0.07	0.03	
CO_2	0.53	1.57	1.14	
Comb H_2O	5.06	7.16	11.15	14.07
H_2O @ 120°C	0.66	0.63	0.30	

a few million to more than a billion tons of alunite-bearing rock. Deposits containing 25 to 50% alunite are the most promising sources for the production of alumina.

Most of the large disseminated alunite deposits of the United States occur in altered acidic to intermediate Tertiary volcanic rocks in the western part of the country (Fig. 3). The best defined are four large alunite deposits located about 30 miles west of Milford, UT, in the Wah Wah Mountains. These deposits occur in hydrothermally altered intermediate to silicic flows and ignimbrites of mid-Tertiary age. The alunite deposits best described in the literature are found in the area around Marysvale, UT (Callahan, 1973 and Kerr et al., 1967). The Marysvale alunite deposits occur in altered intermediate volcanic rocks of Oligocene age and were worked briefly during World War I for potassium and during World War II for alumina.

Several alunite deposits in Colorado have been noted in the literature (Thoenen, 1941). The largest of these appears to be at Red Mountain near Lake City, CO (Fig. 3). This deposit occurs in altered latitic subvolcanic and intrusive rock. Another alunite deposit (Fig. 3) occurs near Rico, CO, in a quartz monzonite stock intruding sedimentary rocks of Pennsylvanian age.

A very large alunite deposit on Red Mountain near Patagonia, AZ (Fig. 3), occurs in altered rhyolitic and latitic volcanic rocks (Corn, 1975). In Nevada (Fig. 3) a number of alunite occurrences have been identified in a linear belt of altered volcanics between Reno and Las Vegas. The largest deposit in this belt is west of Goldfield (Ashley and Keith, 1973).

Alunite mineralization varies from erratic to quite uniform. Variations within a deposit affect the mining and processing and are of major concern. Factors contributing to the uniformity of alunite mineralization are porosity and permeability of the host rock, strength and consistency of the mineralizing fluids, original host-rock composition, structural controls, and fracturing and brecciation of the host rock. In many deposits quartz, kaolin group minerals, amorphous silica, iron oxides, and cristobalite or tridymite are intimately mixed with the alunite. The relative amounts of these minerals can vary significantly within a distance of a few centimeters.

The size and configuration of alunite deposits vary widely from tabular and elongate to essentially circular. In many places the deposits are zoned with the alunite and other minerals distributed zonally. Dimensions of these zones can vary greatly.

In Russia alunite is mined at Azerbaijan and processed to alumina and potash at the town of Kirovabad. This alunite to alumina operation produces approximately 200,000 tpy and is being expanded to 500,000 tpy (Apps et al., 1979). The plant produces potassium sulfate and sulfuric acid as coproducts and reportedly vanadium as a byproduct.

There are two types of processing for treating alunite, a caustic leach and an acid leach. The caustic leach process involves crushing and grinding, roasting, leaching in a weak alkaline solution and crystallizing the potassium sulfate, and treating the residue in a modified Bayer circuit to produce cell grade alumina. In the acid process the roasted ore is leached with sulfuric acid and a potassium alum is crystallized. The potassium alum is roasted at a suitable pressure and temperature to form a basic alum, which can then be calcined to produce potassium sulfate and alumina (Fleischer, 1944).

Alunite is a nonbauxitic ore of alumina that is being used commercially today in Russia. Bauer (1980) reported potential alunite resources of over 8 billion metric tons in the United States with inferred reserves of at least 2.5 billion metric tons. It appears that the US has sufficient reserves and resources for alunite to be considered as a viable potential nonbauxite source of alumina.

Dawsonite and Nordstrandite

Dawsonite [$Na\ Al\ (OH)_2\ CO_3$], nahcolite ($Na\ H\ CO_3$), and halite ($Na\ Cl$) are mixed with high grade oil shale in the lower part of the Parachute Creek member of the Eocene Green River formation in the northern part of the Piceance Creek basin in Rio Blanco County, Colorado. Smith (1980) reported that dawsonite is always accompanied by nordstrandite ($Al\ (OH)_3$) in the Parachute Creek member. Dawsonite and nordstrandite occur as tiny matrix minerals distributed through the oil shale. In addition to alumina, oil, soda ash, and sodium bicarbonite will be produced.

The alumina resource represented by the dawsonite and nordstrandite in oil shale has been evaluated by Beard, et al. (1974). Deposit thicknesses reach more than 305 m (1,000 ft) around the basin's saline depositional center. Total dawsonite in place was estimated as 19 billion tons representing 6.5 billion tons of extractable alumina. The depth of burial of the beds containing the alumina resources is 458 to 610 m (1500 to 2000 ft). The dawsonite and nordstrandite content in this oil shale ranges from 1 to 25 wt % averaging about 10 wt %. The fine-grained mineral matrix of dawsonite oil shale comprises dawsonite, dolomite, quartz, potassium and sodium feldspar, and small amounts of pyrite/pyrrhotite, calcite, and illite. Dawsonite, a white or colorless acid-soluble mineral, generally occurs as stubby to elongate crystals 2 to 20 mμ long disseminated in the oil shale.

Smith and Young (1975) summarized the optimum procedures for extracting alumina from oil shale. The process involves heating to produce shale oil by drying out the organic matter and decomposing the dawsonite and nordstrandite to produce extractable alumina by using a caustic leach process. Smith (1980) believes that the production of alumina from dawsonite-bearing oil shale has enough advantages to offset the low concentrations of the extractable alumina.

The oil shale will be used eventually to produce oil, and alumina from the spent shale is a readily available byproduct that gives the US the availability of a domestic alumina resource. Many mining, processing, and environmental problems must be worked out before this resource can be utilized.

Anorthosite

Anorthosite is an igneous rock characterized by a relatively high alumina content ranging from about 25 to 30%. It is composed principally of plagioclase feldspar and generally less than 10% of clinopyroxene and orthopyroxene. Minor quantities of potassium feldspar and quartz may be present. Accessory minerals in anorthosite include iron and titanium oxides, apatite, zircon, and sphene. Anorthosites are found as large Precambrian massifs or as layers in segregated igneous intrusions. These two types are distinguished by the difference in composition of the dominant plagioclase. The massif-type anorthosite contains the plagioclase (anorthite) varying in composition between An_{35} and An_{80}, whereas in the layered intrusive type it varies between An_{65} and An_{80}.

Anorthosites are dispersed throughout the United States as shown in Fig. 3. Anorthosite bodies may range in size from small stringers and pods measured in meters to enormous intrusions which underlie areas of more than 1500 km^2. The major anorthosite deposits in the United

Table 5. Major Anorthosite Deposits in the United States

Name	Location: County	Location: State	Area, km^2	Al_2O_3 Content	References
San Gabriel Mountains	Los Angeles	California	215	27	Carter and Silver, 1972
N.E. Delaware Piedmont		Delaware			Thompson, 1975
Boehls Butte	Shoshone Clearwater	Idaho	150	29	Hietanen, 1963
Keweenawan Series	Cook Lake	Minnesota	520	30	Green, 1972
Duluth Complex	Lake	Minnesota			Davidson, 1972
Stillwater Complex	Stillwater Sweet Grass	Montana	18	30	Bowes et al., 1973
Bitterroot Range	Adams, Valley	Montana	72.5		Berg, 1969
Adirondacks					
Main Body	Essex, Franklin	New York	3300	26	Buddington, 1969, 1972
Thirteenth Lake	Hamilton, Warren		235	26	
Carthage	Jefferson		2	26	
Rand Hill	Clinton		15	25	
Wichita Mountains	Kiowa Commanche Jackson	Oklahoma	40	28	Scofield, 1975
Honey Brook	Chester	Pennsylvania	40	27	Crawford et al., 1971
Roseland	Nelson, Amhurst	Virginia	$<$ 60		Herz, 1969
Mineral Lake		Wisconsin	$<$ 120		Olmsted, 1969
Laramie Range	Albany, Laramie	Wyoming	650	27	Hagner, 1951

Table 6. Some Major Nepheline Syenite Deposits in the United States

Name	Location: County	Location: State	Area, km^2	Al_2O_3 Content	References
Selawik Hills		Alaska	200	22	Miller, 1972
Little Rock	Puluski	Arkansas	600	18–19	Gordon et al., 1958
McClure Mt.	Fremont	Colorado	50		Parker and Hildebrand, 1962
Highwood Mts.	Great Falls	Montana	400	18	Larsen et al., 1941
Lincoln County		Oregon	250	19	Snavely and Wagner, 1961
Terlingua	Big Bend Brewster	Texas	500	18–19	Yates and Thompson, 1959

States are listed in Table 5 along with the area in square kilometers and the references describing them.

The alkaline nature of anorthosite dictates that an alkaline process be used for the recovery of alumina. The lime-soda-sinter process was tested by the US Bureau of Mines through the pilot plant stage during 1952 to 1954 and found to be feasible for full-scale production of cell-grade alumina from anorthosite (St. Clair et al., 1959). As a first step in the lime-soda-sinter process, limestone and a soda solution (NaOH and Na_2CO_3) are added to alumina containing rock and ground to −200 mesh. The resulting slurry is dried and sintered at 1290°C. The sinter is then ground and leached with sodium carbonate. The soluble sodium aluminate is separated from the insoluble residue. The solution is autoclaved with lime to precipitate excess silica and then washed in carbon dioxide containing flue gas to precipitate aluminum hydroxide. The aluminum hydroxide is separated from the solution and calcined at 1230°C to produce cell-grade alumina.

Nepheline Syenites

Nepheline syenites are coarsely crystalline igneous rocks composed of alkali feldspars and nepheline. Nepheline syenites belong to a family of alkaline rocks noted for higher concentrations of alkali metals and deficiencies of calcium, magnesium, and iron; those containing nepheline or other feldspathoids are also deficient in silica and enriched in aluminum.

Nepheline syenite occurrences are small, rarely exceeding a few square miles in area. They are broadly distributed in the Appalachian Mountains, the Ouachita Mountains, and the western United States (Fig. 3). Table 6 lists the principal nepheline syenites, the areal extent, their Al_2O_3 content, and the references.

Nepheline syenites are comprised of sodic-potassic feldspars and nepheline which make up between 80 and 90% of the rock. The remainder is made up primarily of alkaline pyroxene and accessory minerals including sphene, pyrochlore, magnetite, etc. Although nepheline contains 32 to 34% Al_2O_3, nepheline syenites have considerably less, ranging between 17 and 23%.

Nepheline syenite would be treated by the same lime-soda-sinter process that would be used for anorthosite.

SUMMARY

Bauxite reserves in the world are plentiful and can sustain continued growth of the production of aluminum. However, the United States has limited and inadequate bauxite reserves and must import 95% of its yearly requirements. If economic and/or strategic conditions change so that foreign bauxite becomes unavailable, the United States is abundantly endowed with a variety of nonbauxite resources for aluminum production. These include high aluminum clays including kaolins from Georgia-South Carolina, kaolins associated with the Arkansas bauxites and Oregon laterites, and fire clays or underclays associated with the coal seams from Missouri to Pennsylvania. Alunite occurrences are large and it is estimated there are over 8 billion metric tons of this resource. Dawsonite resources are very large with estimates of over 19 billion tons. Anorthosite and nepheline syenite occurrences are widely dispersed across the United States. When oil from shale is produced the dawsonite may be a viable source of aluminum. However, unless a dire emergency occurs bauxite from foreign sources will continue to be the major ore of aluminum.

REFERENCES

Allen, V. T., Loofbourow, Jr., J. S. and Nichols, R. L., 1951, "The Hobart Butte High-Alumina Clay Deposit, Lane County, Oregon," Circular 143, US Geological Survey, 11 pp.

Apps, J. A., et al., 1979, "Redistribution of Accessory Elements and Compounds from Mining and Processing of Non-Bauxitic Aluminum Ores, *Uranium, Phosphate, and Aluminum*, Pt. 2, National Academy of Sciences, Lib. of Congress Card Catalog No. 79-90468, pp. 105–169.

Ashley, R. P. and Keith, W. J., 1973, "Occurrences of Alunite in Hydrothermally Altered Rocks and Ores at Goldfield, Nevada," *Abstracts with Programs*, Geological Society of America, Vol. 5, No. 6.

Bauer, C. W., 1980, "Aluminum from Alunite—Domestic Resources," Preprint 80-119, Society of Mining Engineers, 3 pp.

Beard, T. M., Tait, D. B. and Smith, J. W., 1974, "Nahcolite and Dawsonite Resources in the Green River Formation of Colorado," *Guidebook to the Energy Resources of the Piceance Creek Basin, Colorado*, D. K. Murray, ed., Rocky Mountain Association of Geologists, Denver, CO., pp. 101–109.

Berg, R. B., 1969, "Petrology of the Anorthosites of the Bitterroot Range, Montana," *Origin of Anorthosites and Related Rocks*, Y. W. Isachen, ed., Memoir 18, New York State Museum, pp. 387–398.

Bowes, D. R., Skinner, W. R. and Skinner, D. L., 1973, "Petrochemistry of the Stillwater Igneous Complex, Montana," *Transactions*, Geological Society of South Africa, Vol. 76, No. 2, pp. 153–163.

Callahan, E., 1973, "Mineral Resource Potential of Prute County, Utah, and Adjoining Area," Bulletin 102, Utah Geological and Mineralogical Survey.

Carter, B., and Silver, L. T., 1972, "Structure and Petrology of the San Gabriel Anorthosite-Syenite Body, California," Report 1972, Section 2, 24th International Geological Congress, pp. 303–311.

Corn, R. M., 1975, "Alteration-Mineralization Zoning, Red Mountain, Arizona," *Economic Geology*, Vol. 70, pp. 1437–1447.

Crawford, W. A., Robelen, P. G., and Kalmbach, J. H., 1971, "The Honey Brook Anorthosite," *American Journal of Science*, Vol. 271, pp. 333–349.

Davidson, D. M. Jr., 1972, "Eastern Part of Duluth Complex," *Geology of Minnesota, A Centennial Volume*, P. K. Sims and G. B. Morey, eds., Minnesota Geological Survey, pp. 354–360.

Fleischer, A., 1944, "The Klunite Process," *AIME Transactions*, Vol. 159, pp. 267–279.

Gordon, M., Jr., Tracey, J. I., Jr., and Ellis, M. W., 1958, "Geology of the Arkansas Bauxite Region," Professional Paper 299, US Geological Survey.

Green, J. C., 1972, "North Shore Volcanic Group," *Geology of Minnesota, A Centennial Volume*, P. K. Sims and G. B. Morey, eds., Minnesota Geological Survey, pp. 294–332.

Hagner, A. F., 1951, "Anorthosite of the Laramie Range, Albany County, Wyoming," Bulletin 43, Wyoming Geological Survey.

Hall, R. B., 1978, "World Non-bauxite Aluminum Resources—Alunite," Professional Paper 1076-A, US Geological Survey, 35 pp.

Herz, N., 1969, "Roseland Alkalic Anorthosite Massif, Virginia," *Origin of Anorthosites and Related Rocks*, V. W. Isachen, ed., Memoir 18, New York State Museum, pp. 357–367.

Hietanen, A., 1963, "Anorthosite and Associated Rocks in the Boehls Butte Quadrangle and Vicinity, Idaho," Professional Paper 344-B, US Geological Survey.

Husted, J. E., 1974, "Potential Reserves of Domestic Non-Bauxite Sources of Aluminum," AIME Preprint A 74-68, 20 pp.

Kerr, P. F., et al., "Marysvale, Utah, Uranium Area," Special Paper 64, Geological Society of America, 212 pp.

Larsen, E. S. and Burgess, C. H., 1941, "Igneous Rocks of the Highwood Mountains, Montana: Part IV, The Stocks," Bulletin 52, Geological Society of America, pp. 1809–1828.

Miller, T. P., 1972, "Potassium-Rich Alkaline Intrusive Rocks of Western Alaska," Bulletin 83, pp. 2111–2128.

Murray, H. H., 1963, "Industrial Applications of Kaolin," *Clays and Clay Minerals*, Vol. 10, Pergamon Press, pp. 191–198.

Nunn, R. F., et al., 1979, "The Comparative Economics of Producing Aluminum from U.S. Non-Bauxite Ores," TMS Preprint LM-77-15, The Metallurgical Society of AIME, pp. 283–333.

Olmsted, J. F., 1969, "Petrology of the Mineral Lake Intrusion, Northwestern Wisconsin," *Origin of Anorthosites and Related Rocks*, Y. W. Isachen, ed., Memoir 18, New York State Museum, pp. 149–173.

Parker, R. L. and Hildebrand, F. A., 1962, "Preliminary Report on Alkalic Intrusive Rocks in the Northern Wet Mountains, Colorado," Professional Paper 450-E, US Geological Survey.

Patterson, S. H., 1965, "Bauxite Reserves and Potential Aluminum Resources of the World," Bulletin 1228, US Geological Survey, 196 pp.

Patterson, S. H., 1977, "Aluminum from Bauxite; Are There Alternatives?" *American Scientist*, Vol. 65, pp. 345–351.

Peters, F. A., Johnson, P. W., and Kirby, R. C., 1962, "Methods for Producing Alumina from Clay; An Evaluation of the Sulfurous Acid Caustic Purification Process," Report of Investigations RI-5997, US Bureau of Mines.

Peters, F. A., Johnson, P. W., and Kirby, R. C., 1963, "Methods of Producing Alumina from Clay; An Evaluation of Three Sulfuric Acid Processes," Report of Investigations RI-6229, US Bureau of Mines.

Peters, F. A., Johnson, P. W., and Kirby, R. C., 1964, "Methods for Producing Alumina from Clay; An Evaluation of a Nitric Acid Process," Report of Investigations RI-6431, US Bureau of Mines.

St. Clair, H. W., et al., 1959, "Operation of Experimental Plant for Producing Alumina from Anorthosite," Bulletin 577, US Bureau of Mines.

Scotfield, N. L., 1975, "Layered Series in Wichita Complex, Oklahoma," Geol. Bulletin 80, Geological Society of America, pp. 732–736.

Smith, J. W., 1980, "Alumina from Oil Shale," SME-AIME Preprint 80-62, 8 pp.

Smith, J. W. and Young, N. B., 1975, "Dawsonite, its Geochemistry, Thermal Behavior, and Extraction from Green River Oil Shale," *Quarterly*, Colorado School of Mines, Vol. 70, No. 3, pp. 69–93.

Snavely, P. D., Jr., and Wagner, H. C., 1961, "Differentiated Gabbroic Sills and Associated Alkalic Rocks in the Central Part of the Oregon Coast Range," Professional Paper 424-D, US Geological Survey, pp. 156–61.

Thoenen, J. R., 1941, "Alunite Resources of the United States," Report of Investigations 3561, US Bureau of Mines, 48 pp.

Thompson, A. M., 1975, "Anorthosite in the Piedmont of Northern Delaware," Abstract 7 (1), Geological Society of America, pp. 124–125.

Tracey, J. I., Jr., 1944, "The High Alumina Clays of Pulaski and Saline Counties, Arkansas," *Journal*, American Ceramic Society, Vol. 27, pp. 246–249.

US Bureau of Mines, 1967, "Potential Sources of Aluminum," Information Circular 8335.

Yates, R. G., and Thompson, G. A., 1959, "Geology and Quicksilver Deposits of the Terlingua District, Texas," Professional Paper 312, US Geological Survey.

3

Unconventional Deposits in Physically Inaccessible Environments

11

Geochemistry of Deep-Sea Manganese Nodules—Organic Involvement

Walter E. Dean

US Geological Survey, Denver, CO

ABSTRACT

Deep-sea manganese nodules are considered to be potential ores of manganese, nickel, cobalt, and copper. Considerable time, effort, and funds have been applied to the study of the distribution of nodules on the seafloor and the location of economically important areas, but less effort has been spent in understanding the origin of nodules. Most theories that have been proposed to explain the origin of manganese nodules have been based on inorganic processes. It is not by coincidence, however, that the area considered to be most economically feasible is in the zone of high biological productivity in the eastern equatorial Pacific Ocean and underlain by siliceous biogenic sediment. The genesis of manganese nodules and the concentration of many trace elements are related indirectly, and possibly directly, to the activities of organisms.

INTRODUCTION

That nodules, crusts, or concretions rich in iron and manganese, collectively referrred to as manganese nodules or ferromanganese nodules, form extensive deposits on the seafloor was first documented during the Challenger Expedition in the 1870s. Until the mid-1960s, however, manganese nodules were mostly considered as curiosities and there were few serious studies of their origin, distribution, or economic potential. With rapid recent advances in marine technology, the primary interest in manganese nodules has been in certain trace metals, especially cobalt, nickel, and copper, that are concentrated along with iron and manganese in the nodules. Most of the literature on manganese nodules has appeared during the past 15 years as a result of the "rediscovery" of manganese nodules and their economic potential. Considerable time, effort, and funds have been applied to the study of the distribution of nodules on the seafloor, their chemical composition, and the location of the most economically important areas, but less effort has been expended on understanding the origin of nodules. For reviews of the geochemistry and distribution of manganese nodules, the interested reader is referred to the slightly dated but still excellent book by Mero (1965), a more recent review by Heath (1981), and collections of papers dealing with all aspects of manganese nodules in Glasby (1977), Bischoff and Piper (1979), and Cronan (1980). One aspect of manganese-nodule genesis that is given little consideration in any of these publications is the role of organisms. I believe that organic activity is more important than has been previously considered by most geologists and geochemists, and may be necessary for nodule formation. In this paper I will briefly review the geochemistry and distribution of manganese nodules, but the main purpose is to outline the main ways in which organisms may be involved in the formation of nodules and in controlling their composition.

COMPOSITION OF MANGANESE NODULES

Element Concentrations

A compilation of the concentrations of selected elements in manganese nodules, pelagic sediments from the Pacific Ocean, and the upper crust is presented in Table 1. It is evident from Table 1 that a number of elements are enriched in both pelagic sediments and manganese nodules relative to the upper crust, but the greatest enrichment is for cobalt, copper, and nickel which may be concentrated more than 1000-fold in nodules. The area of the world ocean that appears to contain nodules with the highest concentrations of these elements is in the northeastern equatorial Pacific Ocean between the Clipperton and Clarion fracture zones and between 114° and 153° west longitude (Fig. 1) where there is estimated to be more than 2×10^9 metric tons (dry weight) of nodules that contain more than 1.8% Cu + Ni within an area of about 2.5×10^6 km^2 (McKelvey et al., 1979).

Regional Variations

According to Cronan (1977 and 1980) the main variables that produce regional differences in the composition of manganese nodules are rate of sedimentation, type and availability of nuclei, and environment of formation. Sedimentation rate is important because if the sedimentation rate is too high, nodules that form will be buried and eventually dissolved as sediment pore waters become reducing. Most abyssal manganese nodules, therefore, are found in areas underlain by slowly accumulating pelagic red clay, or siliceous biogenic ooze that is relatively free of carbonate and terrigenous material (Horn et al., 1973). Although abyssal nodules usually are found in areas of slow, pelagic sedimentation, there is a dilemma because most pelagic sediments accumulate at rates measured in millimeters per million years (Barnes and Dymond, 1967; Ku and Broecker, 1967, 1969; Bender et al., 1966, 1970; Ku, 1977; Moore et al., 1981).

Table 1. Concentrations of Selected Elements (in Weight Percent) in Manganese Nodules, Pelagic Sediments, and Upper Continental Crust.

	Nodules		Sediments		
Element	A Eastern equator. Pacific	B World Average	C Pacific pelagic clay	D DOMES A+B+C	E Upper contin. crust
Mn	28.8	16.2	0.43	0.75	0.069
Fe	6.6	15.6	5.4	4.8	3.5
Si	7.8	8.6	25.7	23.6	30.5
Al	2.8	3.1	8.8	6.1	7.8
Na	1.9	1.9	0.96	2.8	2.5
K	0.82	0.64	2.2	2.7	2.8
Ca	1.5	2.5	0.50	1.9	2.9
Mg	1.8	1.8	2.1	1.9	1.4
Ti	0.61	0.64	0.47	0.40	0.47
Co	0.23	0.30	0.011	0.010	0.001
Cr	0.020	0.001	0.006	0.005	0.007
Cu	0.99	0.26	0.023	0.051	0.003
Ba	0.32	0.20	0.39	0.33	0.059
Mo	0.048	0.041	0.001	0.002	0.0001
Ni	1.2	0.49	0.021	0.028	0.004
Pb	0.048	0.087	0.003	0.005	0.002
V	0.030	0.056	0.012	0.010	0.010
Zn	0.13	0.071	0.017	0.016	0.007

Sources of data:
A—McKelvey and others, 1979.
B—Cronan, 1976.
C—Bischoff and others, 1979.
D—Bischoff and others, 1979.
E—Wedepohl, 1971.

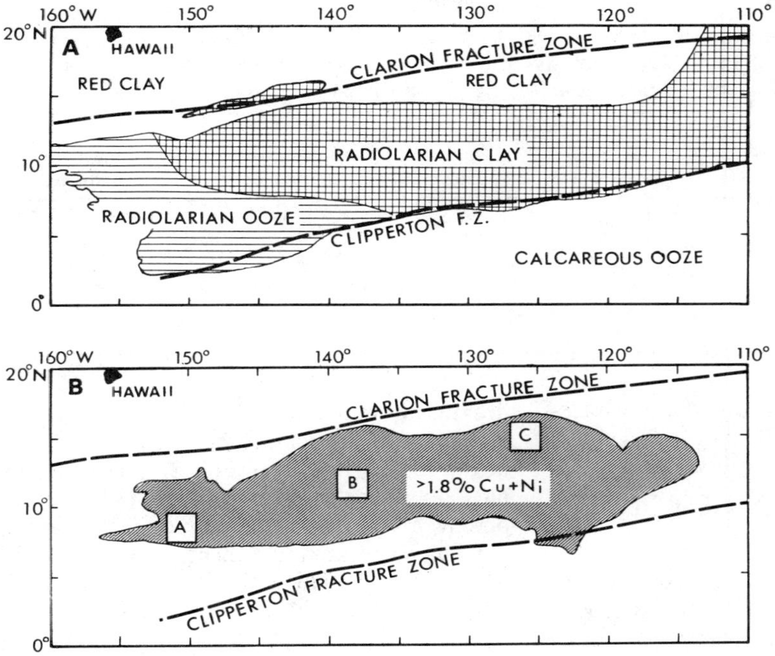

FIG. 1. (A) Distribution of sediment types in the northeastern equatorial Pacific Ocean. From Horn et al., 1973 (B) Distribution of manganese nodules containing more than 1.8% Cu + Ni. Boxes marked A, B, and C refer to the three study areas of the National Oceanic and Atmospheric Administration (NOAA) Deep Ocean Mining Environmental Study (DOMES). From McKelvey et al., 1979.

This suggests that nodules are growing at rates that are faster than the measured radiometric growth rates and (or) that the sediments are accumulating at slower rates or are actually being eroded.

Manganese nodules may have a visible nucleus such as a rock fragment, a volcanic glass fragment, a chip of partly consolidated sediment, a shark tooth, etc., but the nucleus of most nodules is not so obvious, and usually is a micronodule or a fragment of a nodule. The availability of a nucleating surface and the length of time that surface is in contact with seawater may be the most important factors in determining the abundance of nodules and the rate of their formation (Horn et al., 1973; Cronan, 1977; Ku, 1977). The nucleus need not be very large; manganese accretion has been observed on the tests of foraminifera (Dugolinsky et al., 1977), and micronodules (<1mm) may be the nuclei of many abyssal manganese nodules.

The environment in which manganese nodules and crusts form integrates many other variables, each of which may produce variations in nodule abundance and composition. The biological productivity of the overlying waters, the nature of the substrate underlying the site of nodule accumulation, and the redox conditions in bottom waters and sediment interstitial water probably are the most important environmental variables (Cronan, 1977). Redox conditions, or degree of oxidation, affect the oxidation state of the elements and the minerals found in the nodules and in turn are affected by sedimentation rate, bottom-water circulation, bottom-water dissolved-oxygen concentration, and biological oxygen demand. The effect of environment of accumulation on nodule composition is shown by the differences in composition of abyssal and seamount nodules in Table 2. Seamount nodules accumulate in an environment with relatively high bottom-current velocities which means that accumulation of sediment may be inhibited, and that more dissolved oxygen is advected to the bottom-water environment (Cronan and Tooms, 1969; Cronan, 1977). As a result, seamount nodules are composed mostly of δMnO_2 which is a more oxidized phase of manganese oxide that tends to concentrate cobalt (Cronan and Tooms, 1969). Abyssal nodules, on the other hand, are composed mostly of the mineral todorokite, a less oxidized phase that tends to concentrate nickel and copper (Cronan, 1977). Seamount nodules, therefore, tend to be enriched in cobalt, and abyssal nodules tend to be enriched in nickel and copper (Table 2), and

Table 2. Concentrations of Manganese, Iron, Cobalt, Nickel, Copper and Mn/Fe in Manganese Nodules from Pacific Seamounts (1), the Abyssal Seafloor (2), and Active Ridges (3). Data from Cronan (1977). Averages of Three Representative Hydrogenous Manganese Nodules (4) and Three Representative Diagenetic Manganese Nodules (5) are from Bonatti et al. (1972).

Element	1 Pacific seamounts	2 Abyssal	3 Active ridges	4 Hydrogenous nodules	5 Diagenetic nodules
Mn	15.3	16.8	15.5	16.6	36.0
Fe	12.8	17.3	19.2	15.2	3.1
Ni	0.41	0.54	0.31	0.41	0.035
Co	1.5	0.26	0.40	0.37	0.040
Cu	0.062	0.37	0.08	0.14	0.008
Mn/Fe	1.2	0.97	0.80	1.1	11.6

tend to become more enriched in these two elements with increasing age (Cronan and Tooms, 1969; Cronan, 1977).

GENESIS OF MANGANESE NODULES

Inorganic Mechanisms

Most theories that have been proposed to explain the origin of manganese nodules are based on inorganic processes, and have been concerned mainly with sources of manganese and minor elements such as cobalt, nickel, and copper. Genetic classifications of ferromanganese deposits have identified four general types with regard to the sources of elements: (1) hydrothermal deposits, (2) halmyrolic deposits, (3) hydrogenous deposits, and (4) diagenetic deposits (e.g. Bostrom, 1967; Bonatti et al., 1972; Bonatti, 1975; Halbach et al., 1981).

Hydrothermal deposits occur in areas of active volcanism or high heat flow such as near active crustal spreading centers. They may occur as discrete nodules or as ferromanganese crusts on volcanic rock. They are characterized by high concentrations of iron (field I in Fig. 2), rapid rates of accretion, and low concentrations of trace metals (e.g. Bonatti and Nayudu, 1965; Bostrom and Peterson, 1966; Bonatti et al., 1972; Scott et al., 1974; Hekinian and Hoffert, 1975; Moore and Vogt, 1976).

The iron, manganese, and trace metals in halmyrolic deposits are supplied by submarine weathering of volcanic rock. Halmyrolic ferromanganese deposits, usually in the form of crusts on basalt, are common in the southwest Pacific Ocean (Bonatti et al., 1972; Bonatti, 1975).

Hydrogenous deposits result from slow precipitation of iron and manganese oxides and hydroxides from normal seawater. They generally contain more manganese than iron (Mn/Fe ratio of 0.5 to 5.0; field II in Fig. 2), and generally have high trace-metal concentrations as a result of slow accretion and more time for trace-metal scavenging by iron and manganese oxides and hydroxides (Bonatti et al., 1972). Bonatti et al. (1972) considered that hydrogenous nodules were most prevalent in abyssal areas of the oceans where sediment accumulates at a very slow rate, and on parts of topographic highs where little or no sediment accumulates.

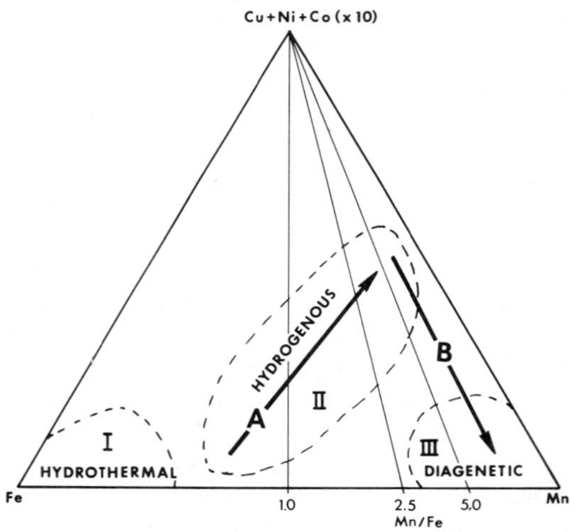

FIG. 2. Triangular diagram showing fields of relative concentrations of manganese, iron, and the sum of Cu + Ni + Co (X10) in hydrothermal (I), hydrogenous (II), and diagenetic (III) manganese nodules. Outlines of fields are from Bonatti et al., 1972. The three vertical lines drawn through the apex for Cu + Ni + Co represent lines of constant ratio of Mn/Fe for values of 1.0, 2.5, and 5.0. Arrow A represents increasing concentration of Cu + Ni + Co with increasing diagenetic influence. See text for discussion.

Diagenetic deposits form by diagenetic reduction of iron and manganese in sediments, followed by upward migration of reduced manganese ions in sediment interstitial waters, and precipitation of iron and manganese oxides and hydroxides at the sediment-water interface. The most important factor that controls the early diagenetic remobilization of manganese is the biological oxygen demand of the sediment, which is determined mainly by the biological productivity of the overlying surface waters. Diagenetic remobilization of iron and manganese is particularly common in areas of continental-margin, hemipelagic sedimentation with relatively high concentrations of organic matter (e.g. Lynn and Bonatti, 1965; Bonatti and Nayudu, 1965; Manheim, 1965; Price, 1967; Glasby, 1972; Ku and Glasby, 1972), but also may be important in affecting growth rates of deep-sea nodules (Halbach et al., 1981; Reyes et al., 1982).

Halbach et al. (1981) found that many Pacific abyssal nodules were of a mixed origin. The upper surfaces of these mixed nodules formed mainly by hydrogenous accumulation of manganese oxides, and the bottom surfaces formed by diagenetic precipitation of manganese. These differences in sources of elements in the upper and lower surfaces of nodules also may explain the common observations that the upper surface usually is smooth and the lower surface usually is lumpy (Raab, 1972; Heath, 1981). Also the lower, lumpy surface usually is enriched in manganese, copper, and nickel relative to the smooth upper surface which is enriched in iron and cobalt (Raab, 1972). The Mn/Fe ratio is the most important criterion that distinguishes hydrogenous from diagenetic nodule growth and concentration of minor divalent cations such as copper and nickel. According to Halbach et al. (1981), the concentration of Ni + Cu in hydrogenous nodules increases with increasing Mn/Fe ratio up to a ratio of about 5.0 (trend A in Fig. 2). In nodules with a Mn/Fe ratio >5.0, the concentration of Ni + Cu decreases as the nodules become richer in manganese derived from early diagenetic reduction in interstitial waters. The manganese presumably is incorporated instead of nickel and copper in lattice sites in todorokite.

Assuming that there is a sufficient supply of elements from one or more sources, the problem still remains as to how these elements become concentrated into nodules or crusts instead of forming metalliferous sediments such as those found near active crustal spreading centers (Bischoff and Rosenbauer, 1977; Bonatti, 1975; Heath and Dymond, 1977). A number of experimental studies have shown that iron oxides and hydroxides will precipitate at pH and Eh values that are much lower than those necessary to precipitate manganese oxides and hydroxides (Hem, 1963 and 1972; Collins and Buol, 1970, 1972; Krauskopf, 1956, 1957; Fig. 3). Under oxidizing conditions (Eh 100-400 mV), a relatively high pH (>8) is required to precipitate manganese which normally tends to remain in solution. Iron, on the other hand, can precipitate under conditions of lower pH and Eh (Fig. 3). Burns and Brown (1972) suggested the following mechanism for the formation of manganese nodules: at pH values greater than about 7.0, colloidal iron hydroxide will precipitate around any negatively charged nucleus. Colloidal iron hydroxide has a positive zero charge potential and is very active. The positively charged iron hydroxide colloids will then adsorb negatively charged hydrolyzed manganous ions which would

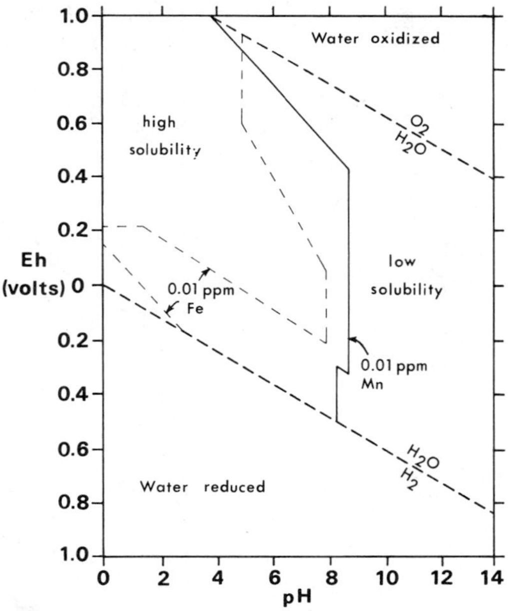

FIG. 3. Simplified pH-Eh diagram showing the comparative solubilities of iron and manganese in the system H_2O-Fe-Mn-S-CO_2, 1 atm. total pressure, 0.01 mg/L each of iron and manganese, and 2000 mg/L each of total CO_2 (as HCO_3^-) and total S (as SO_4^-) [modified from Hem, 1972].

subsequently be oxidized to MnO_2 or a hydrated manganese oxide or hydroxide under conditions of sufficiently high pH and Eh. The negatively charged manganese surface would then attract additional ferric hydroxide colloids, and so on. The precipitation would thus be autocatalytic, and would be one possible explanation for the alternation of iron-rich and iron-poor bands that are characteristic of ferromanganese nodules.

Organic Mechanisms

Inorganic mechanisms for the formation of manganese nodules and crust imply that it should be theoretically possible to form them without the aid of organisms. However, an increasing amount of evidence indicates that organic interactions can accelerate nodule formation in a number of direct and indirect ways, and may indeed be necessary for nodule formation. Possible contributions of organisms to the formation of manganese nodules are shown diagrammatically in Fig. 4.

Direct organic contribution to manganese nodule formation was proposed by Graham and Cooper (1959) and Graham (1959) to explain manganese-rich coatings on foraminifera. Their results suggest that a protein-rich material coats agglutinated arenaceous foraminifera and provides a substrate for other organisms that are capable of extracting iron, manganese, and other trace elements from seawater. Dugolinsky (1976) and Dugolinsky et al. (1977) described abundant and diverse communities of encrusting protozoa attached to manganese nodules from the Pacific Ocean and suggested that these organisms may be important in initiating manganese-nodule growth by concentrating trace elements. Alternatively, the nodules may simply serve as substrates for the organisms to live on. Greenslate et al. (1973) and Greenslate (1974a, 1974b) found that cavities in skeletal debris of planktonic organisms, especially diatoms, apparently were serving as nuclei for incipient manganese-nodule development. Greenslate suggested that bacteria associated with the decay of the planktonic organisms, rather than the organisms themselves, may actually be responsible for the accumulation of the manganese-rich coatings. Wendt (1974) suggested that abundant encrusting foraminifera on manganese nodules may increase the rate of manganese accretion. Accelerated rates of manganese accretion by microorganisms may help to explain the apparent slower growth rate of manganese nodules than the underlying sediments. Most growth rates of manganese nodules are calculated from radiometric decay gradients and are average rates calculated over long periods of time. Dugolinsky (1976) pointed out that the presence of different physical and chemical zones within manganese nodules which give them their characteristic banded appearance is not consistent

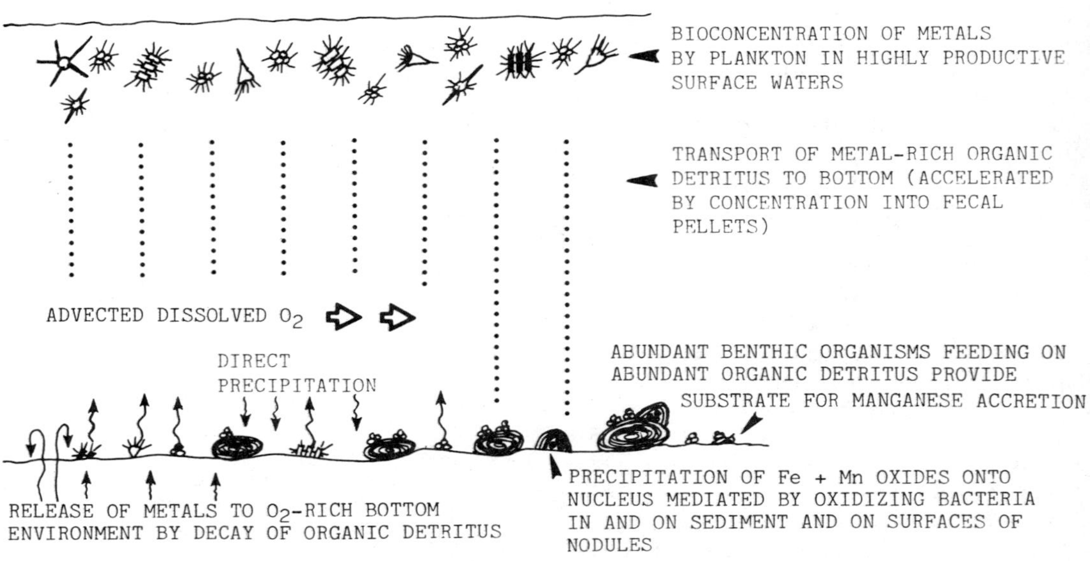

FIG. 4 Possible contributions of organisms to the formation of manganese nodules.

with slow, constant accretion of manganese, and that nodules may undergo prolonged periods of nongrowth or even erosion between periods of rapid accretion. Indeed, detailed accretion-rate studies of marine manganese nodules (Heye, 1978; Sadler, 1980) and freshwater nodules (Moore et al., 1980) show that growth usually is episodic.

The ability of microorganisms to affect the large-scale precipitation of both iron and manganese is well known (Silverman and Ehrlich, 1964; Zajic, 1969). The most extensive investigations on the role of microbes in the origin of manganese nodules have been carried out by Ehrlich and his colleagues (Ehrlich, 1963, 1966, 1968, 1971, 1972, 1975; Trimble and Ehrlich, 1968, 1970; LaRock and Ehrlich, 1975; Arcuri and Ehrlich, 1977). Controlled experiments in an extensive freshwater manganese nodule deposit in Oneida Lake, New York, have demonstrated the importance of microorganisms in the formation of freshwater as well as marine manganese nodules (Harris, 1978; Nealson, 1978; Dean et al., 1981).

The finding of iron and manganese bacteria associated with ferromanganese nodules, and the ease with which bacteria can cause the precipitation of iron and manganese in the laboratory strongly suggest that microorganisms are involved in the formation of the nodules and may be necessary for their growth. Monty (1973) even suggested that manganese nodules from the Blake Plateau are a type of bacterial stromatolite. Ehrlich (1972) concluded that the main role of microorganisms in nodule formation is the enzymatic catalysis of Mn^{2+} oxidation. Manganese-oxidizing bacteria derive energy from Mn^{2+} oxidation that they can use to assimilate organic carbon and thus give them a selective advantage over microorganisms that cannot oxidize manganese (Ehrlich, 1975). Alternatively, precipitation may be caused by adsorption of iron and manganese oxides and hydroxides onto cell surfaces (Zapffe, 1931).

There is therefore a growing list of environments in which manganese (and iron) are concentrated by microorganisms. The effectiveness of microorganisms to concentrate manganese even under the most severe environmental conditions was recently demonstrated by Dorn and Oberlander (1981) who found that microorganisms concentrate manganese to form the black, manganese-oxide rock coating of desert varnish.

In addition to the foregoing direct contributions of microorganisms to the formation of manganese nodules, several indirect contributions also may be possible, and indeed may be essential to the formation of extensive manganese deposits, especially those that are highly enriched in trace metals other than iron and manganese. These indirect mechanisms also may explain why the most metal-rich manganese nodules in the Pacific Ocean are found on a substrate rich in siliceous biogenic debris (radiolarian ooze and radiolarian clay) that occurs in a band directly under the equatorial zone of high biological productivity (Fig. 1). The coincidence of a siliceous biogenic substrate and metal-rich nodules is so striking that there may be a cause and effect relationship (Greenslate et al., 1973; Horn et al., 1973; Lyle et al., 1977). Halbach et al. (1981) considered that siliceous biogenic debris was an important carrier of copper in the deep-sea environment and that this source of copper might explain the enrichment of copper in nodules in the northeastern equatorial Pacific. Both benthic and planktonic microorganisms are known to concentrate iron, manganese, and other trace elements in their cells (Martin and Knauer, 1973; Bostrom et al., 1974; Moore and Bostrom, 1978). Organic detritus containing high concentrations of metals is then transported to the seafloor (Spencer et al., 1978; Martin and Knauer, 1980) where the metals are released by decay to a well-oxygenated bottom-water environment. Oxides and hydroxides of iron and manganese are precipitated either directly (inorganically) or through mediation by other microorganisms, and other trace metals are incorporated by adsorption. Rapid sedimentation of organic detritus is aided by incorporation of detritus into larger fecal pellets (Honjo, 1980, 1982; Brewer et al., 1980). Such a concentration-transport-release mechanism was proposed by Greenslate et al. (1973) and Greenslate (1974a, 1974b) to explain the high concentrations of copper and nickel in manganese nodules in the northeastern equatorial Pacific Ocean manganese-nodule province (Fig. 1). Bostrom et al. (1974) concluded that biogenic debris is one of the most important sources of certain trace elements, especially iron, manganese, copper, nickel, zinc, lead, cadmium, arsenic, mercury, barium, and boron, on the seafloor. The abundant organic detritus also supplies a source of organic carbon to support abundant and diverse communities of benthic microorganisms, some of which (e.g. iron- and manganese-oxidizing bacteria) may aid in the precipitation of iron and manganese oxides, and still others (e.g. benthic foraminifera) may provide additional substrates for microbial activity.

Siliceous biogenic sediments have other properties that may contribute to metal enrichment. They have the highest porosities and water contents of any sediment type in the world ocean. Horn et al. (1973) concluded that metal-bearing solutions could be more easily transmitted in the highly-porous siliceous sediments leading to selective concentration of copper and nickel in nodules in this substrate.

SUMMARY

The most important point that I wish to make in this paper is that there is an intimate association between the concentration of many transition elements into manganese nodules and the activities of microorganisms. Planktic microorganisms can concentrate trace metals. These metals are transported to the seafloor in the organic detritus. This detritus supports abundant and diverse communities of benthic microorganisms that provide additional organic detritus. Siliceous and calcareous skeletons of some microorganisms (e.g. radiolaria, diatoms, and foraminifera) provide substrates for accretion of manganese and iron oxides. Manganese-oxidizing bacteria use the energy derived from oxidation to assimilate organic carbon in organic detritus and, in so doing, precipitate manganese oxides and hydroxides. Decay of organic detritus releases entrapped trace metals, and, at the same time, causes reducing conditions within the sediments. The high porosity of siliceous biogenic sediment facilitates upward diffusion of reduced transition elements which are then oxidized at the sediment-water interface either inorganically or through the mediation of oxidizing-bacteria. Although there is no proof that microorganisms are required for the formation of deep-sea manganese nodules, the mounting evidence suggests that they must play an active role in nodule formation and element concentration.

ACKNOWLEDGMENTS

I would like to thank W. Ghiorse, J.R. Herring, W.S. Moore, and K.H. Nealson for their comments and discussions on the subject of the involvement of organisms in the formation of manganese nodules.

REFERENCES

Arcuri, E.J., and Ehrlich, H.L., 1977, "Influence of Hydrostatic Pressure on the Effects of the Heavy Metal Cations of Manganese, Copper, Cobalt, and Nickel on the Growth of Three Deep-Sea Bacterial Isolates," *Applied and Environmental Microbiology,* Vol. 33, pp. 282–288.

Barnes, S.S., and Dymond, J.R., 1967, "Rates of Accumulation of Ferromanganese Nodules," *Nature,* Vol. 213, pp. 1218–1219.

Bender, M.L., Ku, T.-L., and Broecker, W.S., 1966, "Manganese Nodules—Their Evolution," *Science,* Vol. 151, pp. 325–328.

Bender, M.L., Ku, T.-L., and Broecker, W.S., 1970, "Accumulation Rates of Manganese in Pelagic Sediments and Nodules," *Earth and Planetary Science Letters,* Vol. 8, pp. 143–148.

Bischoff, J.L., and Piper, D.Z., eds., 1979, *Marine Geology and Oceanography of the Pacific Manganese Nodule Province,* Plenum Press, New York, 842 pp.

Bischoff, J.L., and Rosenbauer, R.J., 1977, "Recent Metalliferous Sediment in the North Pacific Manganese Nodule Area," *Earth and Planetary Science Letters,* Vol. 33, pp. 379–388.

Bischoff, J.L., Heath, G.R., and Leinen, M., 1979, "Geochemistry of Deep-Sea Sediment from the Pacific Manganese Nodule Province: DOMES Sites A, B, and C," *Marine Geology and Oceanography of the Pacific Manganese Nodule Province,* J.L. Bischoff and D.Z. Piper, eds., Plenum Press, New York, pp. 397–436.

Bonatti, E., 1975, "Metallogenesis at Oceanic Spreading Centers," *Annual Review of Earth and Planetary Sciences,* Vol. 3, pp. 401–431.

Bonatti, E., and Nayudu, Y.R., 1965, "The Origin of Manganese Nodules on the Ocean Floor, *American Journal of Science,* Vol. 263, pp. 17–39.

Bonatti, E., Kraemer, T., and Rydell, H., 1972, "Classification and Genesis of Submarine Iron-Manganese Deposits," *Ferromanganese Deposits on the Ocean Floor,* D.R. Horn, ed., Lamont-Doherty Geological Observatory, NSF/IDOE, Washington, pp. 149–166.

Bostrom, K., 1967, "The Problem of Excess Manganese in Pelagic Sediments," *Researches in Geochemistry,* Vol. 2, P. Ableson, ed., John Wiley and Sons, New York, pp. 421–452.

Bostrom, K., and Peterson, M.N.A., 1966, "Precipitates from Hydrothermal Exhalations on the East Pacific Rise," *Economic Geology,* Vol. 61, pp. 1258–1265.

Bostrom, K., Joensuu, O., and Brohm, I., 1974, "Plankton: Its Composition and Its Significance as a Source of Pelagic Sediments," *Chemical Geology,* Vol. 14, pp. 255–271.

Brewer, P.G., et al., 1980, "A Sediment Trap Experiment in the Deep Subtropical Atlantic; Isotopic and Elemental Fluxes," *Journal of Marine Research,* Vol. 38, pp. 703–728.

Burns, R.G., and Brown, B.A., 1972, "Nucleation and Mineralogical Controls on the Composition of Manganese Nodules," *Ferromanganese Deposits on the Ocean Floor,* D.R. Horn, ed., Lamont-Doherty Geological Observatory, NSF/IDOE, Washington, pp. 51–62.

Burns, R.G., and Burns, V.M., 1977, "Mineralogy," *Marine Manganese Deposits,* G.P. Glasby, ed., Elsevier, Amsterdam, pp. 185–248.

Collins, J.F., and Buol, S.W., 1970, "Effects of Fluctuations in Eh-pH Environment on Iron and/or Manganese Equilibria," *Soil Science,* Vol. 110, pp. 111–118.

Collins, J.F., and Buol, S.W., 1972, "Patterns of Iron and Manganese Precipitation under Specified Eh-pH Conditions," *Soil Science,* Vol. 110, pp. 157–162.

Cronan, D.S., 1976, "Manganese Nodules and Other Ferromanganese Oxide Deposits," *Chemical Oceanography,* Vol. 5, J.P. Riley and R. Chester, eds., Academic Press, London, pp. 217–263.

Cronan, D.S., 1977, "Deep-Sea Nodules: Distribution and Geochemistry," *Marine Manganese Deposits,* G.P. Glasby, ed., Elsevier, Amsterdam, pp. 10–44.

Cronan, D.S., 1980, *Underwater Minerals,* Academic Press, New York, 362 pp.

Cronan, D.S., and Tooms, J.S., 1969, "The Geochemistry of Manganese Nodules and Associated Deposits from the Pacific and Indian Oceans," *Deep-Sea Research,* Vol. 16, pp. 335–359.

Dean, W.E., Moore, W.S., and Nealson, K.H., 1981, "Manganese Cycles and the Origin of Manganese Nodules, Oneida Lake, New York, U.S.A.," *Chemical Geology,* Vol. 34, pp. 53–64.

Dorn, R.I., and Oberlander, T.M., 1981, "Microbial Origin of Desert Varnish," *Science,* Vol. 213, pp. 1245–1247.

Dugolinsky, B.K., "Chemistry and Morphology of Deep-Sea Manganese Nodules and the Significance of Associated Encrusting Protozoans on Nodules Growth," Ph.D. Thesis, University of Hawaii, 228 pp.

Dugolinsky, B.K., Margolis, S.V., and Dudley, W.C., 1977, "Biogenic Influence on Growth of Manganese Nodules," *Journal of Sedimentary Petrology,* Vol. 47, pp. 428–445.

Ehrlich, H.L., 1963, "Bacteriology of Manganese Nodules. 1. Bacterial Action on Manganese Nodules Enrichments," *Applied Microbiology,* Vol. 11, pp. 15–19.

Ehrlich, H.L., 1966, "Reaction with Manganese by Bacteria from Marine Ferromanganese Nodules," *Developments in Industrial Microbiology,* Vol. 7, pp. 279–286.

Ehrlich, H.L., 1968, "Bacteriology of Manganese Nodules. II. Manganese Oxidation by Cell-Free Extract from a Manganese Nodule Bacterium," *Applied Microbiology,* Vol. 16, pp. 197–202.

Ehrlich, H.L., 1971, "Bacteriology of Manganese Nodules. V. Effect of Hydrostatic Pressure on Bacterial Oxidation of Mn^{2+} and Reduction of MnO_2," *Applied Microbiology,* Vol. 21, pp. 306–310.

Ehrlich, H.L., 1972, "The Role of Microbes in Manganese Nodule Genesis and Degradation," *Ferromanganese Deposits on the Ocean Floor,* D.R. Horn, ed., Lamont-Doherty Geological Observatory, NSF/IDOE, Washington, pp. 63–70.

Ehrlich, H.L., 1975, "The Formation of Ores in the Sedimentary Environment of the Deep-Sea with Microbial Participation: The Case for Ferromanganese Concretions," *Soil Science,* Vol. 119, pp. 36–41.

Glasby, G.P., 1972, "The Geochemistry of Manganese Nodules from the Northwest Indian Ocean," *Ferromanganese Deposits on the Ocean Floor,* D.R. Horn, ed., Lamont-Doherty Geological Observatory, NSF/IDOE, Washington, pp. 93–104.

Glasby, G.P., ed., 1977, *Marine Manganese Deposits,* Elsevier, Amsterdam, 523 pp.

Graham, J.W., 1959, "Metabolically Induced Precipitation of Trace Elements from Sea Water," *Science,* Vol. 129, pp. 1428–1429.

Graham, J.W., and Cooper, S.C., 1959, "Biological Origin of Manganese-Rich Deposits of the Sea Floor, *Nature,* Vol. 183, pp. 1050–1051.

Greenslate, J., 1974a, "Microorganisms Participate in the Construction of Manganese Nodules, *Nature,* Vol. 249, pp. 181–183.

Greenslate, J., 1974b, "Manganese and Biotic Debris Associations in Some Deep-Sea Sediments, *Science,* Vol. 186, pp. 529–531.

Greenslate, J., Frazer, J.Z., and Arrhenius, G., 1973, "Origin and Deposition of Selected Transition Elements in the Sea Bed," *Inter-University Program of Research on Ferromanganese Deposits of the Ocean Floor,* National Science Foundation, International Decade for Ocean Exploration (NSF/IDOE), Phase I Report, pp. 13–20.

Halbach, P., Hebisch, U., and Scherhag, C., 1981, "Geochemical Variations of Ferromanganese Nodules and Crusts from Different Provinces of the Pacific Ocean and Their Genetic Control," *Chemical Geology,* Vol. 34, pp. 3–17.

Harris, T.G., 1978, "The Role of Bacteria in the Fixation of Co, Fe, and Mn in Lake Oneida, N.Y., *EOS,* American Geophysical Union, Vol. 59, p. 412.

Heath, G.R., 1981, "Ferromanganese Nodules of the Deep Sea," *Economic Geology,* 75th Anniversary Vol., pp. 736–765.

Heath, G.R., and Dymond, J., 1977, "Genesis and Transformation of Metalliferous Sediments from the East Pacific Rise, Bauer Deep and Central Basin, Northwest Nazca Plate, *Bulletin,* Geological Society of America, Vol. 88, pp. 723–733.

Hem, J.D., 1963, "Chemical Equilibria and Rates of Manganese Oxidation," Water-Supply Paper 1667-A, US Geological Survey, 64 pp.

Hem, J.D., 1972, "Chemical Factors That Influence Availability of Iron and Manganese in Aqueous Systems," *Bulletin,* Geological Society of America, Vol. 83, pp. 443–450.

Hekinian, R., and Hoffert, M., 1975, "Rate of Palagonitization and Manganese Coating on Basalt Rocks from the Rift Valley in the Atlantic Ocean near 36°50′ N," *Marine Geology,* Vol. 19, pp. 91–109.

Heye, D., 1978, "Changes in the Growth Rate of Manganese Nodules from the Central Pacific in the Area of a Seamount, as Shown by the Io Method," *Marine Geology,* Vol. 28, pp. M59–65.

Honjo, S., 1980, "Material Fluxes and Modes of Sedimentation in the Mesopelagic and Bathypelagic Zones," *Journal of Marine Research,* Vol. 38, pp. 53–97.

Honjo, S., 1982, "Seasonality and Interaction of Biogenic and Lithogenic Particulate Flux at the Panama Basin," *Science,* Vol. 218, pp. 883–884.

Horn, D.R., Horn, B.N., and Delach, M.N., 1973, "Copper and Nickel Content of Ocean Ferromanganese Deposits and Their Relation to Properties of the Substrate," *The Origin and Distribution of Manganese Nodules in the Pacific and Prospects for Exploration,* M. Morgenstein, ed., Valdevia Manganese Exploration Group, University of Hawaii, and NSF/IDOE, pp. 77–83.

Krauskopf, K.B., 1956, "Factors Controlling the Concentrations of Thirteen Rare Metals in Seawater," *Geochimica et Cosmochimica Acta,* Vol. 9, pp. 1–32.

Krauskopf, K.B., 1957, "Separation of Manganese from Iron in Sedimentary Processes," *Geochimica et Cosmochemica Acta,* Vol. 12, pp. 61–84.

Ku, T.-L., 1977, "Rates of Accretion," *Marine Manganese Deposits,* G.P. Glasby, ed., Elsevier, Amsterdam, pp. 249–267.

Ku, T.-L., and Broecker, W.S., 1967, "Uranium, Thorium and Protactinium in a Manganese Nodule," *Earth and Planetary Science Letters,* Vol. 2, pp. 317–320.

Ku, T.-L., and Broecker, W.S., 1969, "Radiochemical Studies on Manganese Nodules of Deep-Sea Origin," *Deep-Sea Research,* Vol. 16, pp. 625–637.

Ku, T.-L., and Glasby, G.P., 1972, "Radiochemical Evidence for the Rapid Growth Rate of Shallow-Water, Continental Margin Manganese Nodules," *Geochimica et Cosmochimica Acta,* Vol. 36, pp. 699–703.

LaRock, P.A., and Ehrlich, H.L., 1975, "Observations of Bacterial Microcolonies on the Surface of Ferromanganese Nodules from Blake Plateau by Scanning Electron Microscopy," *Microbial Ecology,* Vol. 2, pp. 84–96.

Lyle, N. Dymond, J., and Heath, G.R., 1977, Copper-Nickel-Enriched Ferromanganese Nodules and Associated Crusts from the Bauer Basins, Northwest Nazca Plate," *Earth and Planetary Science Letters,* Vol. 35, pp. 55–64.

Lynn, D.C., and Bonatti, E., 1965, "Mobility of Manganese in Diagenesis of Deep Sea Sediments," *Marine Geology,* Vol. 3, pp. 457–474.

Manheim, F.T., 1965, "Manganese-Iron Accumulations in the Shallow Marine Environment," Occasional Publications No. 3, University of Rhode Island, pp. 217–276.

Martin, J.H., and Knauer, G.A., 1973, "The Elemental Composition of Plankton," *Geochimica et Cosmochimica Acta,* Vol. 37, pp. 1639–1653.

Martin, J.H., and Knauer, G.A., 1980, "Manganese Cycling in Northeast Pacific Waters," *Earth and Planetary Science Letters,* Vol. 51, pp. 266–274.

McKelvey, V.E., Wright, N.A., and Rowland, R.W., 1979, "Manganese Nodule Resources in the Northeastern Equatorial Pacific," *Marine Geology and Oceanography of the Pacific Manganese Nodule Province,* J.L. Bischoff and D.Z. Piper, eds., Plenum Press, New York, pp. 747–762.

Mero, J.L., 1965, *The Mineral Resources of the Sea,* Elsevier, Amsterdam, 321 pp.

Monty, C., 1973, "Les Nodules de Manganese sont des Stromatolithes Oceaniques," *C.R. Academie Science,* Series D, Vol. 276, pp. 3285–3288.

Moore, C., and Bostrom, K., 1978, "The Elemental Composition of Lower Marine Organisms," *Chemical Geology,* Vol. 23, pp. 1–9.

Moore, W.S., et al., 1980, "Growth Rates of Manganese Nodules in Oneida Lake, New York," *Earth and Planetary Science Letters,* Vol. 46, pp. 191–200.

Moore, W.S., and Vogt, P.R., 1976, "Hydrothermal Manganese Crusts from Two Sites near the Galapagos Spreading Axis," *Earth and Planetary Science Letters,* Vol. 29, p p. 349–356.

Moore, W.S., et al., 1981, "Fluxes of Metals to a Manganese Nodule: Radiochemical, Chemical, Structural, and Mineralogical Studies," *Earth and Planetary Science Letters,* Vol. 52, pp. 151–171.

Nealson, K.H., 1978, "The Isolation and Characterization of Marine Bacteria that Catalyze Mn Oxidation," *Environmental Biogeochemistry and Geomicrobiology,* Vol. 3, pp. 847–858.

Price, N.B., 1967, "Some Geochemical Observations on Manganese Oxide Nodules from Different Depth Environments," *Marine Geology,* Vol. 5, pp. 511–538.

Raab, W., 1972, "Physical and Chemical Features of Pacific Deep Sea Manganese Nodules and Their Implications to the Genesis of Nodules," *Ferromanganese Deposits on the Ocean Floor,* D.R. Horn, ed., Lamont-Doherty Geological Observatory, NSF/IDOE, Washington, pp. 31–49.

Reyss, J.L., Marchig, V., and Ku, T.-L., 1982, "Rapid Growth of a Deep-Sea Manganese Nodule," *Nature,* Vol. 295, pp. 401–403.

Sadler, P.M., 1980, "The Significance of Time Scale for the Rate of Accretion of Marine Manganese Nodules and Crusts," *Marine Geology,* Vol. 35, pp. M27–32.

Scott, M.R., et al., 1974, "Rapidly Accumulating Manganese Deposits from the Median Valley of the Mid-Atlantic Ridge," *Geophysical Research Letters,* Vol. 1, pp. 355–358.

Silverman, M.P., and Ehrlich, H.L., 1964, "Microbial Formation and Degradation of Minerals," *Advances in Applied Microbiology,* Vol. 6, pp. 163–206.

Spencer, D.W., et al., 1978, "Chemical Fluxes from a Sediment Trap Experiment in the Deep Sargasso Sea," *Journal of Marine Research,* Vol. 36, pp. 493–523.

Trimble, R.B., and Ehrlich, H.L., 1968, "Bacteriology of Manganese Nodules. III. Reduction of MnO_2 by 2 Strains of Nodule Bacteria," *Applied Microbiology,* Vol. 16, pp. 695–702.

Trimble, R.B., and Ehrlich, H.L., 1970, "Bacteriology of Manganese Nodules. IV. Induction of an MnO_2-Reductase System in a Marine Bacillus," *Applied Microbiology,* Vol. 19, pp. 966–972.

Wedepohl, K.H., 1971, *Geochemistry,* translated from the German by E. Althaus, Holt, Rinehart, and Winston, New York, 231 pp.

Wendt, J., 1974, "Encrusting Organisms in Deep-Sea Manganese Nodules," *Pelagic Sediments On Land and Under the Sea,* K.J. Hsü and H.C. Jenkyns, eds., International Association of Sedimentologists, Special Publication 1, pp. 437–447.

Zajic, J.E., 1969, *Microbial Biogeochemistry,* Academic Press, New York, 345 pp.

Zapffe, C., 1931, "Deposition of Manganese," *Economic Geology,* Vol. 26, pp. 799–832.

12

Metalliferous Sediments of the East Pacific

Cyrus W. Field
Department of Geology
Oregon State University
Corvallis, OR

Robert O. Rye
Branch of Isotope Geology
U.S. Geological Survey
Denver, CO

Jack R. Dymond
School of Oceanography
Oregon State University
Corvallis, OR

Joseph F. Whelan
Branch of Isotope Geology
U.S. Geological Survey
Denver, CO

Ronald G. Senechal
Department of Geology
Oregon State University
Corvallis, OR

ABSTRACT

Metalliferous sediments are present throughout much of the ocean basins and are locally prominent in the sedimentary column on and near the Nazca plate of the east-southeast Pacific Ocean. They typically have anomalously high concentrations of the transition metals, and those on parts of the East Pacific Rise, the Bauer Deep, and other basinal depressions may contain more then 12% iron, 4% manganese, 0.1% each of copper and nickel, and geochemically significant amounts of zinc, cobalt, molybdenum, arsenic, beryllium, silver, and possibly gold. Statistical interpretations of chemical data indicate that most metals in sediments on the East Pacific Rise were derived from a hydrothermal component related to submarine volcanism, whereas those in basinal sediments accumulated from both hydrothermal and hydrogenous components of sedimentation. However, concentrations of the metals are controlled more by the absence of biogenic and detrital components than by nearness to centers of hydrothermal activity.

Stable isotope abundances of oxygen (mostly 10.0 to 22.0 permil) in carbonate-free sediment, carbon (−0.1 to 2.1 permil) and oxygen (mostly 29.7 to 30.8 permil) in carbonate, and sulfate-sulfur (17.8 to 21.0 permil) in bulk sediment indicate that these constituents accumulated by low-temperature deposition from seawater. However, depletion of ^{18}O in a few samples of sediment from the East Pacific Rise implies equilibration with heated seawater near hydrothermal vents. Although hydrothermal sulfate-sulfide assemblages of the Andean Cordillera are apparently enriched in ^{34}S (2 to 13 permil) relative to magmatic sulfur (0 permil), this heavy sulfur is interpreted to be a local contaminant of the fluids derived from nearby marine evaporites and not an indicator of remobilized metalliferous sediments through subduction and anatexis.

The total quantity of metals contained in these sediments is enormous, but their principal value is based on that of the aggregate polymetallic assemblage. Because the concentration of each metal is low relative to that of the commerical ore, and because the most economically important metals (iron and manganese) will be difficult to extract, the metalliferous sediments are likely to remain an impressively large but unused resource into the future.

INTRODUCTION

Oceanographic investigations of the past two decades have documented the widespread occurrence of metalliferous sediments as a generally minor component of the sedimentary veneer that overlies basaltic crust of the ocean basins. These sediments were first collected near the East Pacific Rise during the Challenger Expedition (1872–1876) and at which time their characteristically high contents of iron and manganese were also recognized. However, it was not until the detailed studies of Bostrom and Peterson (1966) that the anomalously high concentrations of other transition metals and the probable hydrothermal origin of these sediments were fully appreciated. Accordingly, these metal-rich accumulations of sediment are not merely of scientific interest, but they also constitute a potential resource for the future.

Metalliferous sediments are a small but important fraction of the sedimentary column that mantles ocean-floor basalts on the Nazca plate of the southeast Pacific Ocean. This plate and its convergent boundary with the adjoining South American plate to the east, as given by Fig. 1, represents one of the most geologically dynamic regions of lithosphere on Earth. The Nazca plate occupies about 16×10^6 km^2 and its margins are defined by the Peru-Chile Trench to the east, the East Pacific Rise to the west, the Galapagos Rift to the north, and the Chile Rise to the south. The centrally located and north-trending "fossil" Galapagos Rise (Fig. 1) subdivides the plate into the eastern Peru Basin and the western Central Basin, with the Bauer Deep forming a prominent depression in the northwest part of the Central Basin. Other bathymetric features not illustrated include the Nazca Ridge that extends southwest from the continent at about 15° S, the Sala y Gomez Ridge that trends east-west at about 26° S, and local basins and ridges. Average depths to the ocean floor range from 2800 to 3200 m along the East Pacific Rise, 3500 to 4500 m across the plate, and from 5000 to more than 7500 m within the Peru-Chile Trench (Mammerickx and Smith, 1978). For the purposes of regional perspective, Fig. 1 also shows the ages of underlying basaltic oceanic crust based on magnetic anomalies (Herron, 1972), and the distribution of porphyry-type deposits of copper, molybdenum, and tin that were emplaced in the adjacent Andean Cordillera during late Paleozoic to late Cenozoic time (see Field and Dasch, 1981, and references cited therein). Relationships such as the systematically increasing age of oceanic crust away from the East Pacific Rise and the subparallel coincident association of cordilleran mineral deposits to the nearby Peru-Chile Trench serve to highlight the geologically dynamic character of this region. In addition, Sillitoe (1972) has suggested that the metalliferous sediments, formed by submarine hydrothermal activity during ocean-ridge basaltic volcanism, are the potential source of metals in magmatic hydrothermal deposits of the cordillera as a consequence of the transport and subduction of these sediments by seafloor spreading and their subsequent remobilization and incorporation into granitic magmas.

Purpose, Scope, and Methods of Study

In this report we describe and regionally summarize the data for some major oxides and metals contained in 228 samples of sediment collected from the ocean floor on and near the Nazca plate, and we present additional chemical and isotopic data for a smaller subset of these and other samples. Our purpose is threefold: (1) to show the anomalously high concentrations and widespread distribution of metals in these samples, (2) to further clarify their environment of deposition from stable isotope abundance data, and (3) to evaluate the possible contribution of oceanic sulfur to hydrothermal mineral deposits of the cordillera. Although much of the chemical data have been reported previously, our presentation differs in that metal abundances are given in terms of absolute concentrations (percent or parts per million by weight) and not as relative values determined on a carbonate-free and salt-free basis (Dymond and Corliss, 1980; Dymond, 1981). This format offers a more realistic means of approximating the future role of metalliferous sediments as a potential mineral resource.

Locations of most of the samples employed in this study are shown in Fig. 2. The majority are gravity core samples of ocean-floor sediment obtained by personnel of the Hawaii Institute of Geophysics and oceanographic group of Oregon State University during cruises of the R/V Kana Keoki (HIG), R/V Yaquina (OSU), and R/V Wecoma (OSU) during the period from 1972 to 1977. Smaller sets of other samples described in this report include six drill core samples of sediment [Leg 5, Deep Sea Drilling Project (DSDP)] from the northeast Pacific Ocean, nine dredge haul samples of manganese nodules and five dredge haul samples of ocean-floor basalt from widely different locations in the ocean basins, and suites of hydrothermal sulfate-sulfide assemblages from mineral deposits of Moro-

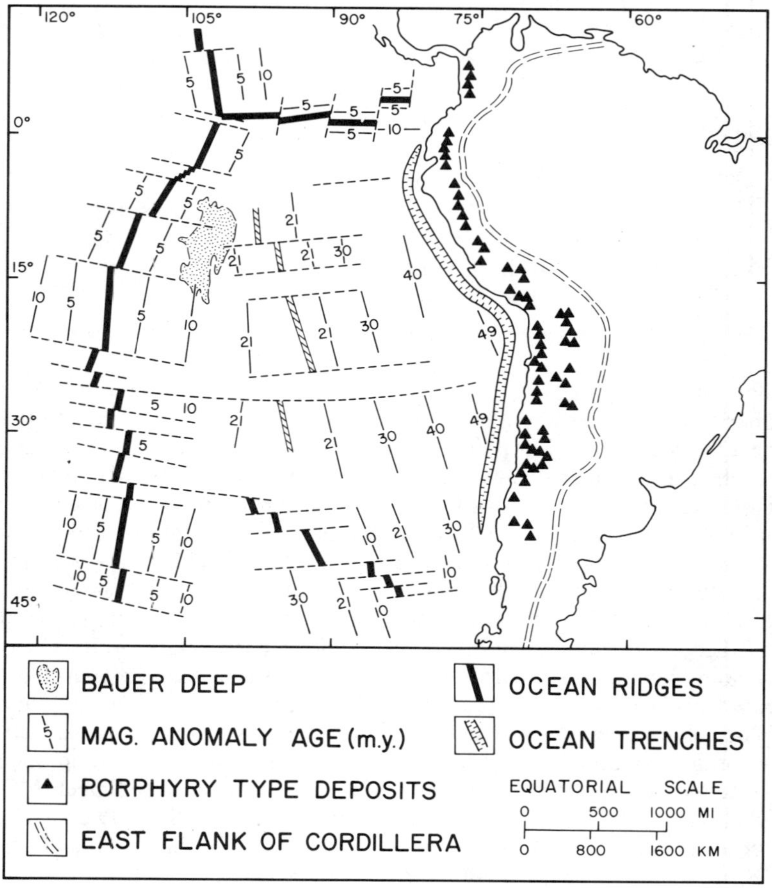

FIG. 1. Tectonic summary of the Nazca and South American plates (ages of oceanic crust after Herron, 1972, and locations of porphyry-type deposits after Field and Dasch, 1981).

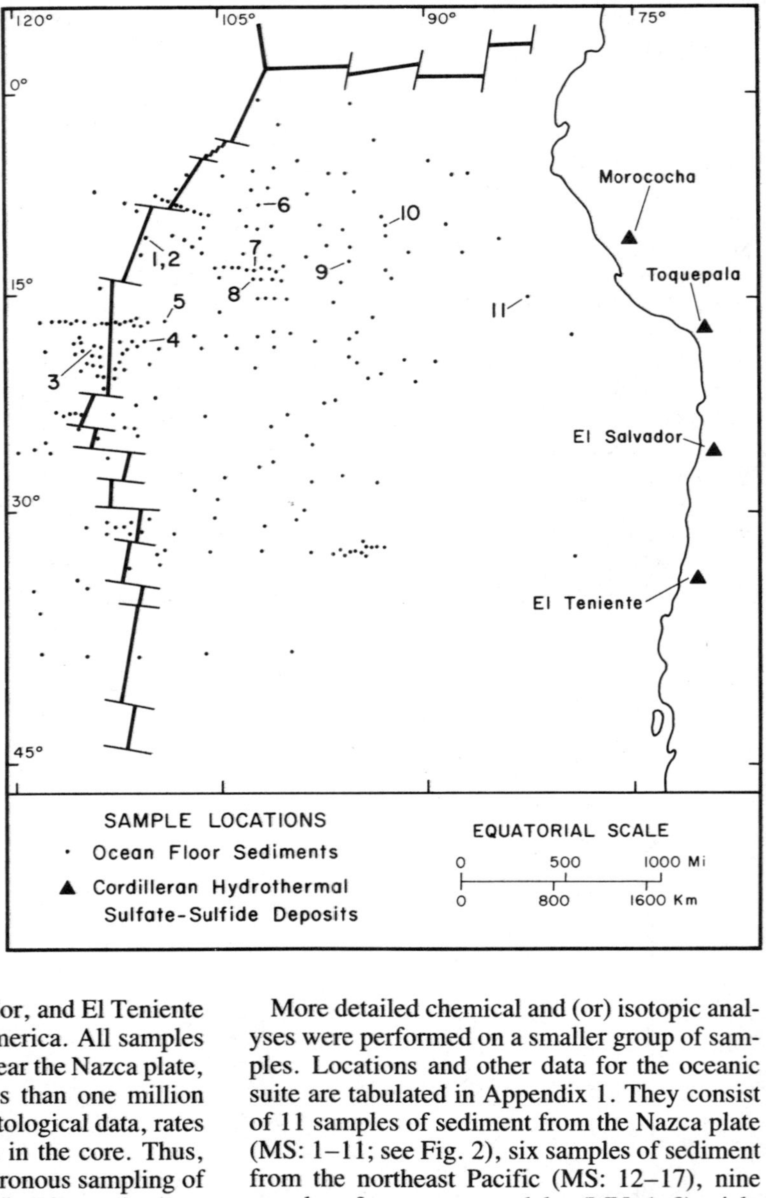

FIG. 2. Locations of samples of sediment from the Nazca plate and hydrothermal sulfate-sulfide assemblages from mines of the Andean Cordillera (numbers refer to piston core samples of sediment for which detailed chemical information is given).

cocha, Toquepala, El Salvador, and El Teniente in the cordillera of South America. All samples of sediment located on and near the Nazca plate, with one exception, are less than one million years in age based on paleontological data, rates of sedimentation, and depth in the core. Thus, they represent a nearly isochronous sampling of sediment from the East Pacific Rise to the eastern part of the plate. These samples were prepared and analyzed for metals (silicon, aluminum, manganese, iron, barium, calcium, cobalt, copper, and zinc) and other constituents ($CaCO_3$ and salt) in the oceanographic laboratories of Oregon State University according to procedures and methods (chiefly atomic absorption spectrophotometry) described previously by Dymond (1981) and Dymond et al. (1973, 1976, 1977).

More detailed chemical and (or) isotopic analyses were performed on a smaller group of samples. Locations and other data for the oceanic suite are tabulated in Appendix 1. They consist of 11 samples of sediment from the Nazca plate (MS: 1–11; see Fig. 2), six samples of sediment from the northeast Pacific (MS: 12–17), nine samples of manganese nodules (MN: 1–9), eight splits from five samples of ocean-ridge basalt (ORB: 1–5), and a reconnaissance suite of hydrothermal sulfate-sulfide assemblages from cordilleran mineral deposits of North and South America. Concentrations of metals (silver, cobalt, copper, molybdenum, nickel, lead, and zinc) and sulfur in the samples of sediment (MS: 1–17) were obtained from AAS and Leco methods of analysis performed by Chemical and Mineralogical Services, Inc., of Salt Lake City,

UT. The data for these and other elements were checked and enhanced with emission spectrographic analyses undertaken by Specomp Services, Inc., of Boise, ID. Isotopic abundances of carbon, oxygen, and sulfur were determined in the samples of oceanic sediment. However, samples of nodules and basalt were analyzed exclusively for oxygen isotopes, and those of sulfates and sulfides in the cordilleran mineral deposits were analyzed only for sulfur isotopes. Sample preparations, extractions, and isotope ratio analyses were largely performed in laboratory facilities at the Branch of Isotope Geology, US Geological Survey, Denver, CO. Standard methods and procedures were used in the treatment and isotopic analyses of these samples (Dymond et al. 1973; Field and Gustafson, 1976; Ohmoto and Rye, 1979). The isotopic data are presented in terms of conventional permil notation, and with reference to appropriate and internationally accepted standards for carbon, oxygen, and sulfur.

Metalliferous Sediments

The results of seismic reflection surveys and deep-sea drilling programs indicate that thicknesses of sediment and sedimentary rock that overlie basaltic crust of the Nazca plate generally increase from west to east. This veneer ranges from variably thin (up to several tens of meters) to absent over the young crest of the East Pacific Rise to more than 150 m in thickness near the Peru-Chile Trench. The sediments are normally composed of three dominant end-members that reflect different sources. These are (1) a detrital component of terrigenous silicates derived from the South American Continent; (2) a biogenic component of opaline silica and (or) calcite derived from the skeletal remains of pelagic microorganisms and minor phosphatic fish debris; and (3) a hydrogenous component of authigenic ferromanganese nodules and silicates derived from seawater and the reconstitution of some biogenic and detrital fractions. The proportions of these three components are a function of their relative accumulation rates, which in turn are largely controlled by proximity to the continent, productivity of pelagic organisms in the surface waters, and depth to the ocean floor (calcite compensation depth about 4300 m). However, a metal-rich hydrothermal component is widespread in small amounts throughout most of the sedimentary column, and it may exceed 10 to 20% and more of the total sediment in basal horizons across the plate, surficial horizons of the Bauer Deep and other basins, and locally elsewhere throughout the stratigraphic section. This unusual component represents the metalliferous sediments, and for which a brief summary follows based on the published investigations of Bostrom and Peterson (1966, 1969), Sayles and Bischoff (1973), Dymond and others (1973, 1976, 1977), Sayles et al. (1975), Field et al. (1976), Bostrom et al. (1976), Heath and Dymond (1977, 1981), Dymond and Corliss (1980), and Dymond (1981).

The metal-rich component of these sediments consists of amorphous to crystalline particles of iron and manganese oxides, hydroxides, and oxyhydroxides. They are normally exceedingly fine-grained and, where relatively uncontaminated by detrital and biogenic components, exhibit a vivid spectrum of brown, red, and yellow-orange colors that indicates a highly oxidized state. The principal components of these sediments include goethite, δMnO_2, psilomelane, todorokite, iron-rich smectite, and amorphous ferromanganese hydroxides and oxyhydroxides. In addition to iron and manganese, significant concentrations of other transition metals have accumulated by sorption processes in these reactive sediments. Except locally, these metal-rich sediments are variably contaminated by the other components of sedimentation that may consist of biogenic opaline silica, calcite, and minor barite and apatite; hydrogenous and (or) authigenic ferromanganese micronodules, iron-rich smectite, and zeolite (phillipsite); and detrital quartz and plagioclase feldspar. Rates of sedimentation vary by more than one order of magnitude across the Nazca plate (Heath and Dymond, 1981; McMurtry and others, 1981) and range from about 10 to 20 m/m.y. along and near the East Pacific Rise to as little as 1 to 2 m/m.y. in the Bauer Deep and more easterly regions. Heath and Dymond (1977, 1981) and Dymond (1981) have grouped the various components of these sediments into "ideal" chemical and mineralogical end members and have applied the methods of normative partitioning and Q-mode factor analysis to a large quantity of analytical data. By doing so, they have documented the important role of hydrothermal processes to the formation of metalliferous sediments. Their results indicate that, on a carbonate-free basis, the hydrothermal component is the dominent source (80% and more) of iron, manganese, copper, nickel, and zinc in sediments on and near the East Pacific Rise. However, both hydrothermal and hydrogenous components are subequally important and are the dominant sources of these metals in sediments of the Bauer Deep and surrounding Central Basin to the east. Stable isotope abundances of carbon, oxygen, and sulfur and the distribution patterns of rare-earth elements collectively point to a seawater provenance for these constituents (Dymond and others, 1973; Field and others, 1976). Although data for the heavy radiogenic isotopes of strontium, thorium, and uranium similarly imply a derivation from seawater, those for lead exhibit a range from marine to mantlelike values (Bender et al., 1971; Dasch et al., 1971; Dymond et al., 1973; Dasch, 1981; Veeh, 1981). The latter values, which are least radiogenic, are suggestive of lead that has been leached from mid-ocean ridge basalts during periods of submarine volcanism and hydrothermal activity.

In summary, the metalliferous sediments are stratiform precipitates of metal-rich oxides that have accumulated at low temperatures on the ocean floor. Geologic relationships and statistical treatment of chemical data suggest that the transition metals are chiefly of hydrothermal origin, whereas the distribution patterns and isotopic data for most other associated constituents indicate that they are of either biogenic and (or) hydrogenous origin, and thus are derived from seawater. The metals presumably represent distal and basinal deposition from hydrothermal fluids vented from sources on mid-ocean ridges such as those recently discovered on the East

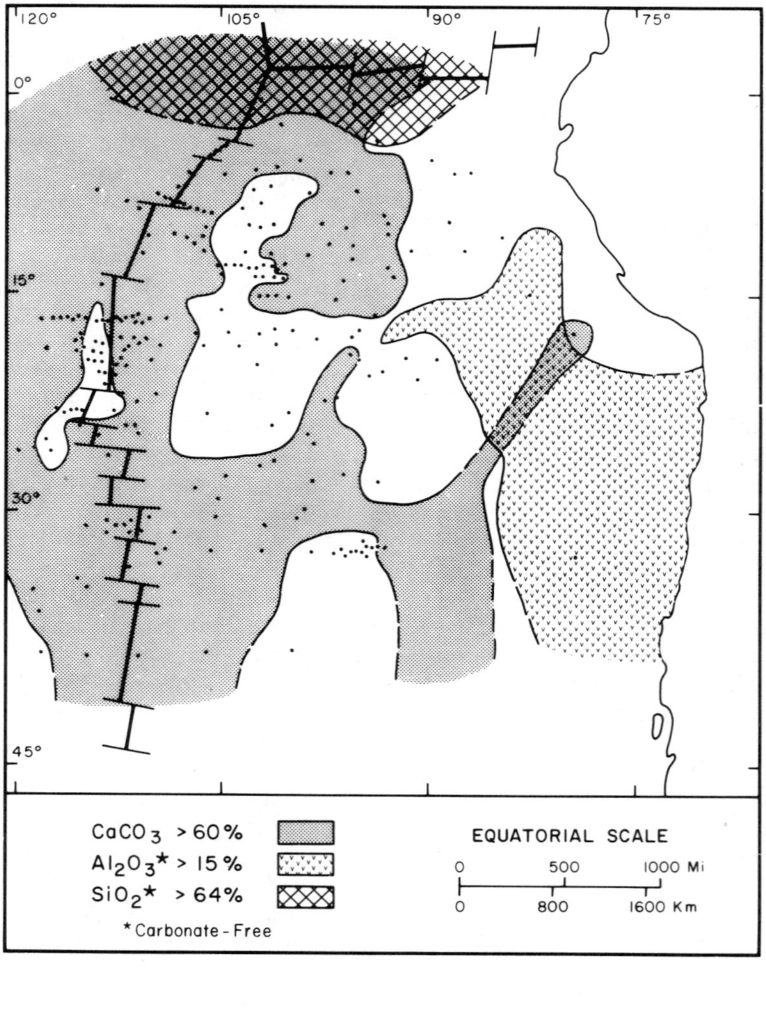

FIG. 3. Chemistry and distribution of principal sediment types (modified after Dymond and Corliss, 1980; Dymond, 1981).

Pacific Rise and elsewhere on active ridge systems of the eastern Pacific Basin (Hekinian and others, 1980).

MAJOR, MINOR, AND TRACE ELEMENT GEOCHEMISTRY

The dominant components of sedimentation on and near the Nazca plate are biogenic calcite and opaline silica and terrigenous detritus swept from the adjacent South American continent. Geographic distributions of these components as defined by $CaCO_3$, SiO_2, and Al_2O_3 are summarized in Fig. 3. It is important to emphasize that not only is $CaCO_3$ generally the most significant of these three components, but also that the abundances of SiO_2 and Al_2O_3 are on a carbonate-free basis. Deposition of carbonate is especially pronounced on the bathymetrically higher parts of the plate that include the Galapagos Rift to the north, the East Pacific Rise to the west, the Sala y Gomez and Nazca Ridges to the south and southeast, and the central north-trending Galapagos Rise. However, it is a minor component of sedimentation at depths below the calcite compensation level (4300 m) that prevail in the Bauer Deep and other basins of the central, southern, and eastern regions. Not surprisingly, the detrital component as designated by Al_2O_3 is most conspicuous in sediment adjacent to the continent, but it becomes diluted with biogenic calcite and silica to the northeast. Relatively large abundances of biogenic SiO_2 that comprise the remains of planktonic diatoms, radiolaria, and other siliceous microorganisms are latitudinally confined to the equatorial zone of high silica productivity.

The regional distributions and contoured abundances of iron and manganese, copper and zinc, and nickel and cobalt in surface sediments on and near the Nazca Plate are illustrated in Figs. 4, 5, and 6, respectively. These transition metals, as previously noted, are largely introduced by the hydrothermal and hydrogenous components of sedimentation. The coincident locations of anomalously high concentrations for each of the metals serve to graphically document their distinct covariant behavior. In addition, most areas of sediment having abundant concentrations of these metals are coincident with the Bauer Deep and other basins that are notable for their lower amounts of biogenic and detrital components. The metal-rich fraction of basinal sediment is derived primarily from nearly subequal contributions of hydrothermal and hydrogenous components. According to the studies of Dymond et al. (1977), Heath and Dymond (1977, 1981), and Dymond (1981), the concentration of metals in these sediments results not from higher absolute rates of accumulation, but rather because of relatively lower accumulation rates of biogenic and detrital components. Thus, the enhanced metal content of the basinal sediments is attributable to distance from the continent (limited detritus), distance from the equatorial zone (limited biogenic silica), and deposition below the calcite compensation depth (limited biogenic carbonate). In contrast, the concentrations of metals in sediments along the East Pacific Rise are generally subdued, in spite of the fact that this area of submarine volcanism is considered to be the source area for the hydrothermal component. Although 80% and more of the metals contained in sediments on and near the rise are ascribed to the hydrothermal component, the rate of sedimentation is much larger than in the basins, and the metal fraction becomes diluted with biogenic components permitted by shallower water depths and (or) proximity to the equator. However, an exception to this generalization is the linear zone of anomalously high metal concentrations located on the west flank of the East Pacific Rise between about 15° and 25°S. Dymond (1981) and Heath and Dymond (1977, 1981) interpret this anomaly to represent a recent east to west change in the submarine flow of hydrothermal fluids off the rise, as deduced from areal and stratigraphic variations in the chemistry of surface and down-core samples of sediment. Their conclusion is further supported by the presence of a mantle-derived hydrothermal plume of ^3He in deep ocean waters above and west of the East Pacific Rise at 15°S (Lupton and Craig, 1981).

Concentrations of some metals, sulfur, and calcium carbonate are listed in Table 1 for a suite of sediment samples from the Nazca plate (MS: 1–11) and northeast Pacific (MS: 12–17). These samples are part of a larger group that was also analyzed for one or more of the stable isotope abundances of carbon, oxygen, and sulfur. Samples from the Nazca plate were selected for regional representation, and all but one (MS-6, a down-core sample from 467–478 cm and

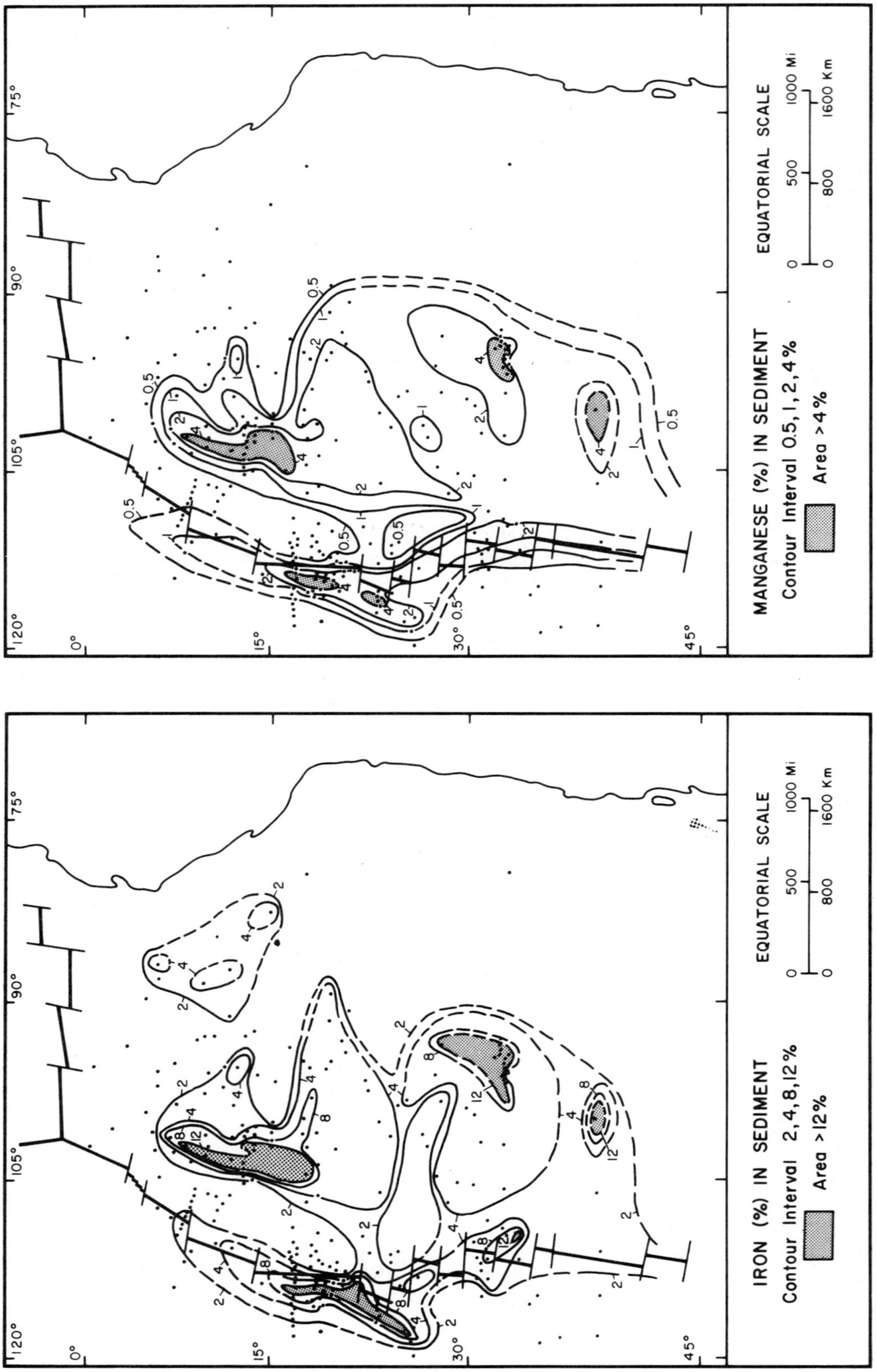

FIG. 4. Abundances and distributions of iron (left) and manganese (right) in sediments of the Nazca plate.

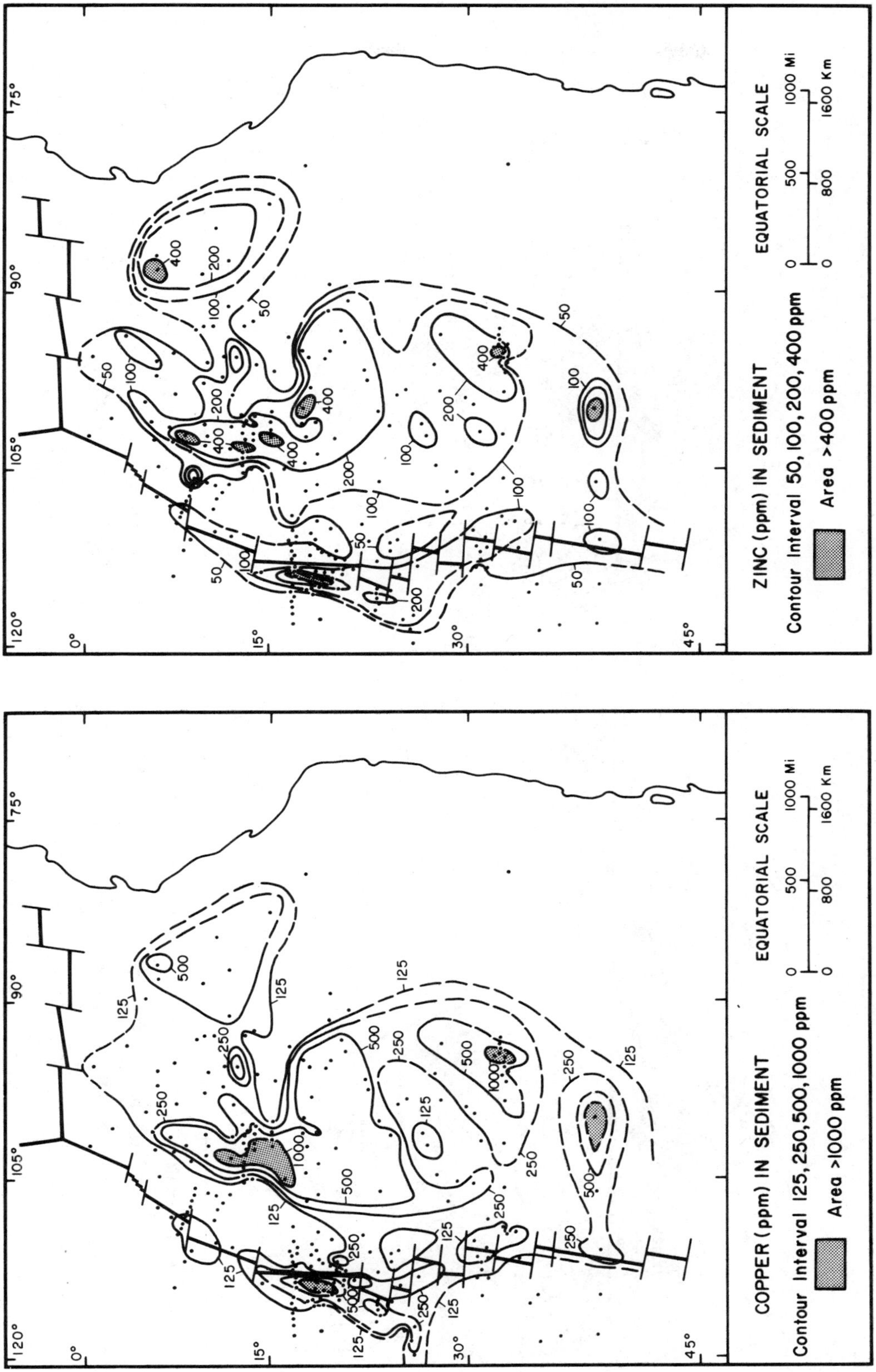

FIG. 5. Abundances and distributions of copper (left) and zinc (right) in sediments of the Nazca plate.

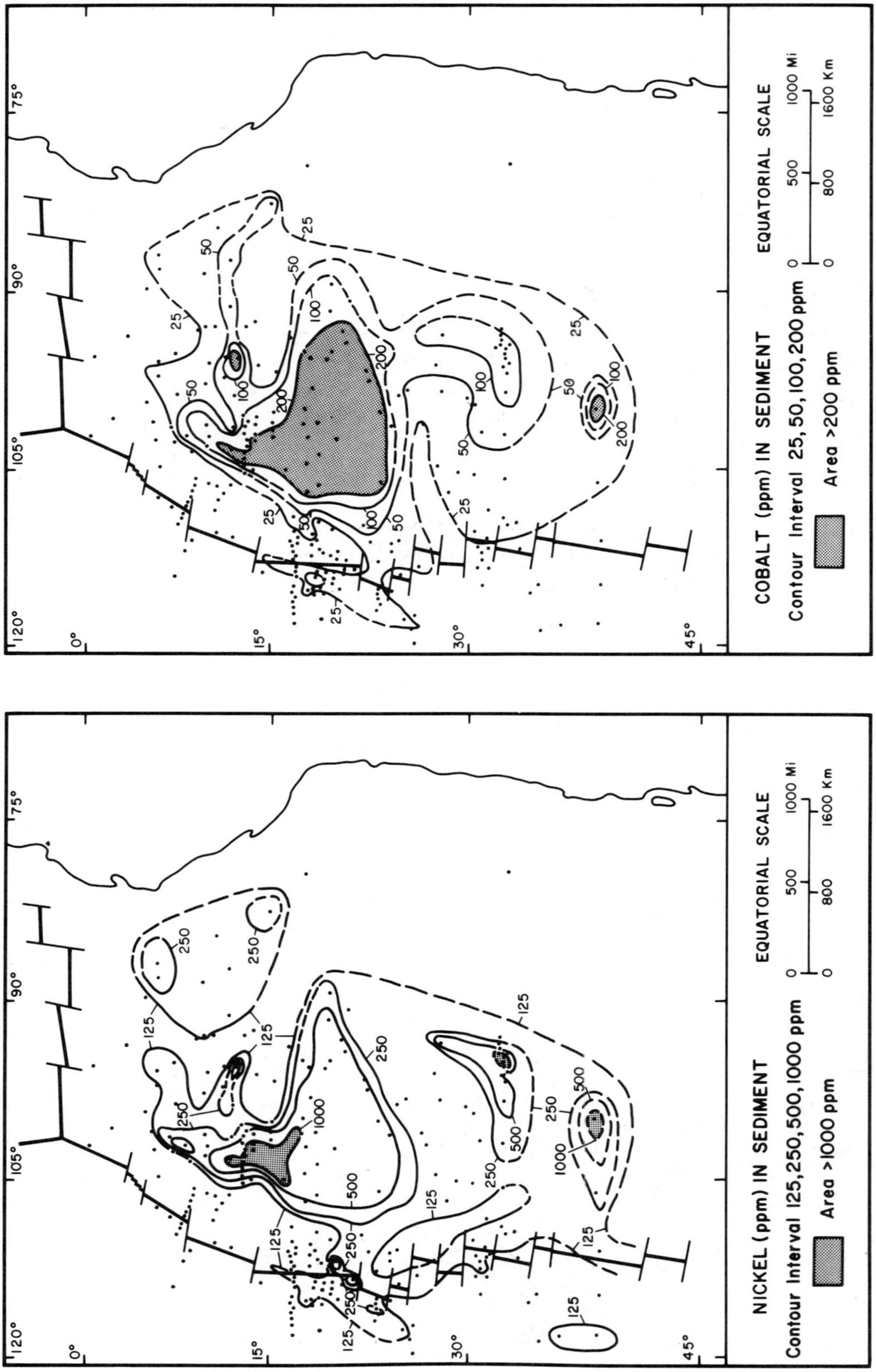

FIG. 6. Abundances and distributions of nickel (left) and cobalt (right) in sediments of the Nazca plate.

probably 5 m.y. in age) are products of recent sedimentation (<1 m.y.). Those from the northeast Pacific are deep-core pairs of samples (MS: 12 and 13, 14 and 15, and 16 and 17) from DSDP sites 37, 38, and 39. Because the stratigraphic location of these sample pairs is close to the underlying basaltic basement, their ages are inferred from magnetic anomaly data to be about 32, 56, and 60 m.y., respectively (see Dymond and others, 1973).

The concentrations of most metals generally exhibit a large range (Table 1), and they are consistent with the regional abundance patterns obtained for the entire Nazca plate sample suite (Figs. 4, 5, and 6). Iron and manganese clearly dominate the metal fraction, and their concentrations show an excellent covariant relationship to those of the other metals (see Dymond, 1981). The generally high Fe:Mn ratios serve to chemically identify the hydrothermal component that is typical of metalliferous sediments, and they contrast markedly with the low ratios and hydrogenous component characteristic of ferromanganese nodules (Dymond et al., 1976; Field et al., 1976). Carbonate, where significantly abundant, effectively masks and obscures the metal-rich contributions of the hydrothermal and hydrogenous components. However, this carbonate "veil," or the contaminating effects of other biogenic and (or) detrital components, can be overcome by utilizing methods of normative partitioning and Q-mode factor analysis. Heath and Dymond (1977, 1981) and Dymond (1981) have demonstrated by application of these methods that up to 95% Fe, 87% Mn, 91% Cu, 76% Ni, and 86% Zn in sediments of the East Pacific Rise have been derived from hydrothermal sources. In sediments of the Bauer Deep, the hydrothermal component provides up to 84% Fe, 70% Mn, 53% Cu, 20% Ni, and 58% Zn, with most of the remainder contributed by the hydrogenous component. The source of metals such as silver, cobalt, molybdenum, and lead is uncertain, but most are probably derived from a combination of both hydrothermal and hydrogenous sources, as determined by proximity to the East Pacific Rise. Concentrations of molybdenum are particularly anomalous, exceeding 300 ppm, and equivalent to byproduct grades in porphyry-type deposits. They greatly exceed values (up to 75 ppm) previously reported by Levashev and others (1975), who also correlated high concentrations of molybdenum with iron and manganese-enriched sediments of the Nazca plate. Although our samples were too small to determine abundances of gold reliably, a composite of samples MS 12–17 provided an economically interesting but analytically questionable value of 0.4 ppm based on atomic absorption spectrophotometry. Emission spectrographic analyses have generally confirmed the concentrations of metals given for samples listed in Table 1, and they also indicate anomalous amounts of other elements such as arsenic (up to 500 ppm), beryllium (up to 7 ppm), and vanadium (up to 1000 ppm) in the more metal-rich sediments. The tabulated data also indicated that the metals are more enriched in samples of sediment from the northeast Pacific than in those of the Nazca plate. The reason for this difference is largely attributable to the absence of carbonate in most samples from the northeast Pacific. However, the cause of this depletion is less certain, and may reflect either rapid accumulation of the metal-rich fraction or a higher calcite compensation depth during early to middle Tertiary time. Concentrations of sulfur do not show any systematic correlations with those of either the metals or carbonate. Sulfides were not identified in any samples of these highly oxidized sediments by visual, optical, or X-ray methods of examination. Thus, sulfur resides in the mineral barite (Dymond and others, 1973), and the association is with barium, which may be of biogenous, hydrogenous, or of hydrothermal origin (Heath and Dymond, 1977 and 1981; Dymond, 1981).

The metallization of deep-ocean sediments takes place by syngenetic hydrothermal-hydrogenous processes, and is temporally distinct from that of cordilleran plutonic rocks which largely result from epigenetic hydrothermal processes. Differences between syngenetic and epigenetic forms of metallization may be ideally detected from the distinctive "signature" patterns for the statistical distribution of metals in their host rocks when the analytical data are plotted on log probability diagrams (see Rose et al., 1979, for a description of this technique). This distinction shown in Fig. 7 is based on the log probability distributions of copper and zinc contained in sediments of the Nazca plate (302 samples, plotted as triangles connected by dashed lines) and in plutonic rocks of the Guichon Creek Batholith in central British Columbia that host porphyry-type mineralization (148 samples, plotted as cir-

Table 1. Concentrations of Some Metals, Sulfur, and Calcium Carbonate in Sediments of the Nazca and Northeast Pacific Plates*

Sample	Location	Fe	Mn	Ag	Co	Cu	Mo	Ni	Pb	Zn	S	$CaCO_3$
MS-1:	Nazca; East Pacific Rise	2.7	0.6	3.0	35	85	1	65	25	50	920	82.8
MS-2:	Nazca; East Pacific Rise	2.4	0.8	2.2	35	90	1	70	25	55	520	80.9
MS-3:	Nazca; East Pacific Rise	22.0	6.7	1.7	70	1130	55	145	65	400	480	29.2
MS-4:	Nazca; East Pacific Rise, east	0.6	0.2	3.6	40	75	8	85	25	25	640	93.1
MS-5:	Nazca; East Pacific Rise, east	0.6	0.2	3.5	40	45	2	50	25	150	400	93.2
MS-6:	Nazca; Bauer Deep	14.0	4.6	5.1	80	930	35	565	145	460	2500	2.2
MS-7:	Nazca; Bauer Deep	9.1	3.3	1.2	140	590	15	575	50	205	1280	28.8
MS-8:	Nazca; Bauer Deep	13.5	5.7	1.3	230	1100	45	1310	75	350	320	8.1
MS-9:	Nazca; Galapagos Rise	7.3	4.0	1.0	240	810	20	780	25	225	1920	14.4
MS-10:	Nazca; central plate	1.1	0.4	2.8	50	135	3	160	25	55	1240	71.1
MS-11:	Nazca; eastern plate	4.7	1.1	0.6	55	415	14	345	30	190	800	0.5
MS-12:	N.E. Pacific; DSDP 37	22.2	7.7	3.4	115	980	305	680	105	600	600	-
MS-13:	N.E. Pacific; DSDP 37	21.6	5.7	3.4	100	860	310	600	90	550	500	-
MS-14:	N.E. Pacific; DSDP 38	27.8	6.5	3.4	130	1190	210	760	100	590	2500	-
MS-15:	N.E. Pacific; DSDP 38	4.8	1.7	2.7	45	210	50	290	40	155	500	76.9
MS-16:	N.E. Pacific; DSDP 39	23.9	5.6	4.8	140	1240	190	620	100	710	600	-
MS-17	N.E. Pacific; DSDP 39	25.0	7.0	4.1	130	1110	260	600	90	690	500	-

*Data for iron, manganese, and $CaCO_3$ in percent; all other elements in parts per million.

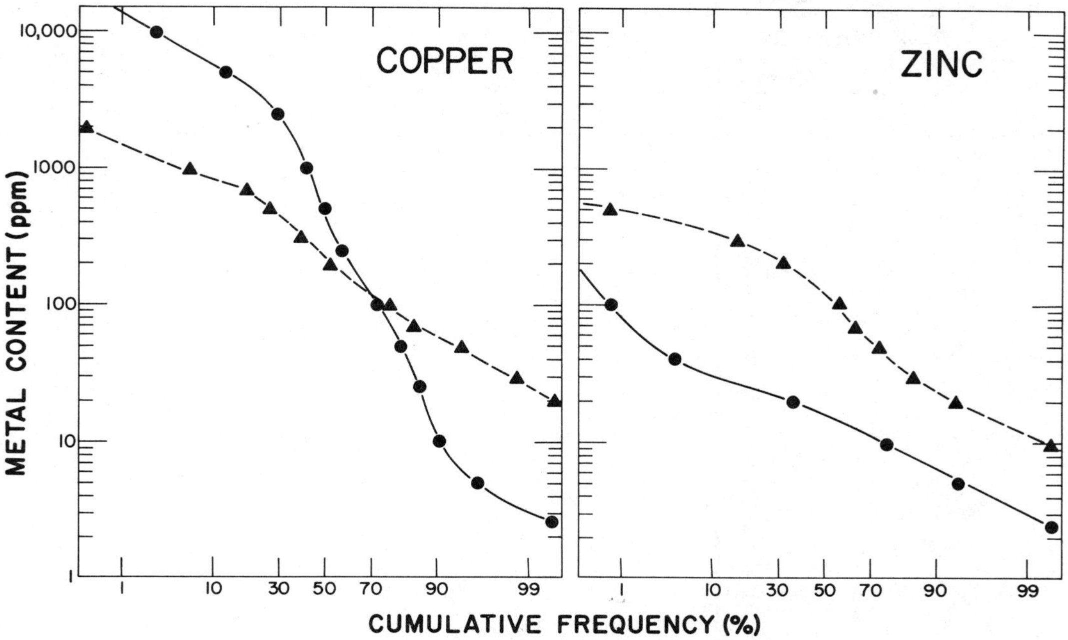

FIG. 7. Log probability distributions of copper and zinc in sediments of the Nazca plate and in plutonic rocks of the Guichon Creek batholith of British Columbia (sediment samples, triangles and dashed lines; pluton samples, circles and solid lines).

cles connected by solid lines; C. W. Field and colleagues, unpublished data). Plotted metal distributions in plutonic rocks of the batholith, and particularly that for copper, define an S-shaped curve that is diagnostic of epigenetic metallization. The curve for copper consists of two end-member populations composed of low syngenetic background values (5 to 10 ppm, and 10% of the sample suite) of primary magmatic origin and anomalously high values (>1000 ppm, and 30% of the sample suite) imposed by later epigenetic hydrothermal mineralization. These end members are connected by intermediate values along the central and steep-sloped part of the curve that represent mixtures of the two populations. The distributional trend for zinc is dominated by a single line of low background values (5 to 40 ppm, and 90% of the sample suite) of syngenetic-magmatic origin, which is consistent with scarcity of this metal in hydrothermal ores of the batholith. In contrast, the log probability distributions of both copper and zinc in sediments of the Nazca plate plot as near-linear trends that are consistent with their presumed syngenetic accumulation from hydrothermal and hydrogenous components in sea water. Reasons for the convex-up bulges on both curves are uncertain, but they probably are related either to dissolution of carbonate in basinal deeps, and (or) to mixed populations of hydrothermal and hydrogenous components.

STABLE ISOTOPE GEOCHEMISTRY

Isotopic abundance of carbon, oxygen, and sulfur in samples of metalliferous sediments, ferromanganese nodules, and basalts are listed in Table 2. Also given are the concentrations of iron and manganese in the sediments, on a carbonate-free and salt-free basis (see Dymond, and others, 1973; Dymond, 1981; Heath and Dymond, 1981), and ferromanganese nodules (Dymond, unpublished data). Isotopic analyses were performed for carbon and oxygen on the carbonate fraction of the sediments, for oxygen on the carbonate-free and salt-free part of the sediments, for oxygen on bulk samples of the nodules and basalts, and for sulfur on bulk samples of salt-free sediment and on concentrates of sulfate and sulfide minerals from hydrothermal ores of the Cordillera. The isotopic data are presented in terms of conventional permil values, and relative to appropriate standards.

Table 2. Isotopic Abundances of Carbon, Oxygen, and Sulfur in Metalliferous Sediments (MS: 1–17), Ferromanganese Nodules (MN: 1–9), and Ocean-Ridge Basalts (ORB: 1–5).*

	Carbonate-free			Bulk sample		Carbonate	
Sample	**Fe, %**	**Mn, %**	**$\delta^{18}O$ ‰**	**$\delta^{18}O$ ‰**	**$\delta^{34}S$ ‰**	**$\delta^{13}C$ ‰**	**$\delta^{18}O$ ‰**
MS-1:	20.0	4.5	5.4		20.6		
MS-2:	18.5	4.0	17.0		20.2	0.7	25.0
MS-3:	34.5	10.4	9.6		20.5		
MS-4:	13.9	5.4	19.4		21.0		
MS-5:	13.3	4.9	-		19.4	−0.1	29.7
MS-6:	14.4	4.7	21.8		20.9		
MS-7:	16.0	5.9	22.0		20.3		
MS-8:	16.6	6.9	10.0		21.0		
MS-9:	10.0	5.5	21.4		20.5		
MS-10:	4.3	1.7	21.9		20.0	0.7	30.8
MS-11:	4.8	1.3	16.4		19.5		
MS-12:	22.2	7.7	15.8		18.5		
MS-13:	21.6	5.7	16.2		19.7		
MS-14:	27.8	6.5	13.2		19.1		
MS-15:	21.1	7.6	11.1		-	2.1	30.4
MS-16:	23.9	5.6	13.9		17.8		
MS-17:	25.0	7.0	12.8		18.9		
MN-1:	13.5	22.0		12.5			
MN-2:	13.0	20.4		10.0			
MN-3:	8.8	20.1		11.0			
MN-4:	12.8	16.7		15.7			
MN-5:	3.8	29.5		10.9			
MN-6:	17.1	11.6		13.6			
MN-7:	21.5	11.1		13.0			
MN-8:	8.9	2.4		11.1			
MN-9:	14.5	16.5		12.0			
ORB-1a:				9.5			
1b:				6.2			
ORB-2a:				5.6			
2b:				6.4			
ORB-3:				6.5			
ORB-4a:				5.5			
4b:				5.3			
ORB-5:				6.2			

*The isotopic data are in permil values relative to traditional standards, and concentrations of iron and manganese are in percent on a carbonate-free and salt-free basis for the sediments.

The isotopic composition of the carbonate fraction in four samples of sediment was measured for carbon and oxygen (Table 2). Data for carbon as given by the $\delta^{13}C$ permil values range from −0.1 to 2.1 and fall within the narrow spread of values (−3 to 3 permil) considered typical of marine limestones (Bender, 1972). Oxygen in the carbonates is significantly enriched in ^{18}O, as expected, because of low temperature equilibration with seawater. Most $\delta^{18}O$ values of the carbonates are isotopically uniform between 29.7 and 30.8 permil, but one sample is relatively depleted in ^{18}O (MS-2; 25.0 permil). All values are in or near the compositional range (26 to 36 permil) of $\delta^{18}O$ in marine carbonates (Garlick, 1972). Isotopic temperature estimates for the more compositionally uniform carbonates, and determined from the fractionation curves summarized by Friedman and O'Neil (1977), suggest that the calcite formed in equilibrium with seawater at 15° to 20°C. However, a somewhat higher temperature of about 45°C is obtained for the more ^{18}O-depleted carbonate. Thus, the isotopic data for carbon and oxygen in carbonates largely support the low temperature origin of this biogenic component that formed from the skeletal remains of planktonic foraminifera and other calcite-bearing microorganisms.

Isotopic abundances of ^{18}O in samples of sediment (carbonate-free and salt-free), ferromanganese nodules, and basalts are illustrated in Fig. 8, and their distributions generally define compositionally discrete populations. Although

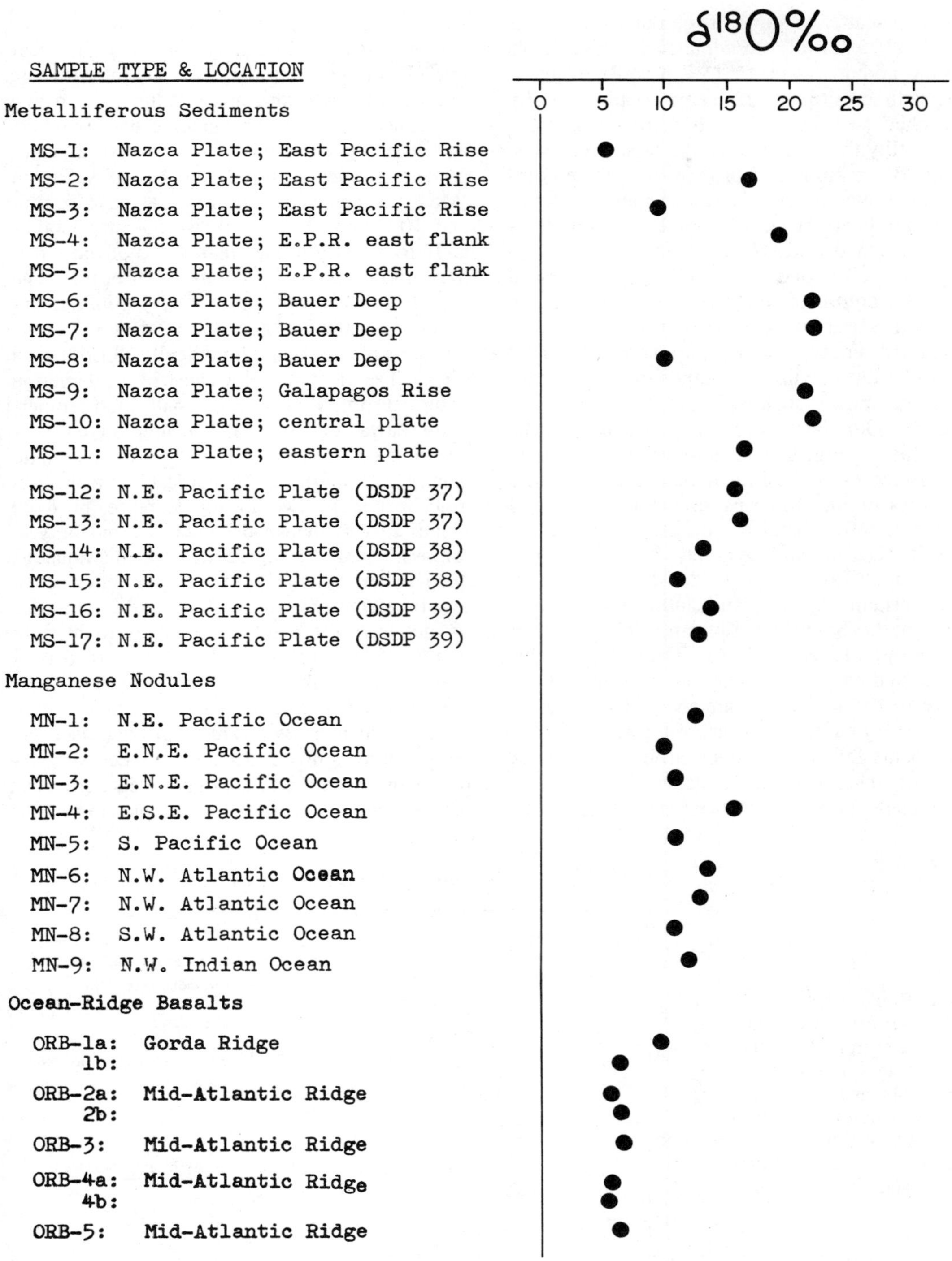

FIG. 8. Distributions of $\delta^{18}O$ permil values in metalliferous sediments, ferromanganese nodules, and ocean-ridge basalts.

there is scatter to some of the data, the basalts (5.3 to 9.5 permil) are depleted in ^{18}O relative to the nodules (10.0 to 15.7 permil), whereas the sediments (5.4 to 22.0 permil) are generally enriched (see Table 2). Much of the isotopic variability that is characteristic of sediments relates to the kinds and proportions of the component minerals, chiefly silicates and oxides of differing provenance in the acid-insoluble fraction, and with each phase having its own $\delta^{18}O$ signature. Dymond et al. (1973) have showed that the confusion imparted by such extreme isotopic variability can be reduced, and interpretations rendered possible, by plotting the $\delta^{18}O$ permil values against a compositionally dependent parameter such as total iron plus manganese. Our data are given in terms of these variables in Fig. 9, and in addition we include the isotopic-compositional data for two other samples of metalliferous sediment (Savin and Epstein, 1970b) and the domains for oceanic basalts (Garlick and Dymond, 1970; Muehlenbachs and Clayton, 1976), detrital clays (Savin and Epstein, 1970b), and authigenic silicates and cherts (Savin and Epstein, 1970a, 1970b; Kolodny and Epstein, 1976). The authigenic silicates and cherts are highly enriched in ^{18}O relative to the iron and manganese-rich nodules. Compositional distributions of the metalliferous sediments fall largely along a tie line between the metal-enriched and ^{18}O-depleted nodules and the metal-deficient and ^{18}O-enriched silicates (Fig. 9). The plotted compositional positions of these carbonate-free and variably metal-enriched sediments implies that they are an intermediate product of low-temperature sedimentation that is defined by oxide and silicate end members and variably contaminated by a detrital component. However, three samples of sediment (MS: 1, 3, and 8) exhibit $\delta^{18}O$ values (5.4, 9.6, and 10.0 permil) that are isotopically moderately to slightly lighter than those of the ferromanganese nodules (10.0 to 15.7 permil). The reason for these apparent ^{18}O-depletions is unknown, but the isotopic effects of two possible mechanisms are diagrammatically illustrated in Fig. 9. The most plausible cause of the depletion is by equilibration of oxides with high-temperature seawater (0 permil). The alternative mechanism, equilibration with oxygen of magmatic waters or more probably that leached from basaltic host rocks, would not decrease the $\delta^{18}O$ values to less than 8 or 9 permil, and only at impossibly high temperatures for this aqueous deep-sea environment. In fact, the net effect of low-temperature equilibration with magmatic oxygen is to shift $\delta^{18}O$ permil values to the right of the nodule-silicate tie line by up to 8 or 9 permil. Although we tentatively ascribe these ^{18}O-depleted sediments to equilibration with heated seawater, we cannot provide isotopic temperature estimates because the necessary experimentally derived fractionation data are lacking. Nonetheless, permissive evidence to support

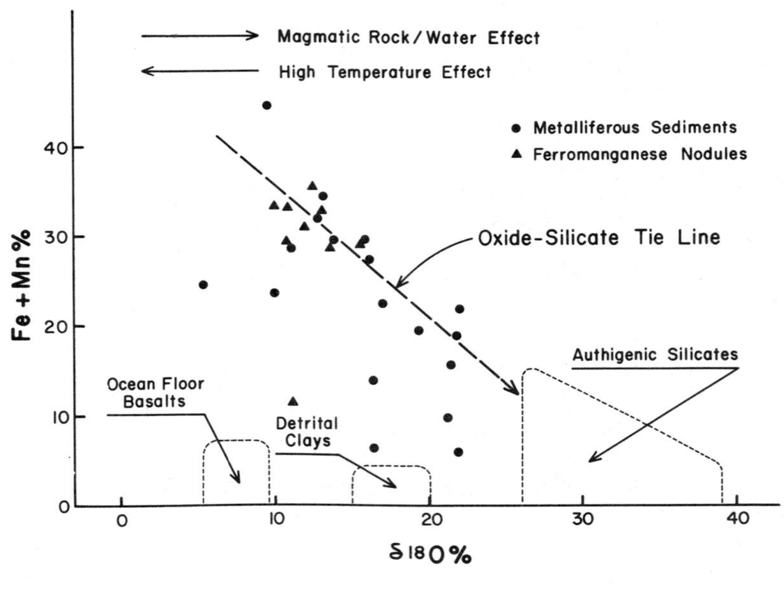

FIG. 9. Distributions of $\delta^{18}O$ permil values vs. percentage iron plus manganese in metalliferous sediments and ferromanganese nodules, and with comparisons to the domains of basalts, detrital clays, and authigenic silicates.

this interpretation includes (1) the location of samples MS: 1 and 3 on the East Pacific Rise; (2) the coincident associations of ^{18}O-depleted sediment (MS-1) and ^{18}O-depleted carbonate (MS-2) in the same piston core and less than 1 m apart; and (3) the presence of vividly colored and sharply defined layers of metalliferous sediment in this core that suggest proximal deposition from a nearby hydrothermal vent.

Metalliferous sediments contain minor quantities of sulfur (see Table 1) that resides in authigenic barite. Abundances of ^{34}S in the sediment samples are graphically portrayed in Fig. 10. The $\delta^{34}S$ permil values are remarkably uniform (17.8 to 21.0 permil) and tightly grouped near the ^{34}S-enriched value for present-day seawater (21 permil). Although Heath and Dymond (1977) and Dymond (1981) have determined that as much as 50% of the barium contained in sediments of the East Pacific Rise has been derived from the hydrothermal component, our $\delta^{34}S$ data clearly imply an oceanic source for the sulfate-sulfur. These data do not provide evidence for any significant contribution of deep-seated magmatic sulfur (0 permil by analogy to meteoritic sulfide-sulfur). It is of interest to note that the sulfur in metalliferous sediments of the northeast Pacific (17.8 to 19.7 permil) is slightly depleted in ^{34}S relative to that in sediments of the Nazca plate (19.4 to 21.0 permil). Because samples from the northeast Pacific are deep-core sediments that range from middle to early Tertiary in age, they are accordingly much older than the core-top samples of recent sediment from the Nazca plate. Thus, the isotopic data suggest an apparent depletion of ^{34}S with increasing age of the sediments. This apparent trend is entirely compatible with that given by the $\delta^{34}S$-age curve for marine evaporites as refined by Claypool and others (1980).

Cordilleran Sulfur

Sillitoe (1972), through compilation and synthesis of a vast amount of data, has elegantly integrated the geologic processes of the ocean basins and continental margins into plate tectonic theory. In essence, he related the coincidence of calc-alkalic magmatism and metallogeny of the Cordillera to seafloor spreading and the eventual subduction of metal-bearing oceanic crust at convergent plate margins. Should this model be correct, the ^{34}S-enriched metalliferous sediments (~20 permil), in contrast to typically magmatic values for hydrothermal sulfides (~0 permil), might serve as an isotopic tracer to document the implied oceanic provenance.

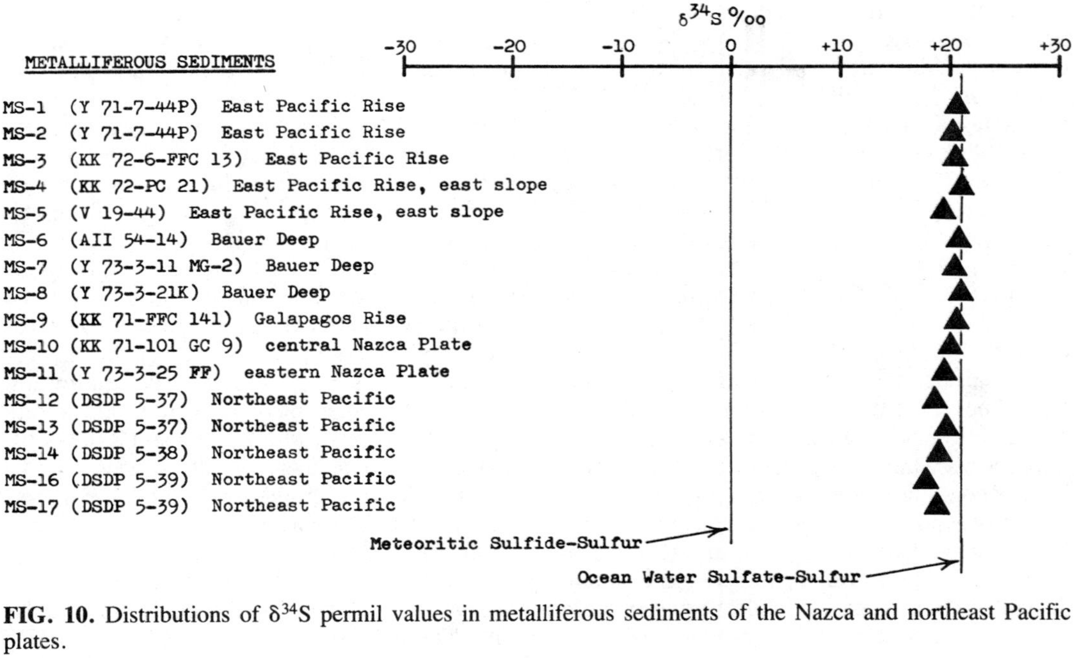

FIG. 10. Distributions of $\delta^{34}S$ permil values in metalliferous sediments of the Nazca and northeast Pacific plates.

Contrary to previous generalizations, Sakai (1968) and Ohmoto (1972) have demonstrated that the isotopic composition of a sulfide is not necessarily identical to the composition of total sulfur ($\delta^{34}S_t$) in the hydrothermal system. The isotopic composition of a sulfide is related not only to $\delta^{34}S_t$, but also to the temperature of deposition and to the acidity and oxidation state of the fluid which exert important controls on the kinds and proportions of oxidized and reduced aqueous species of sulfur in the system. Because of these complexities, the most practical means of estimating $\delta^{34}S_t$ is to determine the $\delta^{34}S$ permil values for coexisting sulfate and sulfide minerals, as the two analyses should bracket the compositional limits of sulfur in the system. Additional information may be obtained from analysis of sulfate-sulfide assemblages that have formed by equilibration over a range of temperatures. These data, when plotted as the $\delta^{34}S$ permil value of each sulfate and sulfide mineral vs. the delta value of the sulfate-sulfide pair, may form linear populations (see Field and Gustafson, 1976). Ideally, the slopes of these lines should define the proportions of aqueous sulfate to sulfide sulfur in the hydrothermal system, and the intercepts of these lines on the ordinate scale should establish the composition of $\delta^{34}S_t$ in the system.

Sulfate occurring as anhydrite and (or) gypsum is uncommonly abundant in many sulfide-bearing hydrothermal deposits of the Andean Cordillera. Illustrated in Fig. 11 are the distributions of $\delta^{34}S$ permil values for sulfate-sulfide assemblages in small suites of samples from hydrothermal deposits of North and South America. The deposits are listed in order of increasing age, not geographically, and the data for El Salvador are a paragenetic composite of a larger suite previously described by Field and Gustafson (1976). The more prominent features depicted by these data include: (1) the pronounced ^{34}S-enrichment of the sulfates relative to the sulfides that is consistent with fractionation theory, (2) the nearly similar fractionation effect between sulfate-sulfide pairs from all of the deposits which is expectable in terms of their largely common parentage in high-temperature porphyry-type environments, and (3) the remarkable sympathetic shifts of the $\delta^{34}S$ permil values in sulfate-sulfide pairs between different deposits. The isotopic variability between these deposits, which is far more than that observed within the individual deposits, intuitively suggests that $\delta^{34}S_t$ is different in each deposit. This inference is further supported by the plotted distributions of the $\delta^{34}S$ permil values of sulfates and sulfides vs. their respective delta values for the individual deposits as shown in Fig. 12. Intercepts of the plotted data for Morococha, Toquepala, El Salvador, and El Teniente, all located in the Andean Cordillera, collectively imply isotopically heavy $\delta^{34}S_t$ that ranges from 2 permil and more at El Salvador to as much as 13 permil at El Teniente.

It is tempting to equate these apparent ^{34}S-enrichments to contributions of isotopically heavy sulfur (and metals?) from metalliferous sediments incorporated in magmas through subduction and anatexis beneath the continent. However, the geologic evidence, small number of samples analyzed, and potential difficulties related to sulfate-sulfide equilibria do not warrant such a bold interpretation at this time. The sulfates and sulfides of Morococha are highly enriched in ^{34}S relative to their counterparts in other pluton-related hydrothermal deposits of the Cordillera. Much of the replacement ore at Morococha is hosted by an evaporite-rich carbonate sequence, and it is geologically probable that much or all of the isotopically heavy sulfur was derived from this source (see Fig. 11). In addition, stratigraphic inference suggests that correlative evaporites may underlie the plutonic-volcanic sequence at Toquepala, and they are known to be present in sedimentary rocks nearby and (or) at depth in the vicinity of El Salvador and El Teniente. Thus, we tentatively attribute these ^{34}S-enriched hydrothermal systems to local near-surface contamination of fluids and possibly of magmas with isotopically heavy sulfur derived from sedimentary evaporites.

It must be emphasized that our interpretations of these data are preliminary and they assume that the isotopic analyses are representative of all sulfur contained in these magmatic hydrothermal deposits and that isotopic equilibrium prevailed between coexisting sulfate and sulfide minerals. According to Ohmoto and Rye (1979) and Ohmoto and Lasaga (1982), aqueous sulfate-sulfide equilibrium in hydrothermal systems is likely to be attained at temperatures above 350°C, and also may be enhanced by other factors such as residence time and (or) rate of cooling, acidity, and sulfur content of the fluids. The data of fluid inclusions, stabilities of min-

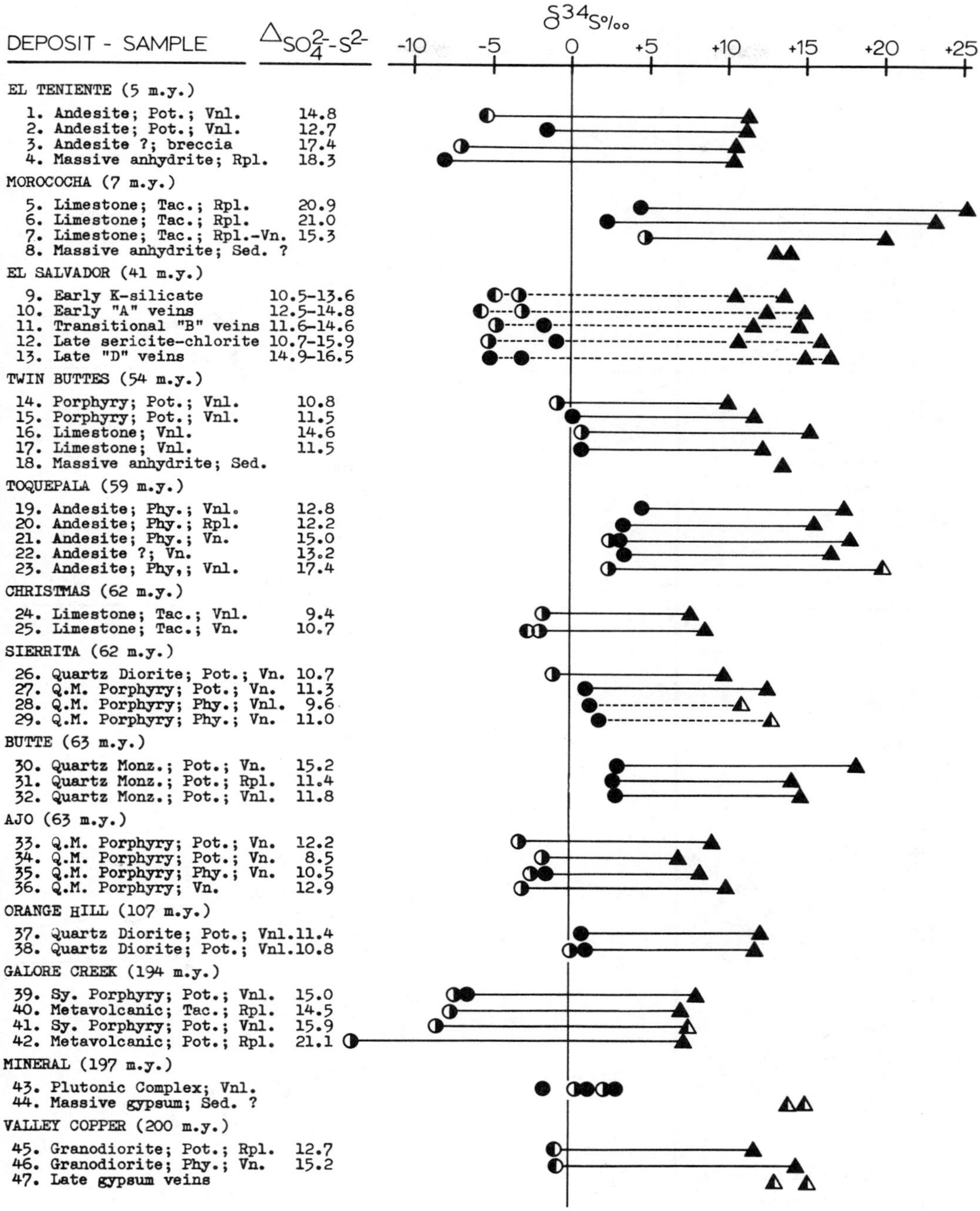

FIG. 11. Distributions of $\delta^{34}S$ permil values in hydrothermal sulfate-sulfide assemblages of the American Cordillera (Sulfates as triangles: anhydrite, closed; gypsum, shaded-left. Sulfides as circles: pyrite, closed; chalcopyrite, shaded-right; and bornite, shaded-left).

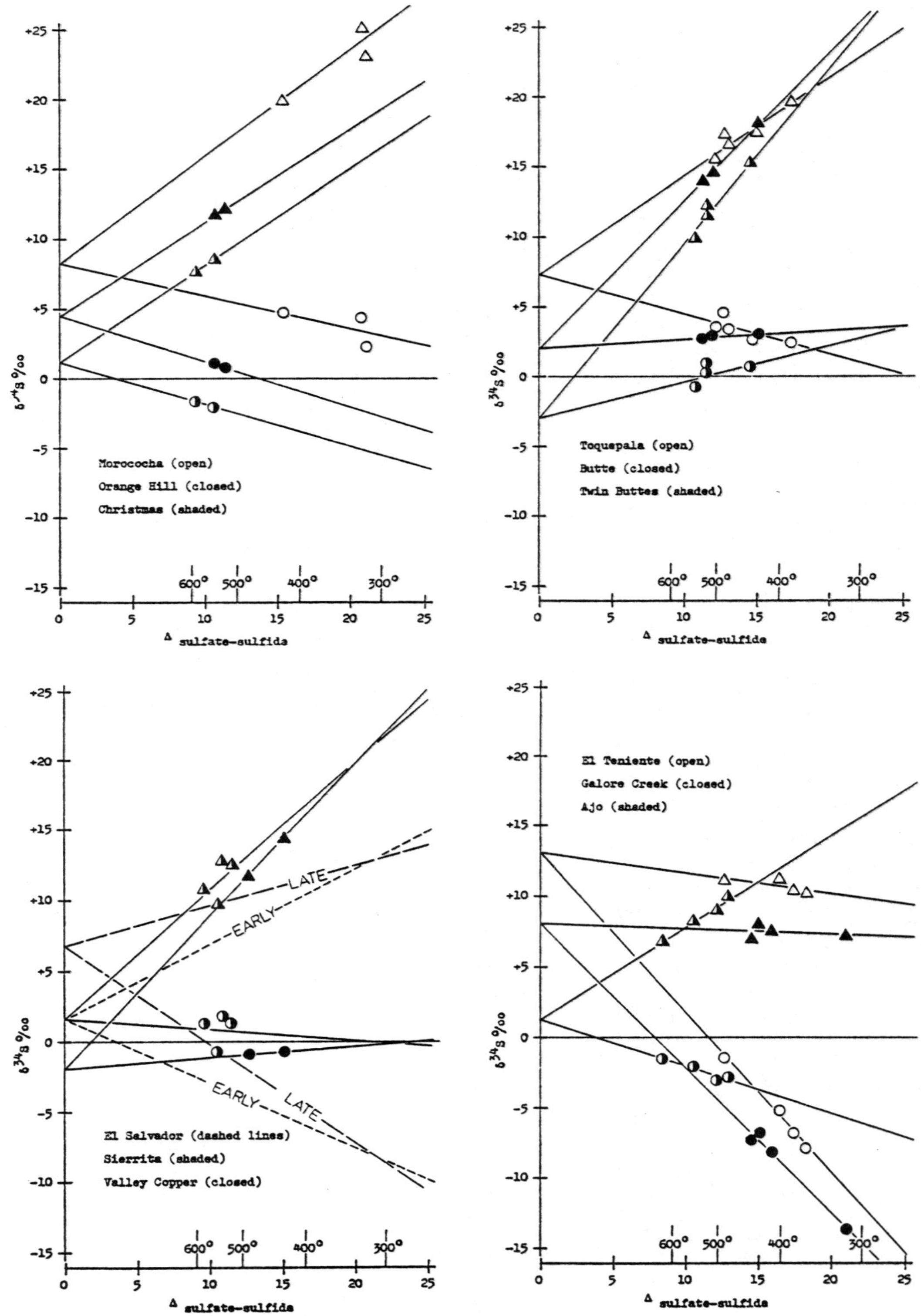

FIG. 12. Estimates of $\delta^{34}S_t$ permil values in hydrothermal systems of the American Cordillera based on plots of sulfate and sulfide $\delta^{34}S$ permil values vs. delta (Δ) values of the associated sulfate-sulfide assemblage.

eral assemblages, and the geometry of mining districts collectively suggest that these conditions are most probable in and near the deeper zones of porphyry-type mineralization associated with hydrothermal systems. Moreover, the isotopic temperature estimates for these sulfate-sulfide mineral pairs range largely from 400 to 600°C (Fig. 12). Equilibrium is likely for these samples unless high-temperature fluid residence times were extremely short, and unless high-temperature remobilization of sulfide and sulfate minerals occurred (Shelton and Rye, 1982). Our studies of this isotopic phenomenon are continuing with more extensive suites of samples from several of these deposits.

RESOURCE POTENTIAL OF METALLIFEROUS SEDIMENTS

It is apparent that the metalliferous sediments, from considerations of their wide distribution on the Nazca plate and elsewhere, and of their anomalously high concentrations of transition metals (Figs. 4, 5, and 6, and Table 1), contain an enormous reservoir of metals that constitute a resource of potentially great importance. For example, surficial sediments of the Bauer Deep (about 5×10^5 km^2) alone contain millions to billions of tons of most transition metals. Although tonnages such as these and most other resource estimates are likely to be impressive, they are meaningless if costs and (or) available technology preclude recovery.

Field et al. (1981) made a preliminary economic appraisal of metalliferous sediments and compared this potential resource to manganese nodules of the ocean basins and to fissure-replacement and porphyry types of ores in the Andean Cordillera. Their results, derived from a combination of factual data and speculative assumptions, indicated that the metal-rich sediments would remain an unused resource into the foreseeable future, relative to manganese nodules and to traditional sources of ore on the continents. The principal reasons for this unfavorable economic analysis may be inferred from the basic data relating to concentrations, quantities, and values of metals contained in these sediments as are summarized in Table 3 (modified from Field et al., 1981, Table 1).

There are three major difficulties relevant to the extraction of metals from these sediments, and they are exclusive of other uncertainties such as ownership of the seafloor, existing technology, and the costs of capitalization and operation. First, although the total value of contained metals ($67.47/t, based on 1980 prices) compares more than favorably with those of ores mined from other large-scale operations, it is based on the aggregate contributions of eight different metals of which not one is present in ore-grade concentration. Second, in addition to the submarginal concentration of each metal, extractive inefficiencies would suggest that the quantity and value of metals recovered should be reduced by at least 10%. Third, relatively high concentrations of iron and manganese contribute more than 60% to the total value of contained metals in the sediment, but metallurgical problems caused by their mutual association would more than likely render their recovery impractical. For these and other reasons, the metalliferous sediments do not presently appear

Table 3. Estimates of Concentration, Quantity, and Value of Metals Contained in Metalliferous Sediment*

Metal	Concentration, %	Quantity, kg/t	Value, $/t
Iron	15.0	150.0	9.00
Manganese	5.0	50.0	32.00
Cobalt	0.02	0.2	11.02
Copper	0.11	1.1	2.48
Molybdenum	0.015	0.15	2.98
Nickel	0.10	1.0	7.60
Silver	0.0004	0.004	2.07
Zinc	0.04	0.4	0.32
Total value			67.47
Total value exclusive of Fe + Mn			26.47

*Modified from Field et al., 1981; metal prices of August 1980.

to be an economically viable source of minerals. Moreover, because of continuing inflation and the generally diminishing values of metals since 1980, the broadly unfavorable economic analysis of this resource given by Field et al. (1981) is all the more unattractive today. Regardless of economic and plate tectonic speculations, the metalliferous sediments constitute a decidedly "unconventional" resource with respect to their location, distribution, and genesis in the ocean basins.

ACKNOWLEDGMENTS

We are pleased to have been accorded the opportunity of contributing our review and research on metalliferous sediments to the symposium and volume in honor of Professor Eugene N. Cameron. Most of this effort was supported by funds provided to the School of Oceanography at Oregon State University from the NSF/IDOE Nazca Plate Project. Corporate donations assisted the senior author's travel to the Branch of Isotope Geology at the US Geological Survey in Denver, CO. Thoughtful reviews given by La Verne D. Kulm, Werner J. Raab, Alan R. Wallace, and an unnamed reviewer did much to improve a preliminary draft of this manuscript. The symposium convener, James J. Eidel, and editor of this volume, Wayne C. Shanks, are thanked for their helpful assistance in this endeavor. Finally, the senior author assumes responsibility for interpretations of the Cordilleran sulfate-sulfide isotopic data.

REFERENCES

Bender, M. L., 1972, "Carbon Isotope Fractionation," Encyclopedia of Geochemistry and Environmental Sciences, Vol. 4A, R. W. Fairbridge, ed., Van Nostrand Reinhold Co., New York, p. 133–136.

Bender, M., et al., 1971, "Geochemistry of Three Cores from the East Pacific Rise," *Earth and Planetary Science Letters*, Vol. 12, pp. 425–433.

Bostrom, K., and Peterson, M. N. A., 1966, "Precipitates from Hydrothermal Exhalations on the East Pacific Rise," *Economic Geology*, Vol. 61, pp. 1258–1265.

Bostrom, K., and Peterson, M. N. A., 1969, "The Origin of Aluminum-Poor Ferromanganoan Sediments in Areas of High Heat Flow on the East Pacific Rise," *Marine Geology*, Vol. 7, pp. 427–447.

Bostrom, K., et al., 1976, "Geochemistry and Origin of East Pacific Sediments Sampled during DSDP Leg 34," *Initial Reports of the Deep Sea Drilling Project*, Vol. 34: R. S. Yeats, et al., eds., US Government Printing Office, Washington, DC, pp. 559–574.

Claypool, G. E., et al., 1980, "The Age Curves of Sulfur and Oxygen Isotopes in Marine Sulfate and Their Mutual Interpretation," *Chemical Geology*, Vol. 28, pp. 199–260.

Dasch, E. J., 1981, "Lead Isotopic Composition of Metalliferous Sediments from the Nazca Plate," *Nazca Plate—Crustal Formation and Andean Convergence*, L. D. Kulm, et al., eds., Memoir 154, Geological Society of America, pp. 199–209.

Dasch, E. J., Dymond, J. R., and Heath, G. R., 1971, "Isotopic Analysis of Metalliferous Sediment from the East Pacific Rise," *Earth and Planetary Science Letters*, Vol. 13, pp. 175–180.

Dymond, J., 1981, "Geochemistry of Nazca Plate Surface Sediments—An Evaluation of Hydrothermal, Biogenic, Detrital, and Hydrogenous Sources," *Nazca Plate—Crustal Formation and Andean Convergence*, L. D. Kulm, et al., eds., Memoir 154, Geological Society of America, pp. 133–173.

Dymond, J., and Corliss, J. B., 1980, "Chemical Composition of Nazca Plate Surface Sediments," Map and Chart Series, Geological Society of America, MC-34.

Dymond, J., Corliss, J. B., and Heath, G. R., 1977, "History of Metalliferous Sedimentation at Deep Sea Drilling Site 319, in the South Eastern Pacific," *Geochimica et Cosmochimica Acta*, Vol. 41, pp. 741–753.

Dymond, J., et al., 1973, "Origin of Metalliferous Sediments from the Pacific Ocean," *Bulletin*, Geological Society of America, Vol. 84, pp. 3355–3372.

Dymond, J., Corliss, J. B., and Stillinger, R., 1976, "Chemical Composition and Metal Accumulation Rates of Metalliferous Sediments from Sites 319, 320, and 321," *Initial Reports of the Deep Sea Drilling Project*, Vol. 34, R. S. Yeats, et al., eds., US Government Printing Office, Washington, DC, pp. 575–588.

Field, C. W., and Dasch, E. J., 1981 "Epilogue—Geostill Reconsidered," *Nazca Plate—Crustal Formation and Andean Convergence*, L. D. Kulm, et al., eds., Memoir 154, Geological Society of America, pp. 817–824.

Field, C. W., et al., 1976, "Metallogenesis in Southeast Pacific Ocean—Nazca Plate Project," *Circum-Pacific Energy and Mineral Resources*, M. T. Halbouty, J. C. Maher, and H. M. Lian, eds., Memoir 25, American Association of Petroleum Geologists, p. 539–550.

Field, C. W., and Gustafson, L. B., 1976, "Sulfur Isotopes in the Porphyry Copper Deposit at El Salvador, Chile," *Economic Geology*, Vol. 71, pp. 1533–1548.

Field, C. W., Wetherell, D. G., and Dasch, E. J., 1981, "Economic Appraisal of Nazca Plate Metalliferous Sediments," *Nazca Plate—Crustal Formation and Andean Convergence*, L. D. Kulm, et al., eds., Memoir 154, Geological Society of America, pp. 315–320.

Friedman, I., and O'Neil, J. R., 1977, "Compilation of Stable Isotope Fractionation Factors of Geochemical Interest," *Data of Geochemistry*, M. Fleischer, ed., 6th ed., Professional Paper 440-KK, US Geological Survey, pp. KK 1–12.

Garlick, G. D., 1972, "Oxygen Isotope Geochemistry," *Encyclopedia of Geochemistry and Environmental Sciences*, R. W. Fairbridge, ed., Vol. 4A, Van Nostrand Reinhold Co., New York, pp. 864–874.

Garlick, G. D., and Dymond, J. R., 1970, "Oxygen Isotope Exchange Between Volcanic Materials and Ocean Water," *Bulletin*, Geological Society of America, Vol. 81, pp. 2137–2142.

Heath, G. R., and Dymond, J., 1977, "Genesis and Transformation of Metalliferous Sediments from the East Pacific Rise, Bauer Deep, and Central Basin, Northwest Nazca Plate," *Bulletin*, Geological Society of America, Vol. 88, pp. 723–733.

Heath, G. R., and Dymond, J., 1981, "Metalliferous-Sediment Deposition in Time and Space—East Pacific Rise and Bauer Basin, Northern Nazca Plate," *Nazca Plate—Crustal Formation and Andean Convergence*, L. D. Kulm, et al., eds., Memoir 154, Geological Society of America, pp. 175–197.

Hekinian, R., et al., 1980, "Sulfide Deposits from the East Pacific Rise Near 21°N," *Science*, Vol. 207, pp. 1433–1444.

Herron, E. M., 1972, "Sea-Floor Spreading and the Cenozoic History of the East-Central Pacific," *Bulletin*, Geological Society of America, Vol. 83, pp. 1671–1692.

Kolodny, Y., and Epstein, S., 1976, "Stable Isotope Geochemistry of Deep Sea Cherts," *Geochimica et Cosmochimica Acta*, Vol. 40, pp. 1195–1209.

Levashev, G. B., et al., 1975, "Mo and W in Sediments in Certain Parts of the Pacific Ocean," *Geochemistry International*, Vol. 12, pp. 138–145.

Lupton, J. E., and Craig, H., 1981, "A Major Helium-3 Source at 15°S on the East Pacific Rise," *Science*, Vol. 214, pp. 13–18.

Mammerickx, J., and Smith, S. M., 1978, "Bathymetry of the Southeast Pacific," Map and Chart Series, Geological Society of America, MC-26.

McMurtry, G. M., Veeh, H. H., and Moser, C., 1981, "Sediment Accumulation Rate Patterns on the Northwest Nazca Plate," *Nazca Plate—Crustal Formation and Andean Convergence*, L. D. Kulm et al., eds., Memoir 154, Geological Society of America, pp. 211–249.

Muehlenbachs, K., and Clayton, R. N., 1976, "Oxygen Isotope Composition of the Oceanic Crust and Its Bearing on Seawater," *Journal of Geophysical Research*, Vol. 81, pp. 4365–4369.

Ohmoto, H., 1972, "Systematics of Sulfur and Carbon Isotopes in Hydrothermal Ore Deposits," *Economic Geology*, Vol. 67, pp. 551–578.

Ohmoto, H., and Lasaga, A. C., 1982, "Kinetics of Reactions Between Aqueous Sulfates and Sulfides in Hydrothermal Systems," *Geochimica et Cosmochimica Acta*, Vol. 46, pp. 1727–1745.

Ohmoto, H., and Rye, R. O., 1979, "Isotopes of Sulfur and Carbon," *Geochemistry of Hydrothermal Ore Deposits*, 2nd ed., H. L. Barnes, ed. John Wiley and Sons, New York, pp. 509–567.

Rose, A. W., Hawkes, H. E., and Webb, J. S., 1979, *Geochemistry in Mineral Exploration*, 2nd ed., Academic Press, New York, 657 pp.

Sakai, H., 1968, "Isotopic Properties of Sulfur Compounds in Hydrothermal Processes," *Geochemical Journal* (Japan), Vol. 2, pp. 29–49.

Savin, S. M., and Epstein, S., 1970a, "The Oxygen and Hydrogen Isotope Geochemistry of Clay Minerals," *Geochimica et Cosmochimica Acta*, Vol. 34, pp. 25–42.

Savin, S. M., and Epstein, S., 1970b, "The Oxygen and Hydrogen Isotope Geochemistry of Ocean Sediments and Shales," *Geochimica et Cosmochimica Acta*, Vol. 34, pp. 43–63.

Sayles, F. L., and Bischoff, J. L., 1973, "Ferromanganese Sediments in the Equatorial East Pacific," *Earth and Planetary Science Letters*, Vol. 19, pp. 330–336.

Sayles, F. L., Ku, L. T., and Bowker, P. C., 1975, "Chemistry of Ferromanganoan Sediment of the Bauer Deep," *Bulletin*, Geological Society of America, Vol. 86, pp. 1423–1431.

Shelton, K. L., and Rye, D. M., 1982, "Sulfur Isotopic Compositions of Ores from Mines Gaspe, Quebec—An Example of Sulfate-Sulfide Isotopic Disequilibria in Ore-Forming Fluids with Applications to Other Porphyry-Type Deposits," *Economic Geology*, Vol. 7, pp. 1688–1709.

Sillitoe, R. H., 1972, "A Plate Tectonic Model For the Origin of Porphyry Copper Deposits," *Economic Geology*, Vol. 67, pp. 184–197.

Veeh, H. H., 1981, "Uranium and Thorium Isotopic Investigations in Metalliferous Sediments of the Northwestern Nazca Plate, *Nazca Plate—Crustal Formation and Andean Convergence*, L. D. Kulm, et al., eds., Memoir 154, Geological Society of America, pp. 251–267.

Appendix 1. Sample Numbers, Locations, and Other Data for Metalliferous Sediments, Ferromanganese Nodules, and Oceanridge Basalts

Metalliferous Sediments

Sample number followed by laboratory and (or) shipboard designation, in parentheses; geographic location; coordinates; water depth; and sample interval and (or) depth below sediment surface.

MS-1: (MS-574; Y 71-7-44P); East Pacific Rise; 10°42′S-110°01′W; 3167 m; 349–350 cm.

MS-2: (MS-1793/94; Y 71-7-44P); East Pacific Rise; 10°42′S-110°01′ W; 3167 m; 425–426 cm.

MS-3: (MS-1593; KK 72-6-FFC-13); East Pacific Rise; 18°51′S-113°31′W; 3034 m; 4–8 cm.

MS-4: (MS-1463; KK 72-PC-21); east flank of East Pacific Rise; 18°23′S-110°26′W; 3525 m; 9–19 cm.

MS-5: (MS-609; V 19-44); east flank of East Pacific Rise; 17°01′S-108°52′W; 3621 m; 2–7 cm.

MS-6: (MS-75; AII-54-14); Bauer Deep; 8°22′S-102°14′W; 4124 m; 467–478 cm.

MS-7: (MS-1306; Y 73-3-11-MG-2); Bauer Deep; 13°02′S-102°35′W; 4555 m; 6–11 cm.

MS-8: (MS-1972; Y 73-3-21K); Bauer Deep; 13°37′S-102°32′W; 4410 m; 5–10 cm.

MS-9: (MS-1452; KK 71-FFC-141); Galapagos Rise; 12°34′S-95°43′W; 4832 m; 0–10 cm.

MS-10: (MS-408; KK 71-101-GC9); central Nazca plate; 9°57′S-93°05′W; 4240 m; 5–10 cm.

MS-11: (MS-1311; Y 73-3-25-FF); eastern Nacza plate; 15°09′S-82°48′W; 4870 m; 5–10 cm.

MS-12: (MS-114; DSDP 37-4-3); northeast Pacific plate; 40°59′N-140°43′W; 4882 m; 25–35 cm and 26 m.

MS-13: (MS-115; DSDP 37-4-5); northeast Pacific plate; 40°59′N-140°43′W; 4882 m; 50–60 cm and 29 m.

MS-14: (MS-116; DSDP 38-5-3); northeast Pacific plate; 38°42′N-140°21′W; 5137 m; 40–50 cm and 33.5 m.

MS-15: (MS-151; DSDP 38-6-4); northeast Pacific plate; 38°42′N-140°21′W; 5137 m; 115–125 cm and 45 m.

MS-16: (MS-152; DSDP 39-2-3); northeast Pacific plate; 32°48′N-139°34′W; 4929 m; 40–50 cm and 11.5 m.

MS-17: (MS-153; DSDP 39-2-6); northeast Pacific plate; 32°48′N-139°34′W; 4929 m; 90–100 cm and 16.5 m.

Manganese Nodules

Sample number followed by laboratory designation, in parentheses; geographic location; and coordinates.

MN-1: (DWHD-47); northeast Pacific Ocean; 40°N-155°W.
MN-2: (P3DH-2); east-northeast Pacific Ocean; 22°N-115°W.
MN-3: (V20-D4); east-northeast Pacific Ocean; 14°N-146°W.
MN-4: (RS 24-15); east-southeast Pacific Ocean; 36°S-134°W.
MN-5: (BT 14-3); south Pacific Ocean; 56°S-164°W.
MN-6: (Ch 58-Al); northwest Atlantic Ocean; 34°N-66°W.
MN-7: (5-M); northwest Atlantic Ocean; 34°N-66°W.
MN-8: (E 28-5); southwest Atlantic Ocean; 58°S-57°W.
MN-9: (62-69-10I1); northwest Indian Ocean; 3°N-60°E.

Ocean-Ridge Basalts

Sample number followed by laboratory designation, in parentheses; geographic location; and coordinates.

ORB-1: (1755G); Gorda Ridge; location uncertain; sample 1a is the rim and 1b the center.
ORB-2: (V25-1-T2); Mid-Atlantic Ridge; 25° 24′ N-45° 18′ W; sample 2a is the rim and 2b the interior.
ORB-3: (TW 10-3); Mid-Atlantic Ridge; 22°29′ N-45°03′ W.
ORB-4: (TW 4–15); Mid-Atlantic Ridge; 22°29′N-45°03′W; sample 4a is the rim and 4b the interior.
ORB-5: (TW 4-412); Mid-Atlantic Ridge; 22° 29′ N-45° 03′ W.

13

Economic and Exploration Significance of Red Sea Metalliferous Brine Deposits

W. C. Shanks, III*

Department of Geology and Geophysics,
University of Wisconsin,
Madison, WI

ABSTRACT

Seventeen deeps, which contain hot or cold brines and/or metalliferous sediments, have been discovered in the axial rift zone of the Red Sea. Metalliferous activity results from the unique geological setting of this incipient ocean basin: oceanic crust intruding in a central rift zone flanked by thick evaporite sequences. Brine pools tend to occur in structural depressions where transform faults intersect the spreading axis.

The most significant metalliferous occurrence is in the Atlantis II Deep, which contains approximately 227 Mt (dry weight) of metalliferous sediment which averages 1.04% Zn and 0.28% Cu. Higher potentially economic grades (3–6% Zn, 1% Cu, 50 g/t Ag) occur in the Southwest Basin of the Atlantis II Deep, comprising total metal tonnages of 1.7 Mt Zn, 0.4 Mt Cu, and 5000 t Ag. The Pre-Pilot Mining Test conducted in 1979 raised 15 000 m^3 of mud by a slurry process and demonstrated the feasibility of mining, concentration by froth flotation, and tailings disposal at sea. A Pilot Mining Test is scheduled for 1983–1984.

The metalliferous discoveries in the Red Sea provide a model example of deep marine mineral exploration. Detailed mapping of bottom bathymetry and brine-interface reflectors using a high-resolution narrow-beam echo sounder has been the most effective exploration tool in the Red Sea. Followup coring has discovered additional metalliferous deposits in basins without present-day brines, and potential exists for additional discoveries. Geochemical exploration using heavy metals values in sediments (iron, manganese, zinc, copper, and mercury) and sulfur isotope values of sulfides are considered to be the most promising techniques for discovery of currently inactive deposits. Positive sulfur isotope values in the range 1 to 17 are an indication of hydrothermal sulfides in the Red Sea, and there is considerable contrast from background bacteriogenic sulfides (−10 to −45).

INTRODUCTION

Metalliferous sediments and/or saline brines have now been discovered in numerous seafloor deeps along 900 km of the axial rift zone of the Red Sea (Fig. 1). The Atlantis II Deep is the most interesting of the brine deeps both in terms of intensity of submarine hydrothermal processes and in terms of potential economic importance. This deposit is unconventional mainly by virtue of occurrence in an inaccessible physical setting. However, successful completion of

* Present address: US Geological Survey, Reston, VA.

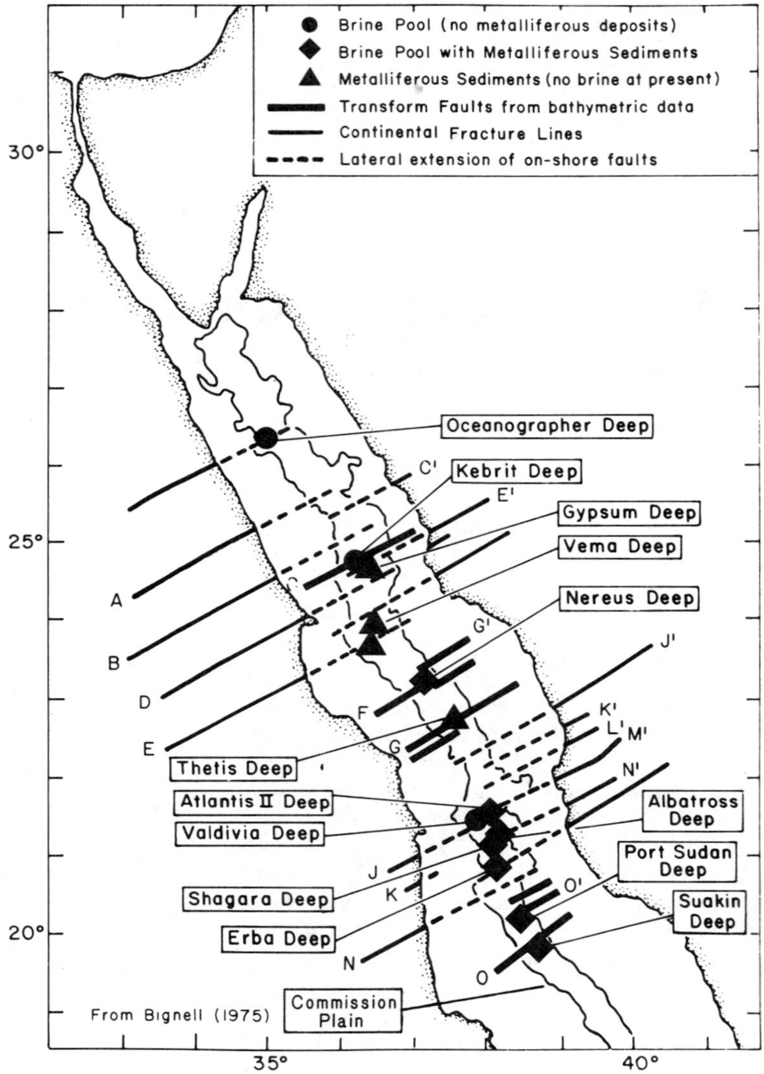

FIG. 1. Location map of Red Sea Deeps with brines and/or metalliferous sediments (after Bignell, 1975). Axial rift zone defined by the 1000-m depth contour. Transform faults and continental fracture line locations from Bignell (1975) with fault identification (A to O) according to Garson and Krs (1978).

a prepilot mining test in 1979 indicates that the Atlantis II Deep may become the first deep-sea base-metal sulfide mine (Mustafa and Amann, 1980). Under the most optimistic scenario, the Atlantis II mine might be established four to six years after completion of a year-long pilot mining test in 1983–1984. If established, mining would produce about 100,000 tpd of wet mud with about 200 tpd contained zinc, and proportional amounts of copper and silver. Expected mine life would be about 15–20 years (Anon., 1981).

In addition to direct economic importance, the Red Sea brines and metalliferous deposits provide significant insight into exploration for seafloor brine-related ore deposits and for analogous ancient deposits. In this paper I will concentrate on the economic potential of Red Sea deposits and seafloor exploration for brine related deposits, but some ramifications of relevance to ancient deposits are extant.

GENERAL STATEMENT

The Red Sea is an embryonic ocean basin which has formed by a two-stage rifting process (Girdler and Styles, 1974). The initial rifting stage occurred in Late Eocene to Oligocene.

Much of the Miocene was characterized by accumulation of thick sequences of evaporites related to subsidence along bounding normal faults. Renewed rifting began about 5 million years (m.y.) ago and continues today. This sequence of events, believed to be fairly typical of continental rifting in arid regions, has produced the unique geologic setting which is of pivotal importance to the formation of submarine brines and associated deposits. This setting includes thick subsurface evaporite sequences (>4 km) of mainly halite and anhydrite in the southern Red Sea. These flank the narrow axial rift zone (10–30 km, Fig. 1) which is a result of the present stage of spreading. Upper Miocene evaporites are overlain by a relatively thin veneer of Pliocene to Recent marine marls.

Oceanic basalts outcrop throughout the axial rift zone, and transform and graben faulting produce numerous structural depressions suitable for brine pools. In particular, significant brine and metalliferous activity occur where major transverse structures intersect the axis of the main Red Sea trough (Fig. 1). Moreover, Garson and Krs (1976) have suggested that the N60°E transform structures in the Red Sea developed along Precambrian fracture lines which cut both Nubian and Arabian shield rocks. However, transforms and associated continental fractures are not continuous across the Red Sea (Fig. 1, Garson and Krs, 1976) due to the left lateral movement of 75–80 km during opening of the Gulf of Aden in Late Cretaceous to Eocene.

Seafloor brine pools occur in Kebrit, Nereus, Valdivia, Albatross, Shagara, and Suakin Deeps (Fig. 1) and in the hot brine area which includes Atlantis II, Chain and Discovery Deeps (Fig. 2). Many of the brines (Table 1) reach salinities similar to the Atlantis II brine, about 25% total dissolved salts, but chemical compositions are quite variable (Bäcker and Schoell, 1972). The Atlantis II brine has uniquely high heavy metal contents (iron, manganese, copper, lead, zinc), is depleted in magnesium and SO_4, and has the highest in situ temperature (62–65°C), (Luck and Nawab, 1981).

Brine formation is clearly related to the flanking evaporite sequences. Oxygen and hydrogen isotope studies (Craig, 1969; Schoell and Faber, 1978) indicate that the brines are a result of evaporite leaching by circulated Red Sea paleowaters. Invariably high Cl/Br ratios in Red Sea brines (Table 1, Schoell et al., 1974) preclude simple derivation from connate brines of evaporative origin.

Subsurface circulation and the evaporite dissolution processes, which convert Red Sea waters to brine, are not precisely understood. However, Bignell (1975) has proposed that northeast striking transform faults (Fig. 1), which cut obliquely across the rift zone and through flanking evaporites, may provide pathways for fluid circulation. Manheim (1974) has proposed that seawater penetrates along the flanks of the Red Sea and circulates in Middle Miocene clastics beneath the capping, impermeable evaporite sequence. Dissolution of basal evaporites then produces dense brine, thus inducing a hydraulic gradient toward the rift zone.

Brine buoyancy due to heating in the rift zone has not been specifically considered, but is clearly important as shown by the detailed hydrography of Atlantis II brines (Schoell, 1976). In particular, Schoell (1976) has shown that new brine vents through the metalliferous sediments of the Southwest Basin (Fig. 2), rises buoyantly to the density interface between the lower (62°C) and upper (50°C) brine layers, and then spreads laterally to other portions of the Deep.

ECONOMIC POTENTIAL OF THE ATLANTIS II DEPOSIT

The Atlantis II Deep is a local tectonic depression within the actively spreading Red Sea rift, and is filled with stratified hot saline brine to a minimum depth of about 2000 m (Fig. 2). Metalliferous sediments have been precipitated in this Deep over the last 25,000 yr, with intense and continuous hydrothermal activity for the last 10,000 yr, forming a complexly layered deposit with a thickness of 5–20 m over an area of 65 km^2. Lithostratigraphic relations are systematic over much of the Deep (Fig. 3, Bäcker and Richter, 1973), but are disrupted and chaotic in the Southwest Basin where new brine currently vents through earlier deposited metalliferous muds. The presently depositing AM zone, the SU zones, and all the zones in Southwest Basin (Fig. 3) are of special economic importance. Zinc and copper occur as sphalerite and chalcopyrite, which are intimately intermixed or finely interlayered with pyrite, iron oxides, iron-rich clay minerals, and anhydrite.

Preussag AG, Hannover, Germany, has played the leading role in geological, geochemical, and

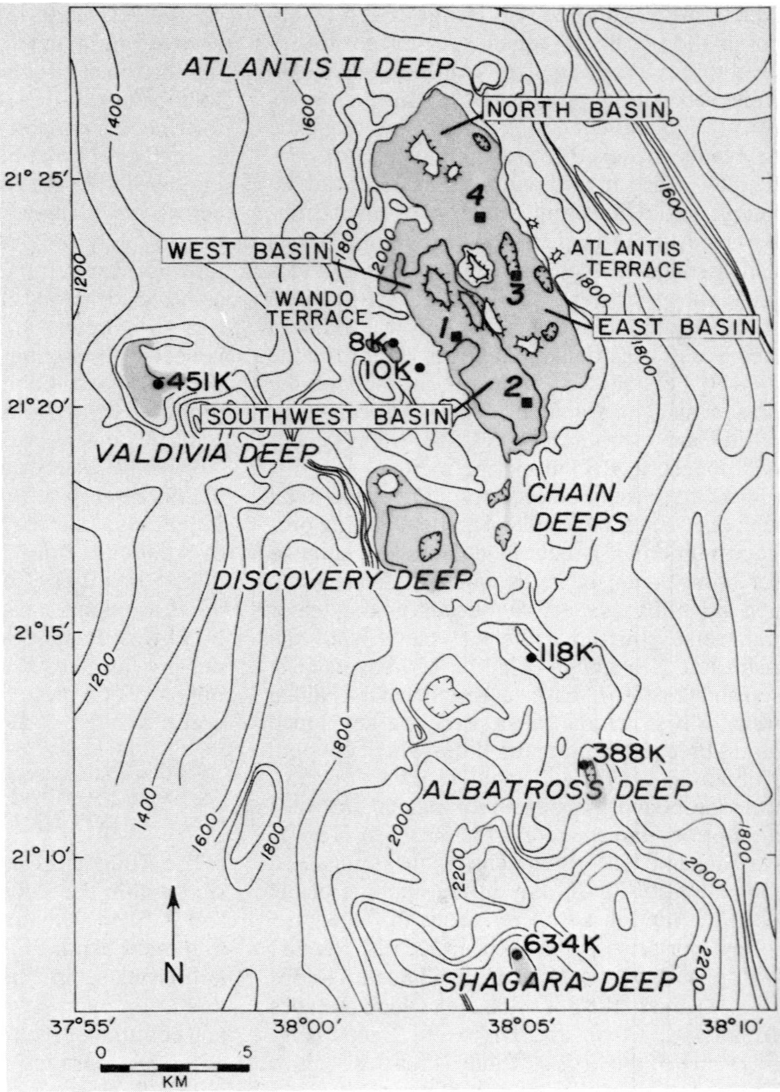

FIG. 2. Bathymetric map for Atlantis II Deep area with core locations and sites (1–4) of Pre-Pilot Mining Test. Shaded areas are brine filled. Bathymetry from Bäcker and Richter (1973).

economic evaluation of the Atlantis II deposit over the last ten years. Since 1976 Preussag has been the main contractor of the Saudi-Sudanese Red Sea Joint Commission. Approximately 650 core samples, including some long cores (>8 m) and some large volume cores, have been taken from the Atlantis II Deep. Recently published estimates of total metal contents in the muds are: 1.7 Mt Zn, 0.4 Mt Cu, and 5000 t Ag (Anon., 1981). An early estimate of overall average grade for dry, salt-free sediment is 1.04% Zn and 0.28% Cu (Bäcker, 1975). However, grades are significantly higher in the Southwest Basin of the Deep, and mining operations would be expected to concentrate on that area. Average grades (dry salt-free basis) of mined material would be approximately 3–6% Zn, 1% Cu, and 50 g/t Ag (Anon., 1981).

Preussag and various subcontractors mined and concentrated approximately 15 000 m^3 of mud from four mine sites in the Atlantis II Deep (Fig. 2) over a three-month period in 1979 (Luck and Nawab, 1981). This prepilot mining test (Table 2) demonstrated effective mining at 2200-m depth using a conventional 5-in. drill string and a suction head with conical vibrating screen (Fig. 4),

Table 1. Chemical Composition of Red Sea Brines (g/Kg)

	T, °C	Na	K	Mg	Ca	SO_4	C1	Ref.
Suakin	24.6	50.6	1.1	1.4	2.2	3.2	84.8	1
Port Sudan	36.2	78	1.5	1.5	1.1	4.0	125	2
Erba	27.9	50	1.0	1.5	1.0	4.1	86.5	2
Shagara	—	62	2.0	2.5	4.8	1.2	113	2
Albatross	24.4	86	2.0	1.0	4.0	1.1	143	2
Valdivia	29.7	95	1.9	1.9	0.8	6.8	140	2
Thetis	22.6	11	0.5	1.3	0.9	3.4	23	2
Nereus	30.2	—	—	1.5	7.9	0.9	130	2
Kebrit	23.3	95	1.5	2.4	7.1	2.2	153	2
Atlantis II	61.5	93	1.9	0.8	5.2	0.8	156	3
Red Sea Water	22.2	12	0.4	1.4	0.5	3.0	21	4

References: (1) Baumann, et al., 1973.
(2) Bäcker and Schoell, 1972.
(3) Brewer and Spencer, 1969.
(4) Craig, 1969.

concentration at sea by froth flotation, and successful disposal of tailings at 400-m depth (Mustafa and Amann, 1980). Minimal environmental impact or contamination of the mine site occurred during this disposal test. Further tests are planned, but monitoring of the disposal plume using the *Valdivia* indicated that the discharge penetrates to 1200-m depth within 20 min, with high solid concentrations at about 800 m. Solids are dispersed to only twice normal Red Sea levels within a few days (Anon., 1981).

The prepilot test was carried out using the drillship SEDCO 445, with suction head location and movement controlled to within 1 m by dynamic positioning of the ship relative to two bottom-tethered acoustic beacons. A submersible radial pump located on the drill string just above the brine interface produced 100 m^3/hr of slurry with solid concentration of 60–100 g/L. Vertical positioning of the head was controlled by an acoustic sediment thickness meter and the pump location relative to the suction head allowed a maximum penetration of 18 m below the mud-brine interface.

Concentration by froth flotation was carried out in the mud mixing tanks of the SEDCO 445. The flotation process was fairly standard and similar to that used for bulk flotation of complex sulfide ores (Mustafa and Amann, 1980; Anon., 1981), except that seawater was used to dilute the slurry to about 30 g/L. Concentrates were produced continuously with 30–35% Zn, 3.5–4.5% Cu, 500–700 g/t Ag, and a few ppm Au at 60–70% recovery rates (Table 2, Weber et al., 1981).

A number of possibilities are currently being tested for metallurgical processing of the concentrates (Weber et al., 1981). Pyrometallurgical processes have been eliminated from consideration, largely because of the 7–12% salts in the filter cake. The most promising methods, being tested in four separate laboratories and differing in detail, involve chloride leaching or oxygen pressure leaching in the presence of chloride followed by solvent extraction and electrowinning. Recoveries in excess of 99% for zinc and 95% for copper and silver have been demonstrated (Anon., 1981).

Table 2. Results of the 1979 Pre-Pilot Mining Test

			Zn content, %		
Site	Time, hr	Quantity mined, m^3	Heads	Tails	Concentrate
1 (W)	21	1800	2.0	0.9	18
2 (SW)	17	1400	5.5	2.4	27
3 (E)	22	1500	3.7	1.7	37
			2.9	1.2	41
4 (N)	32	2800	3.4	1.5	33
	104	8280	3.4	1.4	42

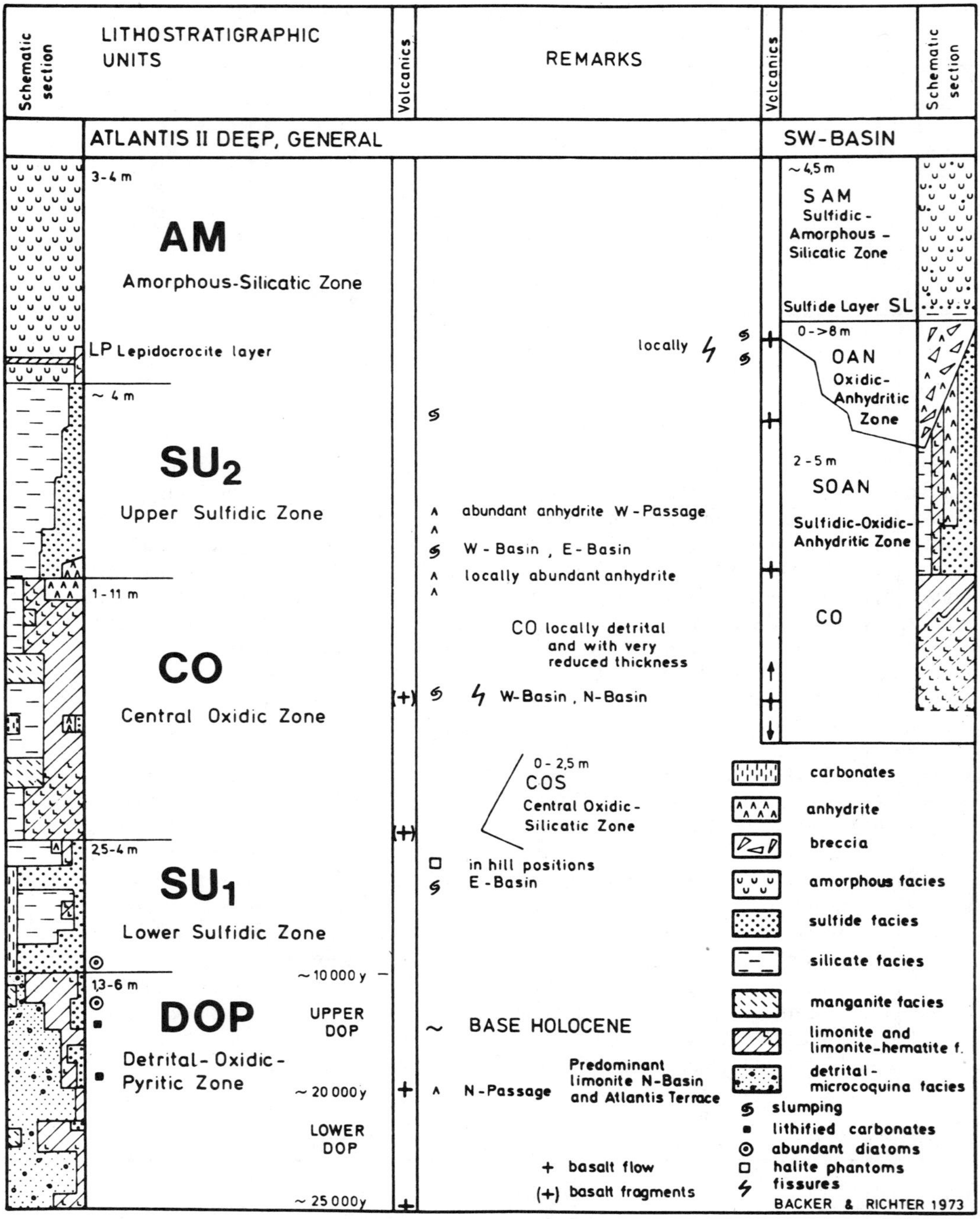

FIG. 3. Lithostratigraphic summary of Atlantis II Deep metalliferous sediments. From Bäcker and Richter (1973).

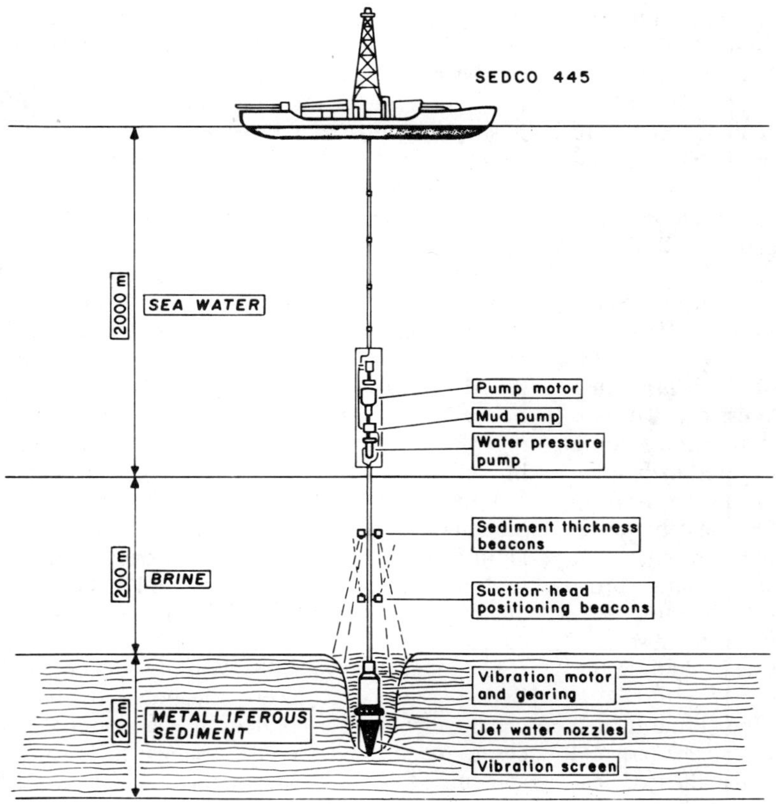

FIG. 4. Schematic drawing of the metalliferous mud mining system. After Mustafa and Amann (1980).

It is interesting to note that the "unconventionality" of the Atlantis II deposit significantly increases the economic potential of this occurrence. If the deposit was lithified and emplaced within a continental land mass, it would be a sheetlike body with dimensions of approximately 6 to 14 km. Expulsion of the 50–90% interstitial brine would leave a bulk deposit thickness of less than 5 m. Mining, crushing, and processing of a lithified equivalent of the Atlantis II deposit would be economic only if metamorphism and deformation produced considerable thickening along fold hinges. However, it is also germane to note that the bulk of Atlantis deposit has developed over only 10–12 thousands years whereas many ancient massive sulfides required millions of years to develop.

Potential advantages of marine mining of the Atlantis II deposit include abundant inexpensive transport and supply by sea, and processing of the unconsolidated muds without crushing and grinding. Tailings disposal can apparently be carried out on site without deleterious effects. Finally, deep-sea mining requires no expensive mine development and demonstrably can be carried on continuously even under heavy sea conditions. A factor which favors the prospect is the Saudi-Sudanese agreement (1974) to hold joint equal and exclusive sovereign rights to economic development of the so-called common zone of the central Red Sea below water depth of 1000 m.

Future prospects for mining the Atlantis II deposit will depend critically on the results of additional metallurgical testing, the year-long pilot mining test scheduled for 1983–84, and on base and precious metal prices. However, present results are moderately encouraging.

EXPLORATION SIGNIFICANCE

Because metalliferous sediments have never been commercially mined from the seafloor, economic exploration for such deposits outside of the Red Sea has not been systematically attempted. The attractiveness of the Atlantis II deposit hinges on its unconsolidated nature and occurrence on the surface of the seafloor. Buried Red Sea type deposits would be uneconomic if removal of significant overburden were necessitated. Thus, seafloor exploration targets are limited to actively forming or recently formed metalliferous muds.

The unique geological setting of the Red Sea is of critical importance. The Atlantis II deposit contains abundant base metal sulfides because of the ability of hot chloride-rich brines to transport heavy metals (Shanks and Bischoff, 1977). The deposit is an unconsolidated mix of sulfides, sulfates, silicates, and oxides because dense brines pool in bathymetric depressions. This anoxic hot brine pool provides a quiescent, sterile environment in which finely layered metalliferous precipitates settle out and are preserved from bacterial degradation, oxidation, and bioturbation. Therefore, the formation of an Atlantis II type deposit requires evaporites to produce brine and igneous activity to heat the brine for effective metal transport. In addition, a subsurface circulation pathway which ends at the seafloor is required.

Global exploration for seafloor brine-related base metal sulfide deposits can thus be targeted to general geological situations like the Red Sea. At present the Red Sea is unique. The Gulf of California is tectonically similar, but significant evaporites have not formed there. The Atlantic Ocean basin was probably very similar to the Red Sea during initial opening. Evaporites did form extensively, but any sulfide deposits which formed would now be deeply buried beneath thick clastic wedges related to the inactive basin margins. In fact, metalliferous sediments containing iron and manganese oxides and some base metal sulfides have been drilled off eastern North America in Cretaceous sedimentary rocks by DSDP hole 105 (Hollister et al., 1972). Another potential area for Red Sea type deposit exploration is in the Gulf of Mexico, where evaporites and subsurface brines occur extensively, but igneous activity is generally not spatially related. Indeed, one hypersaline brine pool has been reported (Shokes et al., 1977) but metallization is insignificant.

Exploration within the Red Sea, by Preussag, has concentrated on the axial rift zone and provides a model example upon which similar exploration programs might be developed. A number of possible exploration tools have been proposed, (Rona, 1978; Cronan, 1980), but in practice detailed bathymetric profiling has been most effective in delineating favorable deeps and in detecting above-bottom reflectors related to suspended matter at brine-seawater interfaces (Bäcker, 1975). In addition, high-resolution narrow-beam echo sounding, combined with simultaneous gravity and magnetic profiling, has provided a wealth of detailed geological and tectonic information on the Red Sea rift zone.

Follow-up sediment sampling in brine-containing deeps and in basins without brine has led to the discovery of additional recently formed metalliferous deposits (Fig. 1; Bäcker and Schoell, 1972) and to an improved understanding of the distribution of metalliferous phases in Recent Red Sea sediments. Recent detrital sedimentation in the Red Sea is dominated by biogenic carbonates (mainly pteropods and foraminifera). Lesser amounts of detrital quartz, feldspars, and clays also occur, and metalliferous components are admixed or interlayered in some areas.

The intensity of metallization of sediments in Red Sea deeps is nicely illustrated by bulk chemical proportions of detrital (aluminum, calcium, magnesium), oxide (manganese, iron), and sulfide (zinc, copper) components (Fig. 5, Schoell et al., 1974). The sediments in many of the Red Sea deeps are dominated by detrital materials with thin layers enriched in metalliferous components. However, significant metalliferous sediments occur in Nereus, Thetis, Gypsum, and Suakin Deeps and in areas flanking the Atlantis II Deep (Discovery, Chain, Wando, and Atlantis Terrace). The Atlantis II Deep contains the only sediments sampled thus far which are dominated by the sulfide component and which attain average grades and tonnages of potential economic levels (Fig. 5).

The careful exploration of the Red Sea rift zone by Preussag (Bäcker, 1975) has most likely discovered any currently active brine pools. However, discovery of metalliferous deposits not related to present brine pools is dependent on direct sediment sampling and, considering the enormous area of the rift zone, important deposits may remain undiscovered. In addition, metalliferous deposits which became inactive a few tens of thousands of years ago may still have economic significance even though buried under a few meters of recent sediments and effectively beyond the reach of gravity corers of

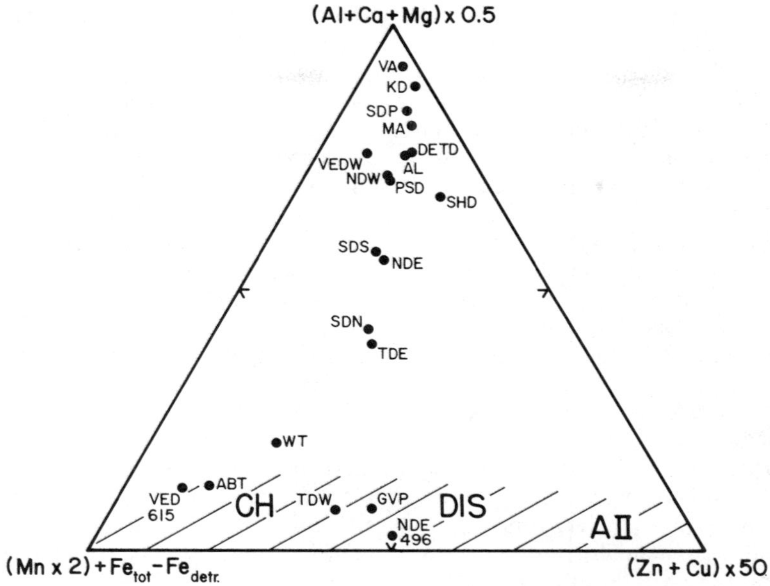

FIG. 5. Relative weight proportions of detrital (aluminum, calcium, magnesium), oxide (iron, manganese), and sulfide (zinc, copper) components in sediments from the Red Sea rift zone. Based on averaged bulk chemical data. From Schoell et al. (1974). Abbreviations: VA—Valdivia Deep; KD—Kebrit Deep; SDP—Suakin Deep Sea plain; MA—Commission Plain, Marsamirit; VED—Vema Deep; ABT—Atlantis Terrace; WT—Wando Terrace; VEDW—West of Vema Deep; NDE—Nereus Deep, East Basin; NDW—Nereus Deep, West Basin; SD—Suakin Deep; AL—Albatross Deep; PSD—Port Sudan Deep; TDWE—Thetis Deep (West, East Basin); DETD—Depression east of Thetis Deep; GYP—Gypsum Deep; CH—Chain Deep; DIS—Discovery Deep; A—Atlantis II Deep.

grab samplers. Finally, the recent spectacular discoveries of base-metal sulfide deposits on the East Pacific Rise (Hekinian et al., 1980; Speiss et al., 1980) and the Juan de Fuca Rise (Delaney et al., 1982) suggest the possibility of additional Red Sea deposits formed by normal-salinity seawater. Recently discovered dispersed metalliferous occurrences on the Commission Plain (Figs. 1 and 5, Bäcker, 1979) may have formed from buoyant seawater hydrothermal fluids and there may be high grade base-metal deposits related to the vent areas for these fluids.

What are the tools which might be useful in exploration for inactive brine related deposits and for possible seawater related deposits?

Discovery of massive sulfide deposits in the Pacific has required very detailed exploration using submersibles, and a substantial element of luck. However, studies of manganese and helium isotope values in the water column have provided excellent general indications of potential hydrothermal vent areas (Craig et al., 1980). Similarly, Lupton et al. (1977) have measured a 3000-fold ^{3}He enrichment in the Atlantis II brine and Holmes and Tooms (1972) have demonstrated enrichment of iron, manganese, zinc and copper by as much as five orders in the seawater column above the Atlantis II Deep. Helium isotope and trace element studies of Red Sea Deep Water should be sensitive methods capable of locating buoyant hydrothermal seawater vent areas. Follow-up investigations utilizing bottom photography, deeply towed instrumental surveys, and manned submersible studies could then be used to pinpoint sulfide deposits.

Continued search for inactive sulfide deposits formed by venting of dense brines has, perhaps, a greater potential for success in the Red Sea. Both geophysical and geochemical exploration have potential applicability. Rona (1978) has suggested that a large residual magnetic low spatially related to the Atlantis II Deep is a result of hydrothermal alteration of underlying oceanic crust by venting hot brine. Similar magnetic anomolies might locate inactive deposits, but very detailed local surveys would be required. In addition, dense brines may travel significant distances along bottom topography before pooling in suitable deeps. Thus, hydrothermally altered rocks in vent areas may not be directly related to sulfide deposits.

Bignell et al. (1976) have demonstrated the usefulness of geochemical exploration in the Red Sea using surficial sediment samples. Dispersion halos of manganese, copper, zinc, and mercury can be recognized for up to 10 km away from Atlantis II and Nereus Deeps. These dispersion halos significantly enlarge the exploration target and are potentially very useful in exploration for inactive as well as active deposits. However, the geochemical anomolies are significantly greater in the area around the Nereus Deep even though Nereus Deep contains much less intensely metallized sediments, far below potentially economic grades and volumes. Thus, numerous geochemical halos may exist in the Red Sea trough which are unrelated to significant mineralization. Once such anomolies are located, additional geochemical techniques which might distinguish significant and insignificant signals would be extremely useful.

Sulfur isotope studies of sulfide phases in sediments have considerable potential for exploration in the Red Sea, and may be more definitive than other geochemical indicators. Shanks and Bischoff (1980) and Kaplan et al. (1969) have shown that hydrothermal sulfides in the Atlantis II Deep have sulfur isotope values which cluster about +6, and range from 1 to 14. Marine bacteriogenic sulfides in the Red Sea and elsewhere give negative and erratic sulfur isotope values, ranging from about −10 to −45 (Table 3). Thus there is significant contrast between hydrothermal and bacterial values, and sulfur isotope studies on metalliferous layers in otherwise normal detrital sediments might be extremely useful in determining proximity to hydrothermal centers.

Systematic sulfur isotope and geochemical studies have been carried out on specially selected metalliferous samples from the Red Sea Deeps and the area around the Atlantis II Deep (Table 3). Specific core locations with respect to brine pools are indicated in Figs. 2 and 6, and sulfur isotope values are plotted against carbonate carbon content in Fig. 7. Carbonate carbon content is useful as an indicator of the relative proportion of detrital material in a given sample. Normal detrital samples in the Red Sea rift zone have about 6% carbonate carbon, and are thus about 50% carbonate minerals with the balance consisting of clays and other silicates. Iron-oxide-containing samples contain less of the detrital component due to the dilution effect. Hydrothermal samples from the Atlantis II Deep have very low carbonate carbon contents (Fig. 7), due to very rapid accumulation of hydrothermal precipitates (50–150 cm/10^3 yr according to Garson and Krs, 1976; Shanks and Bischoff, 1980). Most samples with high carbonate carbon contents have typical marine bacteriogenic sulfur isotope values, but it is encouraging that a number of samples from the DOP zone (Fig. 3) of the Atlantis II sediments and from other areas have high carbonate content and still retain hydrothermal sulfur isotope signatures.

The potential usefulness of sulfur isotope studies in determining proximity to hydrothermal activity in the Red Sea is dependent on two factors: dispersion of hydrothermal sulfides and the meaning of positive sulfur isotope values in terms of significant metallization. Examination of samples from the Atlantis II area (Fig. 2, Table 3) is instructive. Sulfides with positive sulfur isotope values have been analyzed in samples from cores 8K and 10K from the Wando Basin (or Terrace), from core 118K (10 km south of the Atlantis II Deep), and from core 634K from the Shagara Deep. Wando samples are typical of sulfide and silicate zone Atlantis II sediments, 118K samples are iron-oxide rich, and Shagara samples cannot be macroscopically distinguished from normal detrital sediments. Thus, there is a large range in the metalliferous component, but all of these samples have positive sulfur isotope values. Dispersion of hydrothermal sulfides occurs to at least 18 km from the Atlantis II Deep, due either to brine overflow from the main pool or to peripheral venting sites related to the same brine plumbing system. This assertion is supported by the presence of brine in the Albatross Deep which is chemically identical to the Atlantis II lower brine (Table 1).

Table 3. Sulfur Isotope Values of Red Sea Sulfides

Deep	Core	Depth, cm	C^{14} Age	$\delta^{34}S$*	Mineralogy‡	
					Major	Minor
Kebrit	508K	10– 20		−44.5	Ct	Q,Ar
Gypsum	502PT	60– 86		+21.19† 20.37†	Gy,Bs	
	520P	100– 20		+20.07† +20.11†	Gy,Bs	
Vema	525P	31– 36		−33.46	Ct	Py
Nereus	529P	54– 59		—	Ct	
	486K	11– 18		−19.15	Ct	Q,K,Py
		60– 65		−25.73	Go	K,I,M,Py?
Thetis	618K	122–132		+12.10 +11.21	Mt	Py,Sp
	536K	18– 22		−30.16	Ct,Q	D,K,Py
		48– 52		−34.11	Ct,Q	Py,K
		148–152		—	Ct,Q	Ar,K
	534P	85– 95		—	Hm	—
		150–160	6,700y	—	Hm	K,Mt
		190–200		+ 9.88	Hm,Mt	K,I
Hadarba-Hatiba	611PT	419–433		—	Ct	Q,Ar,D
Valdivia	451K	14– 22		−17.68	Ct	Ar,Q,D,Py
Wando	10K	38– 53		− 0.92 − 0.04 − 0.15	Py,Sp	Ct,K,Q,D
		53– 67		+17.19 +14.98	M	K
		186–192	14,940y	−31.25	Q,Ct	Py,K
	8K	55– 86		+ 3.25		
AII vicinity (from Kaplan et al., 1969)	118K	110–120		−27.7		
		160–168		−31.8		
		170–190		−31.8		
		385–387		+10.3		
		387–400		+11.4		
Albatross	388K	30– 42		—	Ct,Q	Py
		60– 70		−26.39	Ct	Q,K,Py
		280–290		−25.64	Ct	Q,K,I,M,Py
Shagara	634K	130–140	3,725y	− 1.08	Ct	Ar,D,K,Q,Py
		180–184		+ 3.35	Ct	Ar,Py
		340–350	3,140y	+ 5.90	Ct	Py,K,D,Q
Saukin	329K	610–615		—	Ct	Q
	332K	280–815		—	Ct,K,Q	I,M
		805–815		−32.00	Ct,Q	K,I,M,Py,Ar
	334K	51– 56		−23.98	Ct,Q	Ar
		350–360		+18.98 +15.99	Ct,Go	Q,I,M,K
	112K	100–120		−27.44	Ct,Q	M,I,K,Py
(from Gant, 1980)	BC76	16– 17		+ 0.71	M,Ct	
		22– 23		−18.17	Ct	
		30– 32		−29.63 −29.18	Ct	Py
		44– 46		—	Go,Ct	

*Brackets indicate duplicate analyses of separate splits of the sample.
†Sulfates.
‡Abbreviations: Ar—aragonite, Bs—bassanite, Ct—calcite, D—dolomite, Go—goethite, Gy—gypsum, Hm—hematite, I—illite, K—kaolinite, M—montmorillonite, Mt—magnetite, Py—pyrite, Q—quartz, Sp—sphalerite.

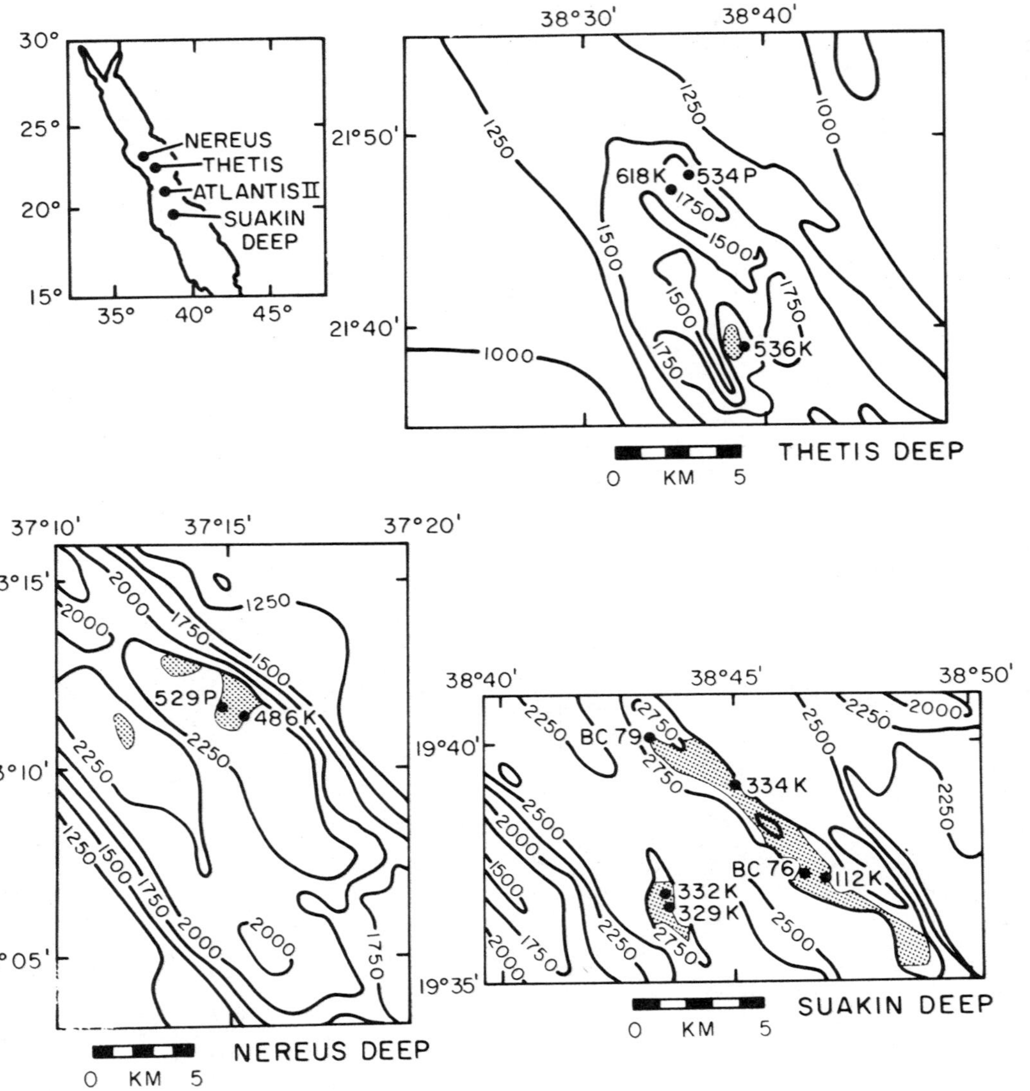

FIG. 6. Bathymetric maps of Thetis, Nereus, and Suakin Deeps showing locations of cores utilized in this study. From Bäcker and Schoell (1972). Stippled areas are brine filled.

Hydrothermal sulfides were not detected in core 388K from Albatross, probably due to poor core placement with respect to the brine pool.

Shanks and Bischoff (1977, 1980) have concluded that the sulfides in the Atlantis II Deep sediments have formed due to a sulfide reduction process in heated subsurface brine. Subsurface brine temperatures in excess of 200°C are probably required to produce the characteristic "hydrothermal" sulfur isotope values which average +6. Thus positive sulfur isotope values in sediments of the Red Sea rift zone indicate high temperature brine activity, as opposed to low temperature brines derived from paleowaters which have not interacted with hot rift zone intrusives. Indeed, outside of the Atlantis II area, hydrothermal sulfides are found only in the Thetis and Suakin Deeps (Table 3, Figs. 6 and 7). Thetis has no brine at present, and Suakin brine has oxygen and hydrogen isotope values which

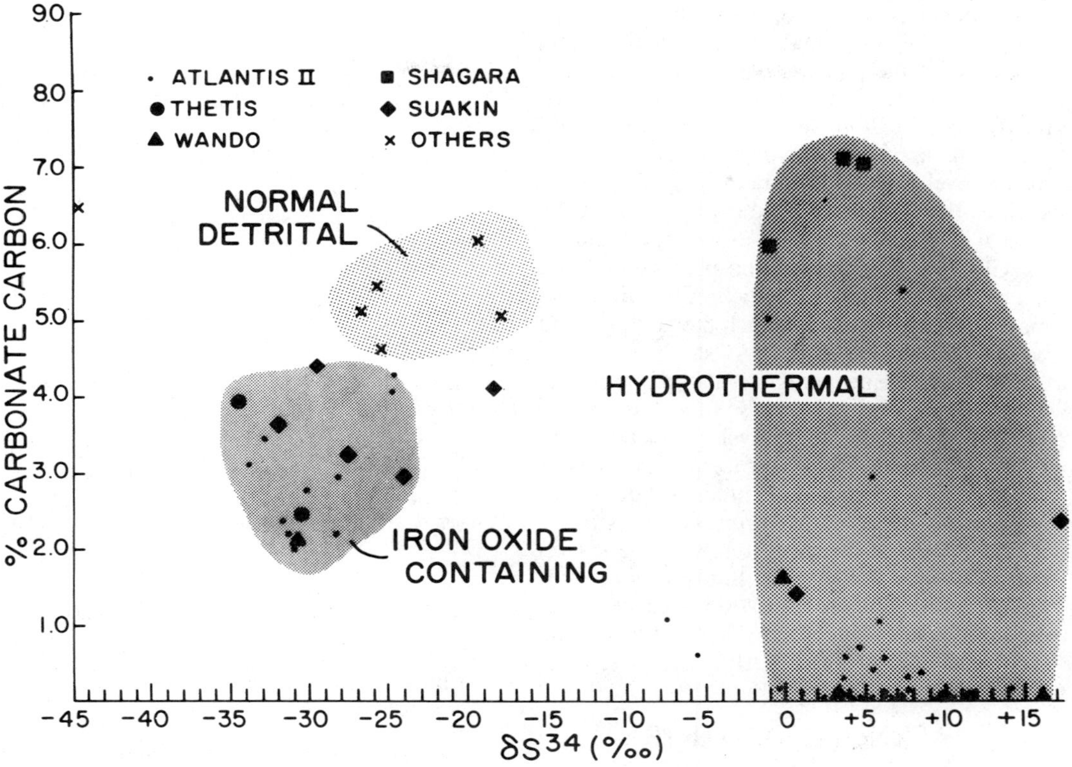

FIG. 7. Plot of carbonate carbon content (weight percent) vs. sulfur isotope value of sulfide minerals (CDT scale). Carbonate carbon content is a general indicator of the proportion of detrital material in a given Red Sea sample. Hydrothermal sulfides have positive isotope values and contain sulfide derived from hot brine venting. Detrital samples have negative sulfur isotope values typical of marine bacteriogenic sulfides.

differ from the paleowater-derived brines and are not amenable to unambiguous interpretation (Schoell and Faber, 1978).

The presence of sulfides with positive isotope values in the Red Sea does not ensure ore-grade accumulation of metalliferous muds. Positive isotope values occur in some metalliferous layers of the Atlantis II DOP zone (Fig. 3) and the Suakin sediments are very similar to the DOP zone (Baumann et al., 1973; Gant, 1980). Thin nontronitic and oxide layers in these sediments indicate only intermittent hydrothermal activity. However, positive sulfur isotope values appear to be a prerequisite for significant hydrothermal deposition and are thus a very positive geochemical indicator. The most significant discovery outside of the Atlantis II area is the Thetis Deep, which contains sediments comprised almost entirely of magnetite with minor but significant base metal sulfides with positive isotope values (Table 3). The Thetis Deep area certainly should be carefully evaluated for more extensive deposits.

SUMMARY AND CONCLUSIONS

The metalliferous sediments of the Atlantis II Deep represent the only known deposit in the Red Sea with reasonable potential for development as a seafloor base metal sulfide ore deposit. Atlantis II muds are estimated to contain 1.7 Mt Zn, 0.4 Mt Cu, 5000 t Ag, and a small gold byproduct. A pre-pilot mining test conducted in 1979 has demonstrated effective mining of the muds by slurry methods, concentration by froth flotation at sea, and tailings disposal without apparent dire consequences.

The Atlantis II deposit is the most attractive thus far discovered in the Red Sea, but the overall grade is low and the economies of mining remain very questionable. Discovery of additional deposits in the Red Sea remains a possibility, especially inactive deposits without associated brine pools.

Exploration for active or inactive seafloor brine deposits in the Red Sea and elsewhere should proceed at approximately three levels of refinement. Level 1 targets exploration in general areas, such as the Red Sea rift, known to contain the critical geological factors: evaporites, igneous activity, and fluid circulation pathways. Level 2 involves detailed bathymetric, gravity, and magnetic profiling to establish the geologic and bathymetric framework. This step should discover active brine pools by above-bottom reflections. Level 3 involves sediment and water sampling, geochemical analysis for metallic element dispersion, and sulfur isotope analysis of sulfides. Level 3 should be directed at discovery of inactive deposits in favorable bathymetric areas, especially if magnetic lows or geochemical anomolies are detected. Sulfur isotope studies of sulfide minerals provide a geochemical exploration tool which may explicitly indicate proximity to hydrothermal sulfide deposits.

ACKNOWLEDGMENTS

I gratefully acknowledge the cooperation of H. Bäcker, Preussag AG, in providing samples and published information on the current status of economic investigations.

This study was supported by National Science Foundation Grants OCE7805481 and OCE8018644.

REFERENCES

Anon., 1981, "Two Thousand Meters Under the Sea," *Mining Magazine,* pp. 114–122.

Bäcker, H., 1975, "Exploration of the Red Sea and Gulf of Aden during the M.S. VALDIVIA Cruises 'Erzschlamme A' and 'Erzschlamme B'," *Geol. Jarbuch,* D 13, pp. 3–78.

Bäcker, H., and Richter, H., 1973, "Die rezente hydrothermal-sedimentare lagerstatte Atlantis II tief im Roten Meer," *Geol. Rundschau,* Vol. 62, pp. 697–741.

Bäcker, H., and Schoell M., 1972, "New Deeps with Brines and Metalliferous Sediments in the Red Sea," *Nature, Physical Science,* Vol. 240, pp. 153–158.

Bäcker, H., 1979, "Metalliferous Sediments of the Red Sea, Occurrence, Exploration and Resource Assessment," *Proceedings,* International Seminar on Offshore Mineral Resources, Orleans, France, BRGM Document No. 7, pp. 317–338.

Baumann, A., Richter, H., and Schoell, M., 1973, "Suakin Deep: Brines and Hydrothermal Sediments in the Deepest Part of the Red Sea," *Geol. Rundschau,* Vol. 62, pp. 684–697.

Bignell, R.D., 1975, "Timing, Distribution and Origin of Submarine Mineralization in the Red Sea," *Transactions,* Institution of Mining & Metallurgy, Vol. 84, pp. B1–B6.

Bignell, R.D., Cronan, D.S., and Tooms, J.S., 1976, "Metal Dispersion in the Red Sea as an Aid to Marine Geochemical Exploration," *Transactions,* Institution of Mining and Metallurgy, Sec. B, Vol. 85, pp. B274–B278.

Brewer, P.G. and Spencer, D.W., 1969, "A Note on the Chemical Composition of the Red Sea Brines," *Hot Brines and Recent Heavy Metal Deposits in the Red Sea,* E.T. Degens and D.A. Ross, eds., Springer-Verlag, New York, pp. 174–179.

Craig, H., 1969, "Geochemistry and Origin of Red Sea Brines," *Hot Brines and Recent Heavy Metal Deposits in the Red Sea,* E.T. Degens and D.A. Ross, eds., Springer-Verlag, New York, pp. 208–242.

Craig, H., et al., 1980, "Geochemical Studies of the 21°N EPR Hydrothermal Fluids, EOS," *Transactions,* American Geophysical Union, Vol. 61, p. 992.

Cronan, D.S., 1980, *Underwater Minerals,* Academic Press, New York, 361 pp.

Delaney, J.R., et al., 1982, "Massive Zinc and Iron-Rich Sulfide Deposits Associated with Hot Springs, Juan de Fuca Ridge," *Abstracts,* American Association of Petroleum Geologists, Annual Convention, p. 43.

Gant, J., 1980, "Mineralogical and Geochemical Studies of Brine-Related Metalliferous Sediments in the Suakin Deep, Red Sea," M.S. Thesis, University of Wisconsin-Madison, 178 pp.

Garson, M.S., and Krs, M., 1976, "Geophysical and Geological Evidence of the Relationship of Red Sea Transverse Tectonics to Ancient Fractures," *Bulletin,* Geological Society of America, Vol. 87, pp. 169–181.

Girdler, R.W., and Styles, P., 1974, "Two Stage Red Sea Floor Spreading," *Nature, Physical Science,* Vol. 247, pp. 7–11.

Hekinian, R., et al., 1980, "Sulfide Deposits from the East Pacific Rise Near 21°N: A Mineralogical and Geochemical Study," *Science,* Vol. 207, pp. 1433–1444.

Hollister, C.D., et al., 1972, "Site 105—Lower Continental Rise Hills," Initial Reports, Deep Sea Drilling Project, Vol. 11, pp. 219–312.

Holmes, R., and Tooms, J.S., 1972, "Dispersion from a Submarine Exhalative Ore Body," *Proceedings,* 4th International Geochemical Exploration Symposium, Institution of Mining & Metallurgy, London, pp. 193–201.

Kaplan, I.R., Sweeney, R.E., and Nissenbaum, A., 1969, "Sulfur Isotope Studies on Red Sea Geothermal Brines and Sediments," *Hot Brines and Recent Heavy Metal Deposits of the Red Sea,* E.T. Degens and D.A. Ross, eds., Springer-Verlag, New York, pp. 474–498.

Luck, K., and Nawab, Z., 1981, "Metalliferous Sediments in the Atlantis II Deep—Recovery and Selected Special Problems, *Conference Report,* International Conference for Ocean Mining, Dusseldorf, pp. 103–115.

Manheim, E.T., 1974, "Red Sea Geochemistry," Initial Reports, Deep Sea Drilling Project, Vol. 23, pp. 975–998.

Mustafa, Z., and Amann, H., 1980, "The Red Sea Prepilot Mining Test 1979," *Proceedings,* 12th Annual Offshore Technology Conference, Society of Petroleum Engineers, Dallas, Paper No. 3874.

Nawab, Z., and Luck, K., 1979, ''Test Mining of Metalliferous Mud from the Red Sea Bottom,'' *Meerestechnik,* Vol. 10, No. 6, pp. 181–187.

Rona, P.A., 1978, ''Criteria for Recognition of Hydrothermal Mineral Deposits in Oceanic Crust,'' *Economic Geology,* Vol. 73, pp. 135–160.

Schoell, M., Bäcker, H., and Baumann, A., 1974, ''The Red Sea Geothermal Systems, New Aspects on Their Brines and Associated Sediments,'' International Association on the Genesis of Ore Deposits (IAGOD), Varna Meeting.

Schoell, M., 1976, ''Heating and Convection within the Atlantis II Deep Geothermal System of the Red Sea,'' *Proceedings,* 2nd United Nations Symposium on the Development and Use of Geothermal Resources, Vol. 1, pp. 583–590.

Schoell, M., and Faber, E., 1978, ''New Isotopic Evidence for the Origin of Red Sea Brines,'' *Nature,* Vol. 275, pp. 436–438.

Shanks, W.C., III, and Bischoff, J.L., 1977, ''Ore Transport and Deposition in the Red Sea Geothermal System: A Geochemical Model,'' *Geochimica et Cosmochimica Acta,* Vol. 41, p. 1507–1519.

Shanks, W.C., III, and Bischoff, J.L., 1980, ''Geochemistry, Sulfur Isotope Composition, and Accumulation Rates of Red Sea Geothermal Deposits,'' *Economic Geology,* Vol. 75, pp. 445–459.

Shokes, R.F., et al., 1979, ''Anoxic, Hypersaline Basin in the Northern Gulf of Mexico,'' *Science,* Vol. 196, pp. 1443–1446.

Spiess, F.N., et al. (RISE Project Group), 1980, ''East Pacific Rise: Hot Springs and Geophysical Experiments,'' *Science,* Vol. 207, pp. 1421–1432.

Weber, H., et al., 1981, ''Metallurgical Treatment of Red Sea Concentrates,'' *Conference Reports,* International Conference for Ocean Mining, Dusseldorf, pp. 116–122.

4

Unconventional Exploration Concepts

14

Potential for High-Grade Shallow-Marine Manganese Deposits in North America

W.F. Cannon and E.R. Force

US Geological Survey
Reston, VA

ABSTRACT

Many large high-grade manganese deposits occur in marine-transgressive sedimentary sequences deposited on stable platforms, although the manganese mineralogy varies from carbonate to oxide facies and host lithologies vary from carbonate- to clay-dominated sections. Examples include the Nikopol and related deposits of the Soviet Union and Groote Eylandt, Australia.

Recent work has shown that high sea-level stands are synchronous with stratified and stagnant oceans. We suggest that during transgressions and consequent anoxic events, the anoxic deep waters rich in dissolved manganese were dilute ore-forming fluids. Manganese was precipitated on the oxidized side of oxidation and alkalinity interfaces on shallow shelves. Where the precipitate was preserved, manganese accumulated into ore deposits.

Sedimentary manganese deposits have not been widely explored for in North America, nor are they always obvious in outcrop. Large areas of North America are underlain by rocks that we consider broadly favorable for such manganese deposits, and the disparity between the high potential but insignificant reserves of the United States is difficult to reconcile. Important undiscovered deposits seem likely. An area of Cretaceous rocks in South Dakota is presented as a possible location for a United States deposit analogous to foreign deposits described herein.

INTRODUCTION

Manganese is a metal essential to modern industrial society and has been sought extensively for more than a century. In spite of long exploration efforts, North America has always been deficient in economic manganese deposits.

Manganese is a relatively common element and can be redistributed and concentrated in nature by many geologic processes. In other parts of the world, manganese ores have formed in a great diversity of host rocks and in a wide range of physical and chemical environments. Similar rocks and environments are widespread in North America, but major manganese deposits have not been found in them. Does North America lack such deposits, or do undiscovered deposits exist? In this paper, we present our reasons for believing that deposits of one major type, shallow marine sedimentary deposits, remain to be discovered in North America. Using analogies between some well-studied deposits elsewhere and the geology of parts of North America, we have identified some terranes that we consider may contain deposits and that we believe have not been adequately explored for manganese.

Manganese deposits presently of major importance in world markets can be roughly classed in two categories: (1) primary shallow-marine sedimentary deposits, in which manganese was concentrated largely during sedimentation; and (2) supergene deposits, in which tropical weathering of manganiferous rocks has produced economic concentrations. Deep-sea manganese nodules, volcanogenic manganese deposits, and other deposit types are not yet of great economic importance and are not treated in this paper.

Supergene deposits are restricted to the present-day tropics or to stratigraphic horizons where zones of ancient tropical weathering products are preserved. Supergene deposits account for about 25% of current world production but supply about 50% of United States consumption. The deposits of Brazil, India, and Gabon are examples of supergene deposits, some enriched to economic grades from primary sedimentary deposits; such deposits are not likely to have formed in the temperate climates of most of North America.

Primary sedimentary deposits such as those at Nikopol (USSR), the Kalahari field (Republic of South Africa), Groote Eylandt (Australia), and Molango (Mexico) can be very large; these four deposits alone account for more than 80% of world reserves and about 75% of world resources. Unlike supergene deposits, primary deposits are not restricted to tropical latitudes and have a wide geographic and stratigraphic distribution. As such primary deposits are more likely to occur in North America than are supergene deposits, we will concentrate our discussions on the primary deposits, emphasizing what we consider to be some frontier areas for exploration in North America.

CONTROLS ON MANGANESE DEPOSITION

Transgression

A review of sedimentary manganese deposits from around the world has revealed a striking association of these deposits with transgressive marine rock sequences deposited in shallow water on stable platforms. In fact, deposits accounting for more than 80% of world resources have this common sedimentologic setting, broadly defined. They occur in marine rocks from within a few meters to as much as several hundred meters stratigraphically above widespread unconformities or above transitions from continental to marine sediments. Within this general transgressive setting, deposits have a wide range of internal mineralogy and textures and are interbedded with a wide range of rock types. In detail, some deposits can form regressive units within transgressive sequences.

The importance of transgression in controlling manganese deposits probably lies with the association of high sea-level stands with ocean anoxic events. The nature of the link between the two phenomena is discussed in a later section, but empirical evidence that they coincide is abundant (Schlanger and Jenkyns, 1976; Fischer and Arthur, 1977; Hallam and Bradshaw, 1979; Arthur and Schlanger, 1979; Jenkyns, 1980). Both occur during periods of worldwide warm climate.

Anoxic Events

Recent work, spurred largely by petroleum exploration as well as by the Deep Sea Drilling Project, has shown that at times in the geologic past, the world's oceans have not been well oxygenated as they are at present, but rather have been anoxic in large part. These periods of widespread ocean anoxia, best documented for the Albian-Aptian and Cenomanian-Turonian stages of the Cretaceous, were times of widespread ocean stagnation and subsequent widespread deposition of black muds rich in organic matter at depths of the oceans below the zone of surface turbulence to a depth of at least 3000 m (Schlanger and Jenkyns, 1976; Degens and Stoffers, 1976; Ryan and Cita, 1977; Arthur and Schlanger, 1979; Jenkyns, 1980; Demaisson and Moore, 1980; Summerhayes, 1981; Weissert, 1981). The chemical character of such oceans must have been vastly different from that of modern oceans because anoxic water has a much higher solubility for many metals than does oxygenated water.

The solubility of manganese in anoxic seawater is about 500 times that of oxygenated sea water. In anoxic seas, an interface between deeper anoxic water and oxygenated water of the surface mixed zone is therefore a sharp gradient for dissolved manganese. This is discussed in greater detail in a subsequent section. Clearly, however, mixing and diffusion processes at the interface are capable of precipitating manganese on the oxidizing side. Preservation of the precipitate can occur at the intersection of the in-

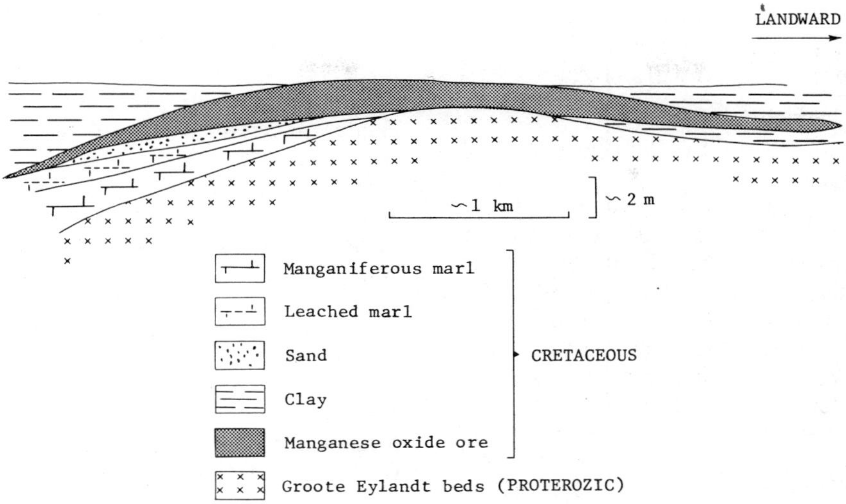

FIG. 1. Schematic cross section of Groote Eylandt, Australia, deposits, compiled from Smith and Gebert (1969, Fig. 3) and Slee (1980, Fig. 16).

terface with the shelf, and we hypothesize that the shallow marine sedimentary manganese deposits of stable platforms are the precipitate, locally modified by other processes. The broadly transgressive nature of these deposits is the only stratigraphic clue linking some of them to anoxic seas; for others there is also an association with black-shale basins (but commonly not at the site of manganese deposition), age of deposition corresponding to known anoxic events, and/or carbon isotope data indicating a nearby reservoir of anoxic water.

DEPOSIT DESCRIPTIONS

The following sections briefly describe some of the world's important manganese deposits and emphasize their common transgressive nature, when viewed on a large scale, and their wide diversity on a smaller scale.

Groote Eylandt, Australia

The recently discovered Groote Eylandt manganese deposit in the Gulf of Carpentaria, Northern Territory, has made Australia a major supplier of manganese. Reserves are more than 250 million tons at 48% Mn (National Materials Advisory Board, 1981). Ore occurs as a layer as thick as 15 m (Fig. 1), averaging 3–4 m (Smith and Gebert, 1969), and consists of manganese oxides. Both sedimentary and diagenetic processes have produced ore. Unusually good evidence (rerounded broken pisolites) exists for the sedimentary origin of cryptomelane-pyrolusite pisolites (Smith and Gebert, 1969) that form a considerable part of the deposit. On the other hand, much ore formed by replacement of clay by manganese oxides (Slee, 1980; Ostwald, 1980).

The ore layer is a member of the Albian (Lower Cretaceous) Mullaman beds, which consist of nearly horizontal clays, sands, and marls as thick as 100 m. The marls themselves have cream-colored rhodochrosite-calcite nodules and contain as much as 14% Mn (Smith and Gebert, 1969, p. 589). Ore is found on buried physiographic highs, and different ore textures have been related to different depositional settings in a deeply embayed shoreline environment (Smith and Gebert, 1969). The Mullaman beds contain marine Foraminifera below, in, and above the ore. These beds unconformably overlie Proterozoic sedimentary rocks and clearly formed during a regional transgression. In addition to this regional transgression at the base of the Mullaman beds, a more local transgression apparently separates underlying weathered marl from ore (Slee, 1980, pp. 145–146).*

*B.R. Bolton and L.A. Frakes (written communication, 1982) believe that the ore above this unconformity is regressive in nature, and Early Cenomanian in age, based on rare microfossils.

The Albian was a period of widespread ocean anoxia (Arthur and Schlanger, 1979) and high eustatic sea-level stands (Vail and others, 1977).

Molango, Mexico

The Molango deposit of Mexico is the only major manganese deposit in North America. It lies along the Sierra Madre Orientale in Hidalgo state. The ore is bedded carbonate rocks, consisting of mixtures of kutnahorite, rhodochrosite, and calcite. Reserves include 230 million tons of ore averaging 25% manganese, and resources are estimated at 15 billion tons having grades of 10% or higher (National Materials Advisory Board, 1981). The ore occurs in the Taman formation of Oxfordian/Kimmeridgian (Late Jurassic) age (Tavera and Alexandri, 1972) (Fig. 2) and is exposed as a continuous unit for about 50 km along strike. The Taman formation is as much as 500 m thick and consists mostly of dark-gray calcareous siltstone and shale. The base of the manganiferous unit is about 300–400 m above the base of the formation near the current area of production, but the manganese bed laps onto older units elsewhere and is commonly only a few tens of meters above the base of the formation (Carillo Bravo, 1965). The manganiferous unit is about 25 m thick and where it is being mined contains several meters of ore containing about 30% Mn. It consists of manganiferous carbonate beds interlayered with siltstone and shale.

The Taman formation lies with angular unconformity on the Middle Jurassic Cahausas formation, which consists mostly of continental conglomerate, sandstone, and shale. The late Jurassic was a time of major marine transgression throughout this part of Mexico (Imlay and others, 1948) and worldwide (Vail and others, 1977), and the Taman formation records this marine incursion in the Molango area. The Oxfordian is also a widespread ocean anoxic event (Hallam and Bradshaw, 1979).

Nikopol, USSR

The Oligocene manganese deposits of the Black Sea area include those of the Varna district in Bulgaria (Aleksiev and Bogdonova, 1974) as well as those of the more important Nikopol (Ukraine) and Chiatura (Georgia) districts in the USSR. The latter two districts have been of great historical importance and still contain re-

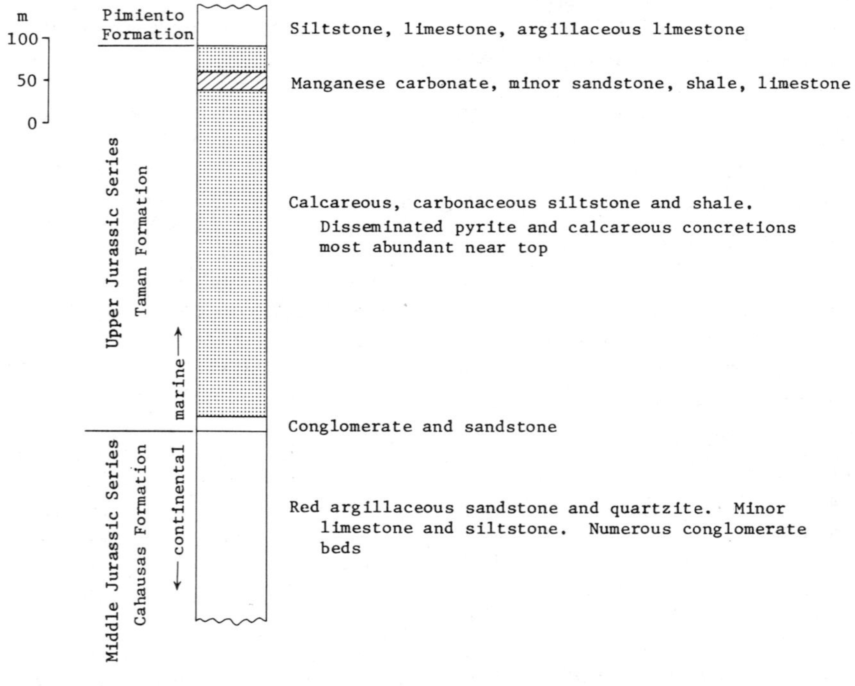

FIG. 2. Stratigraphic column of Jurassic rocks in the vicinity of the Molango deposit, Mexico, from data of Carillo Bravo (1965) and Imlay et al. (1948).

serves on the order of 1.7 billion tons of ore. Low manganese grades of 13%–33% are now being worked, but 60 years ago, average grades were as high as 55% (National Materials Advisory Board, 1981).

For simplicity, the largest deposits, those of the Nikopol district (Fig. 3), will be described as being representative of all the deposits of the Black Sea area. A large but mostly redundant literature in English includes works by Gryaznov (1970), Gryaznov and Danilov (1980), Shevchenko (1970), Strakhov and others (1970), Varentsov (1964), and Varentsov and Rakhmanov (1980). The following summary is from these reports. Only specific statements will be referenced.

The ores range from 0.5 to 5 m in thickness and consist of several manganese oxides and carbonates, mostly having nodular textures. Carbonate minerals predominate at the base of the ore and in the downdip, presumably deeper water, southern parts of the deposits. In recent reports, the consensus is that most of the oxides formed from carbonates by Oligocene and Miocene diagenesis or weathering, although some formed at the same time as or even before the carbonate minerals.

The ore layer is enclosed predominantly in clay (mostly above) and glauconitic sand (mostly below) in a gently dipping thin marine lower Oligocene section, which laps onto Precambrian crystalline rocks. A paleogeographic model has been worked out in which the crystalline rocks formed a large island, the deposit being on the south side (Gryaznov, 1970). Ore-bearing clay pinches out islandward into coal-bearing fluvial beds and passes seaward into thicker clay that lacks ore beds. The east-trending paleoshoreline was strongly embayed, and the ore bodies fill north-oriented depressions (see Strakhov and others, 1970, Fig. 8), so that deposition in or just offshore of a group of estuaries has been suggested.

The marine origin of the ore and enclosing rocks is well established because the rocks contain a diverse marine fauna locally including fish, crabs, corals, and other forms that suggest normal salinity, shallow water, and warm temperatures (Varentsov and Rakhmanov, 1980, Fig. 7; Gryaznov, 1970). Pollen indicates a humid-temperate paleoclimate.

Crystalline rock is commonly in contact with and elsewhere closely underlies the manganese ore; the entire Oligocene section is less than 20 m thick, and ore is near the base (Varentsov and Rakhmanov, 1980, p. 340). Pre-Oligocene units intervene locally.

The ore is one of the basal units resulting from a lower Oligocene marine transgression of regional and worldwide extent (Vail and others, 1977). During the Oligocene, Tethys was closing to become a paleo-Black Sea. Conditions in the deeper parts of the basin in the Oligocene are unknown, but probably the paleo-Black Sea was already a silled basin and was anoxic during times of temperate to warm climate.

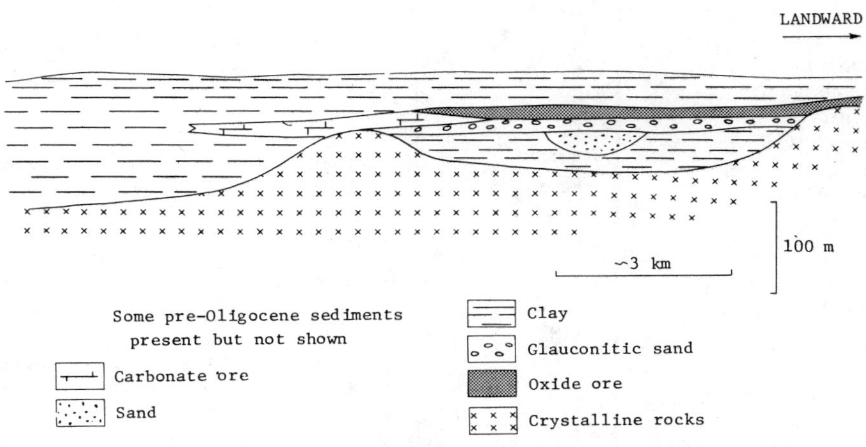

FIG. 3. Schematic cross section of the Nikopol, USSR, deposits, compiled from descriptions and diagrams by Gryaznov (1970, pp. 152–153), Varentsov and Rakhmanov (1980, Fig. 5), and Shevchenko (1970, Fig. 1).

Moanda, Gabon

The Moanda deposit in Gabon is currently the principal manganese supplier for the United States. Mining is from a high-grade supergene deposit, but that deposit has formed by enrichment of manganiferous shales deposited in a transgressive environment. Reserves of high-grade ore (50% Mn) are 194 million tons, but the amount of manganiferous shale containing 10%–25% manganese has not been determined (National Materials Advisory Board, 1981).

The manganiferous shale is a unit about 150 m thick in the Franceville series (Fig. 4), about 1.8 billion years old (Leclerc and Weber, 1980). The Franceville consists of a lower sequence of continental rock composed of coarse-grained feldspathic sandstone, conglomerate, and minor dolomite. These rocks are overlain with slight angular unconformity by marine rocks consisting, from the base up, of about 20 m of conglomerate and 100 m of alternating siltstone, shale, and fine-grained dolomitic sandstone. The manganese-bearing unit lies on these rocks and consists of black shale containing dolomitic, ferruginous, and manganiferous beds. The manganese in the manganiferous beds occurs mostly in carbonate minerals. The unit contains minor sandstone and dolomitic sandstone interbeds. Thus, although the present ore body at Moanda is of supergene origin and is not classed as a sedimentary deposit, the protore is sedimentary, and it formed under marine-transgressive conditions.

Morro do Urucum, Brazil

Manganese ores of the Morro do Urucum deposit in Mato Grosso state, Brazil, consist of beds of manganese oxides, mostly cryptomelane, interbedded with hematitic jasper (Dorr, 1945). Reserves are 45 million tons of ore having a grade of 46% Mn, and the deposit contains about 15 million tons of submarginal resources of comparable grade (National Materials Advisory Board, 1981).

The deposit is in the Band' Alta formation, probably of Cambrian or Ordovician age (Dorr, 1973), which consists of banded hematite and jasper and subordinate beds of manganese ox-

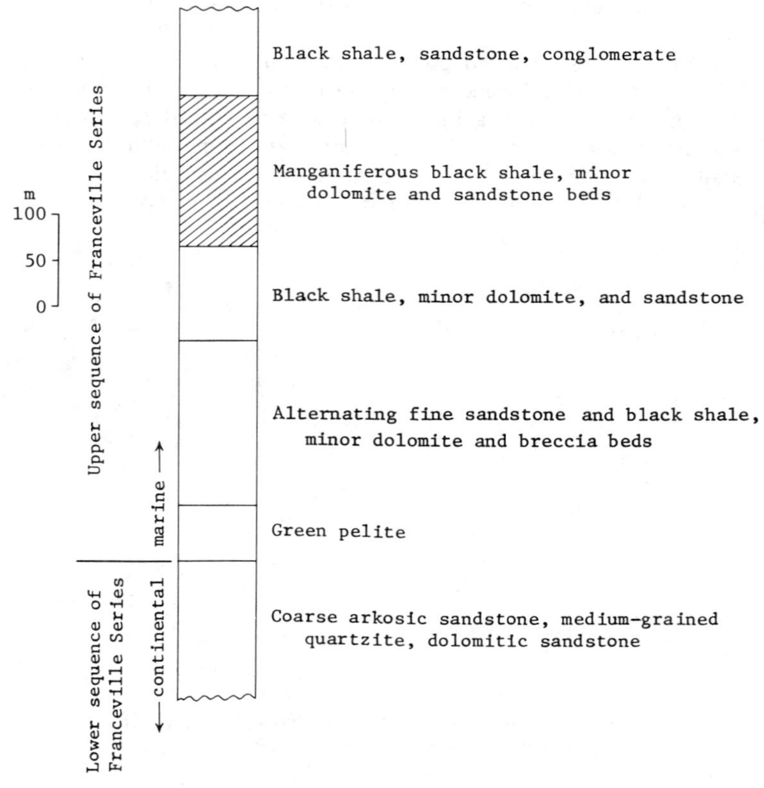

FIG. 4. Stratigraphic column of Precambrian rocks in the vicinity of the Moanda deposit, Gabon, from data of Leclerc and Weber (1980).

ides, siltstone, and sandstone. The manganese beds are near the base of the formation (Fig. 5). The Band' Alta overlies the Corrego dos Pedras formation, which is about 100 m thick and grades from ferruginous arkosic rocks of continental origin near the base to hematitic jasper of marine origin (Dorr, 1945) near the top. The Corrego dos Pedras formation records the transition from continental to marine sedimentation, and the manganiferous beds of the Band' Alta formation are a few tens of meters above the earliest marine rocks.

Imini, Morocco

The manganese deposits of Imini near the Atlas Mountains of Morocco are of importance more because they contain chemical-grade ore than because of their size (3 million tons reserve at 46% manganese, National Materials Advisory Board, 1981).

The ore is in three manganese oxide beds, each about a meter thick, that occupy an east-trending zone 25 km long and as wide as 1 km (Varentsov, 1964, pp. 27–31; Rakhmanov and Tchaikovsky, 1980, pp. 131–139). One bed is near the base, one is within, and the third is

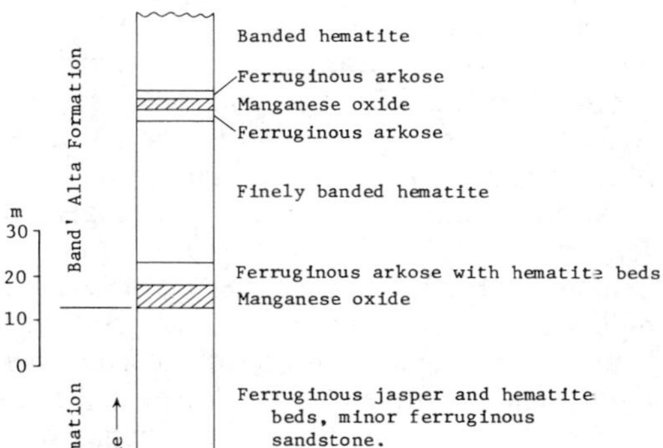

FIG. 5. Stratigraphic column of rocks in the vicinity of the Morro do Urucum deposit, Brazil, from data of Dorr (1945).

near the top of a 20-m-thick dolomite sequence of Cenomanian-Turonian (Late Cretaceous) age (Fig. 6). Each bed is associated with solution-collapse breccia, some manganese-cemented, and laminated insoluble-residue deposits. Hollandite and pyrolusite are common ore minerals; lead-bearing coronadite ore becomes increasingly abundant westward in the belt.

The dolomite is sugary-textured replacement dolomite. Some beds are extremely fossiliferous; locally, manganese oxides replace thick-shelled *Exogyra sp*. Traces of burrowing organisms are so abundant that bedding is difficult to see.

The dolomite is both underlain and overlain by fluvial clastic sediments. The sediments below have locally been bored by marine organisms and filled in by gritty dolomite from the basal dolomite bed. At the top of the dolomite are zoned chert-calcite-manganese oxide nodules. These features record an overstep transgression, marine deposition, then freshwater alteration of an emerged surface. At the time of deposition, the carbonate platform at Imini was adjacent to a black-shale basin to the southwest (Einsele and Weidmann, 1982).

Our work at Imini suggests two controls on oxide mineralization. The first was primary depositional distribution of some manganese-bearing mineral, probably a carbonate, in a strikingly transgressive limestone sequence during an ocean anoxic event. This manganiferous material formed three beds having great continuity parallel to depositional strike. The second control was the replacement of this sequence by dolomite (for limestone) and manganese oxide ore and solution collapse breccia (for the precursor manganese phase). Replacement probably occurred during emergence by mixing ground waters.

Comparison

All the deposits are related to high sea-level stands, and all but Molango seem related to oxidizing conditions. Three (Groote Eylandt, Molango, Imini) formed during periods of ocean anoxia, and local anoxic conditions are likely for Nikopol also. Diagenesis was important in producing ore at Imini and perhaps all the deposits. Depositional settings include embayed shorelines of the open sea (Groote Eylandt), arms of the sea (Imini), and bays in restricted basins (Nikopol). Paleoclimates include humid and arid, tropical, and temperate (but not arctic). Primary manganese mineralogy varies from te-

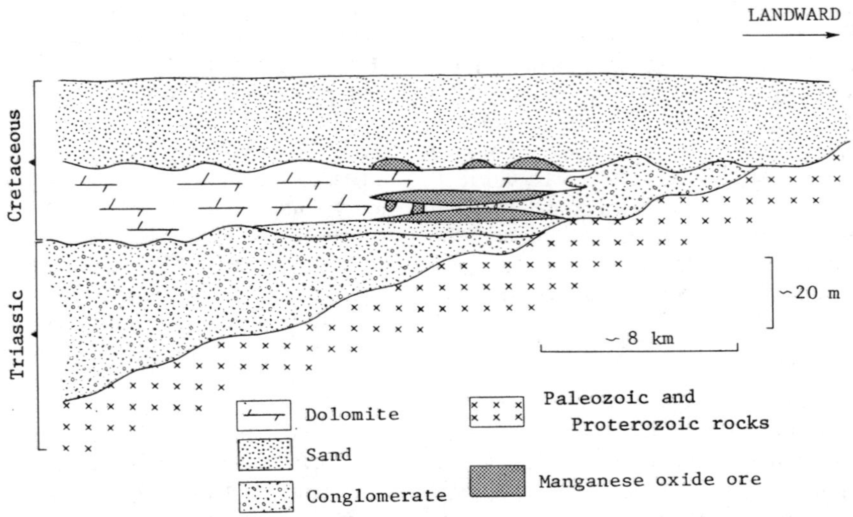

FIG. 6. Schematic cross section of the Imini deposit, Morocco, modified from Varentsov (1964, Fig. 5 and pp. 27–30).

travalent oxides (Groote Eylandt) to bivalent carbonates (Nikopol, Molango). Host rocks are clays (Nikopol, Groote Eylandt) and carbonates (Imini).

RELATIONS AMONG HIGH SEA STANDS, OCEAN ANOXIA, AND MANGANESE DEPOSITION

The unmistakable affinity of manganese deposition for marine transgressive rocks strongly indicates that transgressions somehow produce highly favorable environments for manganese sedimentation both very near the shoreline (at the base of the section) and somewhat farther from shore (in deeper water facies now preserved as much as a few hundred meters above the basal transgressive beds).

Some aspects of transgression may inherently favor manganese deposition. These aspects include drowned irregular topography, which permits heterogeneous geochemistry at the depositional interface, silled anoxic basins, and efficient trapping of coarse sediment at the heads of estuaries. (Modern manganese deposits of silled Black and Baltic Seas owe their existence to Holocene transgression.) Given a heterogeneous bottom, specialized conditions may exist in which manganese can precipitate without other chemical constituents like iron and without clastic dilution. These aspects are consistent with the Nikopol and Groote Eylandt deposits which show deposition on an irregular bottom near a deeply embayed shoreline, with abrupt facies changes.

However, we believe that transgression per se is not as critical as the association of high sea level with anoxic ocean water. This association is empirical (cf. Schlanger and Jenkyns, 1976; Arthur and Schlanger, 1976) but there is also a likely genetic link between high sea level and widespread anoxic conditions (Schlanger and Jenkyns, 1976; Ryan and Cita, 1977; Arthur and Schlanger, 1979; Jenkyns, 1980). For instance, high sea level (a higher ratio of sea surface to land surface) changes the earth's albedo, making it more retentive of solar energy and hence warmer (Barron et al., 1980). The resulting warm climate and probable lower pole-to-equator thermal gradients lessen deep oceanic circulation and diminish the rate of oxygen replenishment to deep oceans. The deep ocean oxygen supply comes mostly from the sinking and equatorward spreading of cold, well-oxygenated polar waters. During warm climatic episodes, not only is such deep circulation diminished, but the oxygen supply is further restricted because oxygen solubility in water decreases as temperature increases.

High sea level also creates extensive shelf seas which are usually very biologically productive and transgressions may also consume areas of high terrestrial productivity. Hence, there are multiple mechanisms that logically can be inferred to (1) diminish oxygen supply to oceans and (2) increase oxygen demand to degrade organic matter, which occur simultaneously during coincident high sea-level stands and warm climate. The result is that conditions favorable to stratified oceans, anoxic below a shallow surface zone that can obtain oxygen from the atmosphere, are better developed during periods of high sea level and warm climate.

This association of high sea levels and anoxic events may combine to create manganese accumulations. When the oxidation-reduction interface in seawater transgresses across a shelf, reduced bottom waters dissolve manganese from

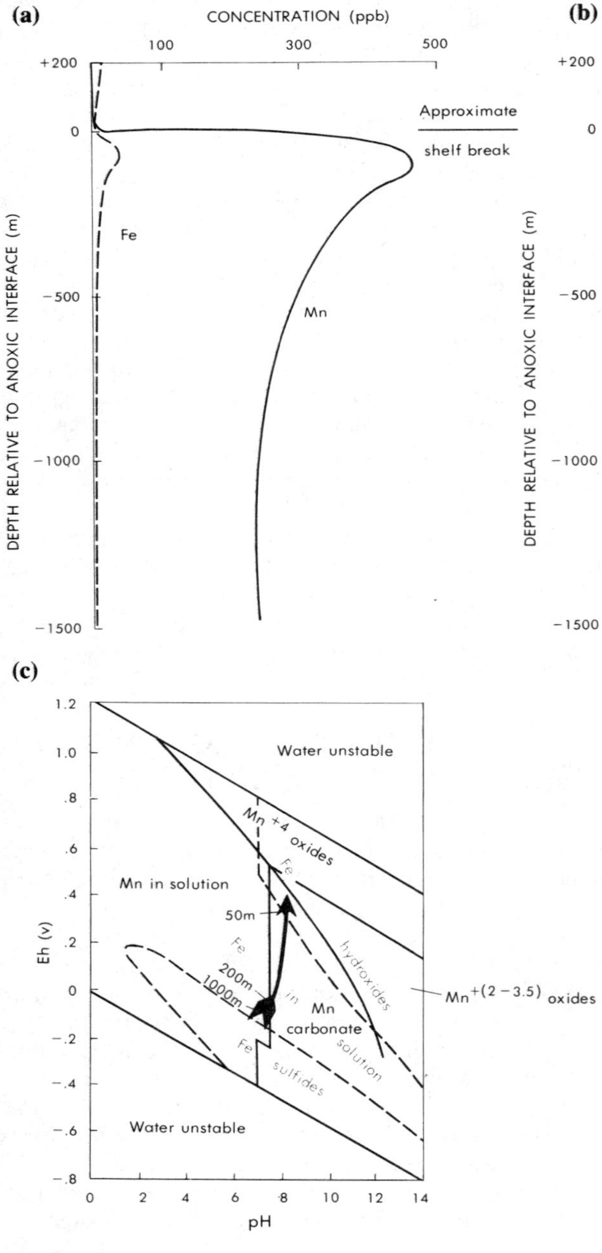

FIG. 7. Geochemistry of Black Sea water in profile and Eh-pH fields. [a] Distribution of dissolved manganese and iron in the Black Sea (from Brewer and Spencer, 1974). [b] Distribution of Eh and pH values in the Black Sea (from data in Richards, 1965; Grasshoff, 1975). [c] Relative stability of manganese and iron compounds at concentrations prevailing in the Black Sea, from data in Hem (1972) and Krauskopf (1957). Arrow shows locus of Black Sea water compositions through depths as marked.

sediments until the solubility limit is reached. This manganese may reprecipitate where the interface intersects the shelf, but because the interface is transgressing, precipitated manganese remains stable only at or beyond the line of maximum transgression of the oxidation-reduction interface. The effect may be much like sweeping previously deposited scraps of deposits together. The sluggishness of the manganese-oxidation reaction may permit precipitation to be well into the oxidized zone rather than entirely at the interface. Hence, the richest deposits may be near the feather edge of transgressive wedges, as they apparently are in Nikopol and related deposits.

Note that in this model the source of the manganese has little bearing on the formation of the deposits or their locations. Manganese from fluvial sources, submarine weathering, and submarine hydrothermal vents all could be incorporated in a large manganese reservoir in the anoxic part of the water column. This is the "ore-forming fluid." Precipitation occurs wherever this fluid is oxidized. Lateral transport of manganese along anoxic layers may be considerable; where this transport proceeds more slowly than local precipitation, then local sources of manganese may influence deposit location.

AN ANALOGY WITH THE BLACK SEA

Our concept of the role played by anoxic waters in the formation of sedimentary manganese deposits may best be illustrated by a look at the Black Sea; the present condition of the upper part of its water column is probably similar to that of the oceans during worldwide anoxic events (Degens and Stoffers, 1976).

The Black Sea is stratified. At a depth of about 200 m, a boundary is present between anoxic deep waters that have Eh -0.15, pH 7.6 and dissolved H_2S, and oxidized surface waters that have Eh > 0, pH 8.0–8.3, and dissolved oxygen (Fig. 7). Black sulfidic mud is currently being deposited on much of the bottom below the boundary.

The dissolved-manganese content of anoxic waters is greater by a factor of about 500 than that of oxidized surface waters (Brewer and Spencer, 1974) (Fig. 7). A manganese-maximum zone just below the base of the oxygenated zone is due to diffusion of dissolved manganese upward into the oxidized zone, precipitation there, and dissolution of the precipitate as it sinks back through the interface.

A high ratio of dissolved manganese to dissolved iron (a ratio about 100 times greater than that of normal seawater) is present in anoxic bottom water (Fig. 7). This is probably due to removal of iron into sulfides on the bottom.

The interface between oxygenated and anoxic water intersects the bottom well up the flanks of the deeper part of the basin (Fig. 8). A shallow-water shelf area is present in the northwest part of the Black Sea. The formation of manganese oxide nodules in one area may be controlled by the circulation pattern of anoxic waters; surface circulation (Fig. 8), which moves water at depths down to about 300 m (Grasshoff, 1975), is apparently causing upwelling of anoxic waters there. Mixing with oxygenated shallow waters could cause precipitation of the dissolved manganese on oxidized substrates and may account for the nodules. The geochemical trend, shown in Fig. 7C, could cause manganese to precipitate as either oxide or carbonate. Paleo-Black Sea deposits at Nikopol contain both. A similar analogy between modern processes and Oligocene deposits of the Black Sea area was presented by Sapozhnikov (1971).

Another modern depositional environment of manganese is the Baltic Sea, where individual silled anoxic basins are rimmed with manganese oxide nodules on oxidized substrates (Manheim, 1965). Some stratified lakes show the same phenomena.

A GENERALIZED EXPLORATION MODEL

On the basis of the foregoing empirical and theoretical considerations, a generalized model can be constructed to aid in predicting favorable sites for manganese concentrations.

Foremost is the occurrence of manganese ore in shallow-water transgressive marine sequences. Such sequences deposited during known ocean anoxic events are probably the most favorable in view of the probable connection between anoxic water and manganese accumulation. In sequences of any age, oxidized shallow marine rocks contemporaneous with nearby black shales are favorable if it can be shown that color of the shale is not due to post-depositional reduction below the sediment-water interface.

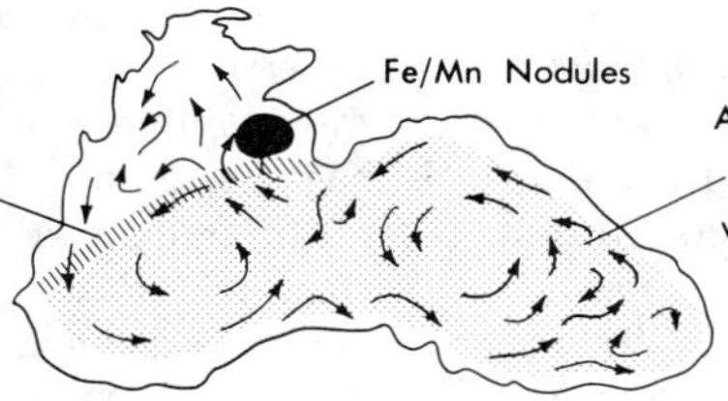

FIG. 8. Relation of surface (0–300 m) currents, anoxic bottom waters (>about 200 m), shelf edge (~200 m), and manganese oxide nodule deposits in plan view of the Black Sea.

Clastic dilution should be minimal. Topographic highs on shelf seas marginal to anoxic oceans may be ideal sites for the formation of ore because at some point during the transgressive history they will stand as isolated promontories projecting through the anoxic/oxic interface. They will thus be sites at which manganese can accumulate without being diluted by large amounts of clastic sediments. Fig. 9 diagramatically summarizes these relationships.

Sequences chosen as exploration targets should then have three attributes:

1) Record a high sea stand.
2) Have evidence of contemporaneous anoxic water nearby.
3) Consist of condensed sequences of fine clastics or chemical sediments.

There are probably scores of these sequences, however, each potentially spread over extensive areas. Until the model can be further refined, the next step may be physical exploration. Fortunately, important sedimentary manganese deposits are large and cover hundreds or even thousands of square kilometers, so they present large targets for exploration. Preliminary exploration could consist of a search for anomalously manganiferous beds at sites spaced along outcrop belts of transgressive rocks, so large regions could be explored relatively efficiently. Chemical analyses are essential because some types of manganiferous sediments are difficult to recognize in the field.

POTENTIAL FOR SHALLOW-MARINE MANGANESE DEPOSITS IN NORTH AMERICA

We suggest two broad regions where shallow-marine transgressive manganese deposits might occur in North America (Fig. 10). These are (1) the outcrop belt of Cretaceous marine rocks extending through the midcontinent region of the United States and Canada and (2) the Cretaceous and Tertiary rocks of parts of the Atlantic and Gulf Coastal Plains of the United States and Mexico. Other areas also exist but are not discussed.

Cretaceous Rocks of the Midcontinent

Cretaceous marine rocks crop out in a broad arc from Kansas to British Columbia and record a major marine transgression over the stable continental interior. They lie unconformably on

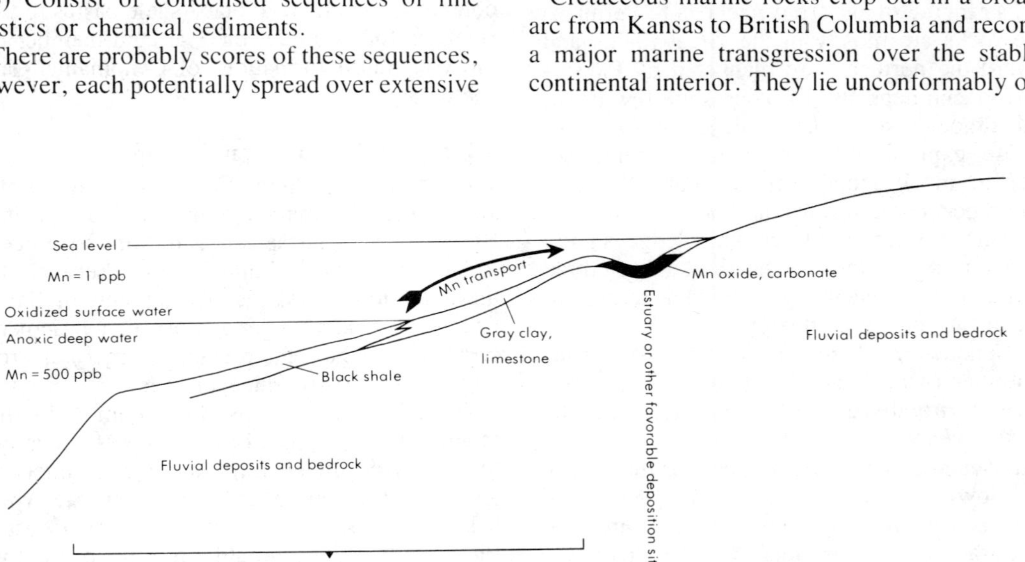

FIG. 9. Schematic cross section of our hypothesized ideal depositional setting for sedimentary manganese deposits.

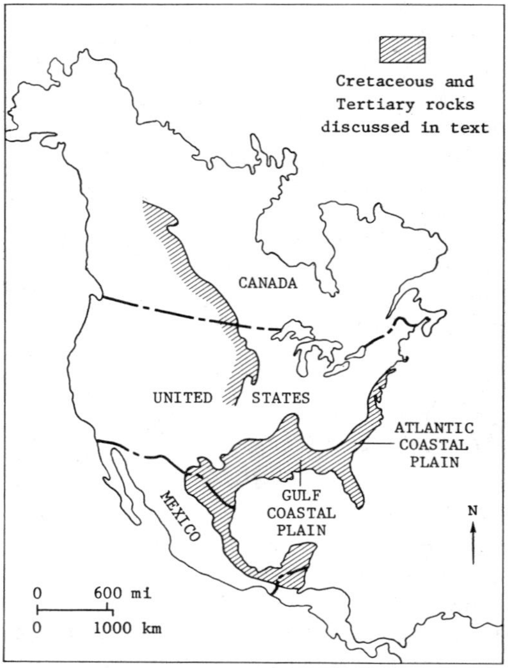

FIG. 10. Map of North America showing areas of Cretaceous and Tertiary rocks believed to be potentially manganese bearing. Highly generalized in Mexico where Tertiary tectonism has produced complicated map pattern.

rocks ranging in age from Archean to Permian. The rocks are mostly in the Great Plains and toward the northeast are blanketed by Pleistocene glacial deposits. Exposures are few there, and although some drill holes (not for manganese exploration) have penetrated the sequence, much remains to be learned of the lithologies, compositions, and facies changes of units in the sequence. Where knowledge is good enough to permit their delineation, at least three "cyclothems," each representing a transgressive event, have been recognized in the Albian, the Cenomanian-Turonian, and the Coniacian-Santonian (Kauffman, 1977); these events took place at virtually the same time as Cretaceous anoxic events.

East-central South Dakota, east of the extensive low-grade Chamberlain manganese deposit, is a particularly favorable target area for high-grade manganese deposits. The Chamberlain deposit is in the Oacoma member of the Pierre shale of Campanian (Late Cretaceous) age and is exposed for more than 250 km along the Missouri River. The underlying Sharon Springs member is a black shale formed in a Santonian anoxic event. Manganiferous nodules averaging about 15% manganese and of early diagenetic origin occur throughout a stratigraphic zone about 10 m thick (Pesonen and others, 1949). The enclosing shale averages about 0.7% manganese. The nodules are unevenly distributed, and although some beds are a nearly continuous pavement of nodules, the overall grade is low, well below economic grades. The stratigraphic section in the area of the deposits consists of the Dakota sandstone (Lower Cretaceous) unconformably overlying Precambrian crystalline rocks and overlain by shale and limestone of the Graneros, Greenhorn, Carlile, and Niobrara formations, all marine. The base of the Pierre shale is here about 300 m above the Precambrian basement (Fig. 11). The Pierre shale lies disconformably on the Niobrara and both overlap the lower units eastward. Although the Chamberlain deposit itself is of little economic interest, we suggest that it may be the low-grade distal part of higher grade deposits that may exist to the east where Pierre shale was deposited in shallow marine environments adjacent to a large peninsula or island of Precambrian Sioux quartzite. In addition, deposits of lower cyclothems are favorable in the same area where near-shore oxidized environments are adjacent to and contemporaneous with black shale basins to the west, as for example, the Hartland member of the Greenhorn formation, of Cenomanian-Turonian age.

Atlantic and Gulf Coastal Plains

The emerged Atlantic Coastal Plain is largely an unexplored province for manganese deposits. However, manganese concentrations have been found elsewhere in similar geologic settings. In the Atlantic Coastal Plain, sediments of Early Cretaceous and younger age overlie a basement complex in which the rocks are largely of Proterozoic Z to early Paleozoic age.

Transgressive deposits are common in the stratigraphic sequence of cover rocks. Locally, these deposits are near the base of the sequence, and, locally, marine units are in contact with basement rocks. Thus, deposits of this Coastal Plain are broadly favorable in terms of the hypotheses of this article. Transgressive near-shore marine rocks of Campanian-Santonian Age (a less documented anoxic event) crop out in South

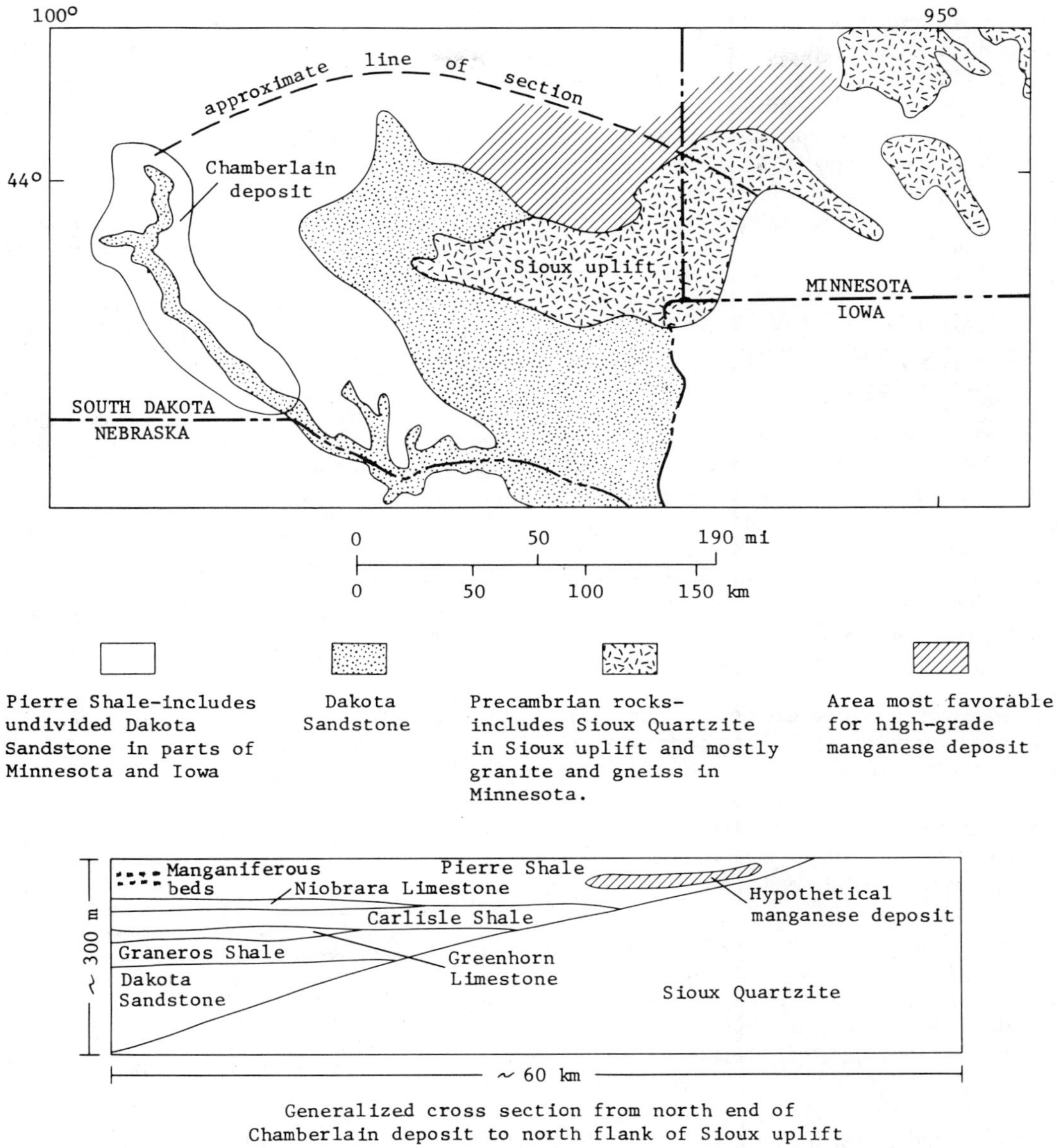

FIG. 11. Map and generalized section of eastern South Dakota, showing location and stratigraphic setting of Chamberlain deposit and area judged favorable for manganese deposits. Section is generalized from subsurface data of Barkley (1953) and Erickson (1954).

Carolina, where they form the lower Peedee-upper Black Creek formational boundary.

In the Gulf Coastal Plain, deposits representing transgressive cycles correlative with several Cretaceous anoxic events can be recognized (Kauffman and others, 1977). Here, the problem may be that facies represented in outcrop by the peak of transgression are chalky deeper water deposits.

COULD MAJOR MANGANESE DEPOSITS REMAIN UNDETECTED IN NORTH AMERICA?

Important considerations in assessing the potential for undiscovered manganese deposits in North America are whether extensive bedded deposits could have escaped detection and whether any such deposits would already have been found either by systematic exploration or by accident. We would like to point out some features that lead us to believe that even very large exposed deposits could have gone undetected.

1) Many of the areas that we consider favorable are regions of very low topography and poor bedrock exposure; hence, discoveries, even accidental discoveries, are not likely. Conversely, some of the known sedimentary deposits around the world are in areas of very good exposure where the ore beds are easily visible in outcrop.

2) Many manganiferous rocks, particularly carbonate rocks and manganiferous shales, are not readily distinguished in the field and may not attract the attention of observers, even geologists. For example, the Molango deposit is a 25-m-thick manganese carbonate bed very well exposed for a strike length of 50 km, yet it was not recognized as a manganese ore until 1960. Even then, it was recognized only after small supergene enrichments on it led exploration geologists to the area specifically to search for manganese. Consider also Varentsov and Rakhmanov's (1980, pp. 331–332) description of the thickest bed of ore in a section at Nikopol: "Abundant brownish nodules and soft earthy masses of manganite emplaced in the grayish white carbonate-clay matrix." Such a rock outside a known manganese field might not attract attention of either exploration geologists or casual observers.

3) Buried manganese ores, even if covered by only a thin veneer of other material, are difficult to detect. They give no strong geophysical response, and geochemical signatures are generally not diagnostic because of the common occurrence of manganese in a great many environments devoid of ore. Although a deposit might produce a geochemical anomaly, geochemical exploration is very inefficient because of widespread anomalies not related to ore deposits.

4) To our knowledge, the areas that we have identified as having potential for sedimentary manganese deposits have not been well explored. The link between manganese deposits and transgressive marine rocks formed during anoxic events does not seem to have been widely recognized, at least in exploration in North America, and we suspect that many favorable areas have not been examined for manganese.

REFERENCES

Aleksiev, B., and Bogdanova, K., 1974, "The Obrochiste Manganese Deposit," *Twelve Ore Deposits of Bulgaria,* International Association on the Genesis of Ore Deposits Symposium 4, pp. 149–159.

Arthur, M.A., and Schlanger, S.O., 1979, "Cretaceous 'Oceanic Anoxic Events' as Causal Factors in Development of Reef-Reservoired Giant Oil Fields," *Bulletin,* American Association of Petroleum Geologists, Vol. 63, pp. 870–885.

Barkley, R.C., 1953, "Artesian Conditions in the Area Surrounding the Sioux Quartzite Ridge," Report of Investigations 72, South Dakota Geological Survey, 68 pp.

Barron, E.J., Sloan, J.L., II, and Harrison, C.G.A., 1980, "Potential Significance of Land-Sea Distribution and Surface Albedo Variations as a Climatic Forcing Factor; 180 m.y. to the Present," *Palaeogeography, Palaeoclimatology, and Palaeoecology,* Vol. 30, pp. 17–40.

Brewer, P.G., and Spencer, D.W., 1974, "Distribution of Some Trace Elements in Black Sea and Their Flux between Dissolved and Particulate Phases," *The Black Sea—Geology, Chemistry, and Biology:* E.T. Degens and R.A. Ross, eds., American Association of Petroleum Geologists, Memoir 20, pp. 137–143.

Carrillo Bravo, J., 1965, "Estudio Geologico de Una Parte del Anticlinorio de Huayacocotla, *Boletin,* Asociacion Mexicana de Geologos Petroleros, Vol. 17, pp. 73–96.

Degens, E.T., and Stoffers, P., 1976, "Stratified Waters as a Key to the Past," *Nature,* Vol. 263, pp. 22–27.

Dorr, J.V.N., II, 1945, "Manganese and Iron Deposits of Morro do Urucum, Mato Grosso, Brazil, Bulletin 946-A, US Geological Survey, 47 pp.

Dorr, J.V.N., 1973, "Iron-Formation and Associated Manganese in Brazil," *Proceedings,* Genesis of Precambrian Iron and Manganese Deposits, UNESCO, Paris, pp. 105–113.

Einsele, G., and Weidmann, J., 1982, "Turonian Black Shales in the Moroccan Coastal Basins, First Upwelling in the Atlantic Ocean?," *Geology of the Northwest African Continental Margin,* Springer Verlag, Berlin, pp. 396–414.

Erickson, H.D., 1954, "Artesian Conditions in East Central South Dakota," Report of Investigations 74, South Dakota Geological Survey, 116 pp.

Fischer, A.G., and Arthur, M.A., 1977, "Secular Variations in the Pelagic Realm, *Deep-Water Carbonate Environments,* H.E. Cook and P. Enos, eds., Society of Economic Paleontologists and Mineralogists, Spec. Pub. 25, pp. 19–50.

Grasshoff, K., 1975, "The Hydrochemistry of Landlocked Basins and Fjords," *Chemical Oceanography,* 2nd ed., Vol. 2, J.P. Riley and G. Skirrow, eds., Academic Press, London, pp. 455–584.

Gryaznov, V.I., 1970, "Genesis of Manganese Ores of the Nikopol Basin and Prediction of Marine Sedimentary Manganese Deposits," D.G. Sapozhnikov,* ed., pp. 146–159.

Gryaznov, V.I., and Danilov, I.S., 1980, "Oxidized Manganese Ores of the Nikopol Manganese Deposit, Ukrainian SSR," I.M. Varentsov and G.Y. Grassely,† eds., pp. 403–416.

Hallam, A., and Bradshaw, M.J., 1979, "Bituminous Shales and Oolitic Ironstones as Indicators of Transgressions and Regressions," *Journal,* Geological Society of London, Vol. 136, pp. 157–164.

Hem, J.D., 1972, "Chemical Factors that Influence the Availability of Iron and Manganese in Aqueous Systems," *Bulletin,* Geological Society of America, Vol. 83, pp. 443–450.

Imlay, R.W., et al., 1948, "Stratigraphic Relations of Certain Jurassic Formations in Eastern Mexico," *Bulletin,* American Association of Petroleum Geologists, Vol. 32, pp. 1750–1761.

Jenkyns, H.C., 1980, "Cretaceous Anoxic Events: From Continents to Oceans," *Journal,* Geological Society of London, Vol. 137, pp. 171–188.

Kauffman, E.G., 1977, "Geological and Biological Overview: Western Interior Cretaceous Basin," *Cretaceous Facies, Faunas, and Paleoenvironments Across the Western Interior Basin,* E.G. Kauffman, ed., Mountain Geologist, Vol. 14, pp. 75–100.

Kauffman, E.G., Hattin, D.E., and Powell, J.D., 1977, "Stratigraphic, Paleontologic, and Paleoenvironmental Analysis of the Upper Cretaceous Rocks of Cimarron County, Northwestern Oklahoma," Memoir 149, Geological Society of America, 150 pp.

Krauskopf, K.B., 1957, "Separation of Manganese from Iron in Sedimentary Processes," *Geochimica et Cosmochimica Acta,* Vol. 12, pp. 61–84.

Leclerc, J., and Weber, F., 1980, "Geology and Genesis of the Moanda Manganese Deposits, Republic of Gabon," I.M. Varentsov and G.Y. Grassely,† eds., pp. 89–109.

Manheim, F.T., 1965, "Manganese-Iron Accumulations in the Shallow-Marine Environment," *Symposium on Marine Geochemistry,* D.R. Schink and J.T. Corliss, eds., University of Rhode Island Narragansett Marine Laboratory, Occasional Publication 3, pp. 217–275.

National Materials Advisory Board, 1981, "Manganese Reserves and Resources of the World and Their Industrial Implications," Board Publication NMAB 374, United States National Materials Advisory, 334 pp.

Ostwald, J., 1980, "Aspects of the Mineralogy, Petrology, and Genesis of the Groote Eylandt Manganese Ores," Varentsov, I.M. and G.Y. Grassely,† eds., pp. 149–182.

Pesonen, P.E., Tullis, E.L., and Zinner, P., 1949, "Missouri Valley Manganese Deposits, South Dakota. Part 1, General Investigations, Stratigraphic Studies, and Tonnage and Grade Estimates," Report of Investigations 4375, US Bureau of Mines, 90 pp.

Rakhmanov, V.P., and Tchaikovsky, V.K., 1980, "Manganese Ores in Activated Platforms," in *International Association on the Genesis of Ore Deposits Symposium 5,* E. Schweizerbartsch Verlagsbuchhandlung, Stuttgart, Germany, pp. 125–141.

Richards, F.A., 1965, "Anoxic Basins and Fjords," *Chemical Oceanography,* 1st ed., Vol. 1, J.P. Riley and G. Skirrow, eds., Academic Press, London, pp. 611–644.

Ryan, W.B.F., and Cita, M.B., 1977, "Ignorance Concerning Episodes of Ocean-Wide Stagnation," *Marine Geology,* Vol. 23, pp. 197–215.

Sapozhnikov, D.G., 1971, "On Manganese Content in Seawater (in Connection with the Problem of Manganese Deposit Formation)," Society of Mining Geology of Japan, Spec. Issue 3, pp. 484–488.

Schlanger, S.O., and Jenkyns, H.C., 1976, "Cretaceous Oceanic Anoxic Events: Causes and Consequences," *Geologie en Mijnbouw,* Vol. 55, pp. 179–184.

Shevchenko, E.S., 1970, "The Ore Content of the Eastern Part of the Nikopol Manganese Ore Deposit," D.G. Sapozhnikov,* ed., pp. 160–169.

Slee, K.J., 1980, "Geology and Origin of the Groote Eylandt Manganese Oxide Deposits, Australia," I.M. Varentsov and G.Y. Grassely,† eds., pp. 125–148.

Smith, W.C., and Gebert, H.W., 1969, "Manganese at Groote Eylandt, Australia," *Proceedings,* 9th Commonwealth Mining and Metallurgical Congress, pp. 585–609.

Strakhov, N.M., et al., 1970, "The Mechanism of Manganese Ore-Formation Processes (Oligocene Ores in the Southern Part of the USSR)," D.G. Sapozhnikov,* ed., pp. 34–57.

Summerhayes, C.P., 1981, "Organic Facies of Middle Cretaceous Black Shales in Deep North Atlantic," *Bulletin,* American Association of Petroleum Geologists, Vol. 65, pp. 2364–2380.

Tavera, I.E., and Alexandri, R., 1972, "Molango Manganese Deposits, Hidalgo, Mexico [Abstract]," *Acta Mineralogica Petrographica,* Vol. 20, pp. 387–388.

Vail, P.R., et al., 1977, "Seismic Stratigraphy and Global Changes of Sea Level," *Seismic Stratigraphy-Applications to Hydrocarbon Exploration,* C. Payton, ed., American Association of Petroleum Geologists, Memoir 26, pp. 49–212.

Varentsov, I.M., 1964, *Sedimentary Manganese Ores,* Elsevier Publishing Co., Amsterdam, 119 pp.

Varentsov, I.M., and Rakhmanov, V.P., 1980, "Manganese Deposits of the USSR (Review)," I.M. Varentsov and G.H. Grassely,† eds., pp. 319–392.

Weissert, H., 1981, "The Environment of Deposition of Black Shales in the Early Cretaceous: An Ongoing Controversy, Special Publication 32, Society of Economic Paleontologists and Mineralogists, pp. 547–560.

*Sapozhnikov, D.G., ed., refers to *Manganese Deposits of the Soviet Union* (English translation), Israel Program for Scientific Translations, Jerusalem, 522 pp., republished from Academy of Sciences of the USSR.

†Varentsov, I.M., and Grassely, G.Y., eds., refers to *Geology and Geochemistry of Manganese,* Vol. 2, *Manganese Deposits on Continents,* E. Schweizerbartsche Verlagsbuchhandlung, Stuttgart, Germany, 513 pp. Proceedings of the 2nd International Symposium on Geology and Geochemistry of Manganese, Sydney, 1976.

15

Conceptual Models of Epithermal Precious Metal Deposits

Byron R. Berger

US Geological Survey, Denver, CO

Paul I. Eimon

Pioneer Nuclear Corp., Amarillo, TX

INTRODUCTION

Epithermal silver-gold deposits form at low to moderate temperatures in near-surface environments. These deposits are found in all rock types, but historically, the most important deposits occur as veins in volcanic rocks (e.g., Creede, Colorado; Pachuca, Mexico) and as replacements in sedimentary sequences (e.g., Mercur, Utah; Carlin, Nevada). Increases in the open-market prices for silver and gold in the last few years have resulted in renewed interest in precious metals and the economic viability of large-tonnage low-grade deposits. These bulk-mineable deposits form as replacements in permeable horizons or as stockworks in the upper parts of otherwise bonanza-type hydrothermal systems and newly recognized systems that erupted onto the surface as hot springs, fumaroles, or geysers. Large tonnage low-grade deposits were an unconventional resource prior to price increases; however, now they constitute one of the main sources of precious metals in the United States and the importance is growing worldwide.

This paper presents conceptual geologic-geochemical models of epithermal precious-metal deposits derived from studies of productive deposits throughout the world. Much of the data presented is not new. However, the interpretation of the data in terms of current concepts of hydrothermal processes is new. This paper deals with idealized and general ore-forming mechanisms, and no attempt is made either to explain the origin of the different chemistries of ore-forming fluids or to examine all of the variations in specific deposits due to local structural controls and host-rock compositions.

GENERAL CHARACTERISTICS OF EPITHERMAL PRECIOUS METAL DEPOSITS

Lindgren (1933) recognized the difficulties in devising a rigorous classification for precious-metal epithermal deposits that clearly distinguished the various types of occurrences. He divided epithermal deposits into categories based on host rock, mineralogy, and elemental associations (Table 1).

Ferguson (1929) divided epithermal deposits into two major categories based on the weight ratio of gold to silver (gold-silver: Au/Ag > 1;

TABLE 1. Characteristics that Classify a Hydrothermal System as Being Epithermal (after Lindgren, 1933)

Depth of formation	Surface to 1000m
Temperature of formation	50° to 300°C
Form of deposits	Thin to large veins, stockworks, disseminations, replacements
Ore textures	Open-space filling, crustification, colloform banding, comb structure, brecciation
Ore elements	Au, Ag, (As, Sb), Hg, [Te, Tl, U], (Pb, Zn, Cu)*
Alteration	Silicification, argillization, sericite, adularia, propylitization
Common features	Fine-grained chalcedonic quartz, quartz pseudomorphs after calcite, brecciation

*[ ] brackets indicate elements seldom present in more than subeconomic concentrations; ( ) parentheses indicate elements often present in economic concentrations but usually less valuable than associated precious metals.

silver-gold: Au/Ag < 1; Table 2). Table 3 gives examples of deposits according to Ferguson's classification.

The geochemistry of epithermal deposits is not as well documented as is the geology of these deposits. Most geochemical research has utilized light stable isotopes which have shown that the hydrothermal fluids were predominantly of meteoric origin (e.g., Bethke and Rye, 1979; O'Neil and Silberman, 1974; Taylor, 1973), either from a single reservoir as in the case of Bodie, California (O'Neil et al., 1973) or from multiple reservoirs as in the case of Creede, Colorado (Bethke and Rye, 1979).

Fluid inclusion studies (e.g., Nash, 1972; Vikre, 1981) indicate that most of the epithermal deposits were formed from dilute alkali-chloride solutions (approximately 0.5-5 wt % NaCl equivalent) over a temperature range of 150°–300°C. Detailed knowledge of the trace elements and zoning patterns associated with different kinds of deposits is limited.

Extensive studies of active thermal spring areas by D.E. White, R. Fournier, and others of the US Geological Survey have led to a better understanding of the fluid chemistry, wallrock alteration, and structural complexities of hydrothermal systems. White (1974) believes that active geothermal areas such as Steamboat Springs, Nevada, and Broadlands, New Zealand, are analogs of epithermal precious metal deposits. Some of the salient features of these hot-springs systems are depicted schematically in Figs. 1 and 2. The alteration at the top of the system consists characteristically of silicification underlain by a zone of mixed-layer illite-montmorillonite (Fig. 1). This clay zone changes with depth into a sericite zone. At Steamboat Springs, the K-feldspar occurs from a depth of 45 to 135 m (Sigvaldason and White, 1962). An important attribute of both the Steamboat and Broadlands systems is that primary quartz

TABLE 2. Common Attributes of Gold-Silver and Silver-Gold Deposits (adapted from Nolan, 1933)

Gold-silver deposits	Silver-gold deposits
Au:Ag : : 1:1 to >10:1	Au:Ag : : 1:1 to >1:100
	Electrum
Native gold	Silver sulfides, sulfosalts, and selenides
Gold tellurides	Base-metal sulfides
Simple sulfides of arsenic, antimony, and mercury	Complex arsenic- and antimony-bearing sulfides
Fluorite and/or alunite	

TABLE 3. Examples of Typical Epithermal Precious-Metal Deposits

A. Gold-silver deposits
1. Veins and replacements in intermediate to acidic volcanic rocks

Goldfield, Nevada	Pearce, Arizona
Round Mountain, Nevada	Cripple Creek, Colorado

2. Veins and replacements in clastic and carbonate sedimentary rocks

Carlin, Nevada	Cinola, British Columbia
Manhattan, Nevada	Baley, Transbaikal
Silver Stack, Quebec	

B. Silver-gold deposits
1. Veins and replacements in intermediate to acidic volcanic rocks

Virginia City, Nevada	Delamar, Idaho
Tonopah, Nevada	Guanajuato, Mexico
Creede, Colorado	Chanarcillo, Chile

2. Veins and replacements in clastic and carbonate sedimentary rocks

Taylor, Nevada	Hardshell, Arizona
Calico, California	Real de Angeles, Mexico

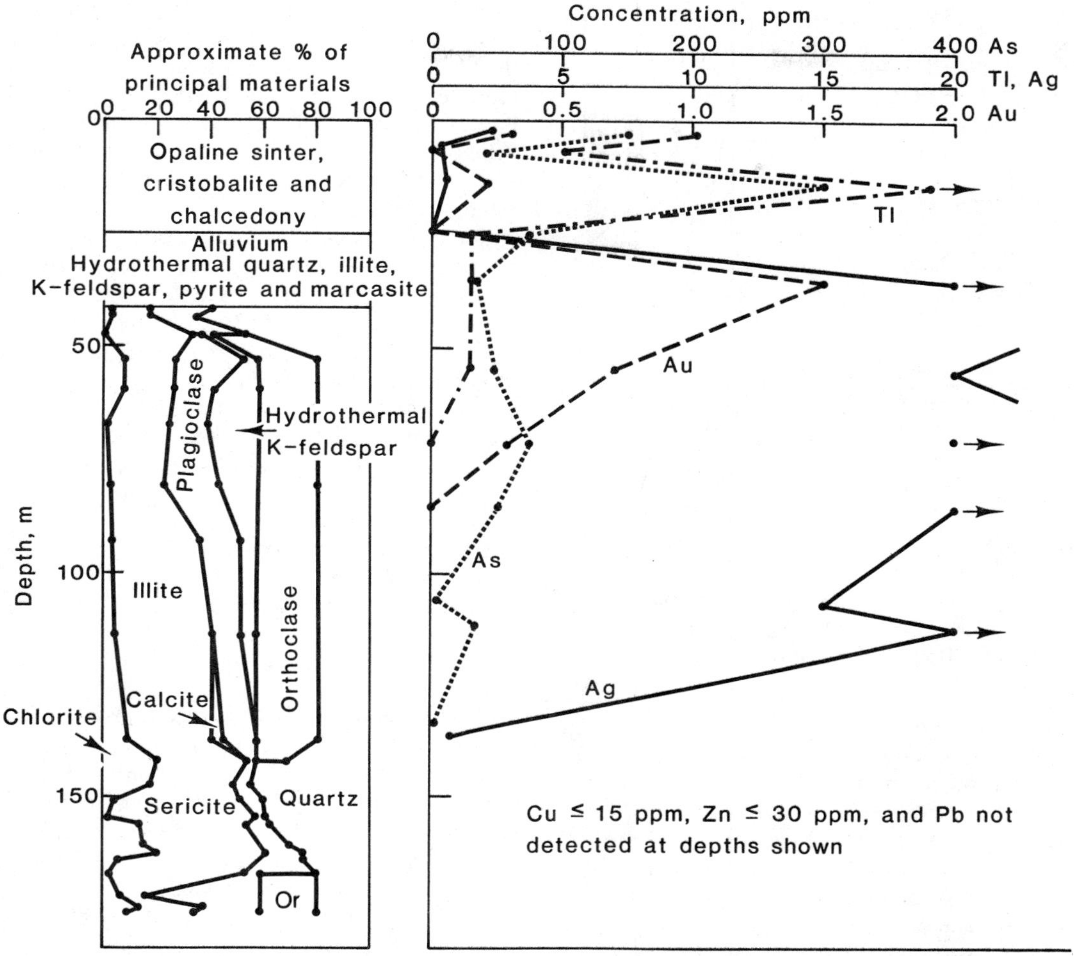

FIG. 1. The spatial relationship of trace-element concentrations and alterations in drill hole GS-5, Steamboat Hot Springs, Nevada (data from Sigvaldason and White, 1962; White, 1980).

and orthoclase are unaltered throughout the drilled intervals. Kaolinite is an important phase only where acid-leaching has occurred or is occurring, and is, therefore, superimposed on top of the other clay alteration patterns mentioned. Precious metals are abundant in the upper parts of the systems (Fig. 1), with base metals generally being more abundant at deeper levels. Depending upon the structural framework, the isotherms in active systems are generally flat surfaces in the broader context of the whole system (Fig. 2). The consequences of this in productive systems are ore horizons at predictable elevations.

CONCEPTUAL MODELS OF EPITHERMAL PRECIOUS METAL HYDROTHERMAL SYSTEMS

Geological and geochemical data from our own research, unpublished reports, and the published literature were compiled for a large number of epithermal silver-gold districts throughout the world to formulate generalized models. Three models are considered in this study:

1) Hot-Springs Deposition Model: The deposition of precious metals was in the near-surface portions of a hydrothermal system that was expressed at the surface as hot springs, fu-

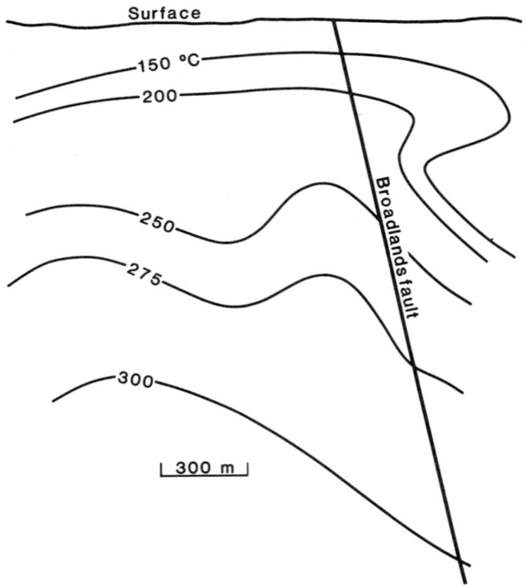

FIG. 2. Cross section of the Broadlands thermal area, New Zealand, showing the generally flat upper surfaces of the isotherms (based on data from Grindley, 1970).

maroles, or geysers; the metallization occurs at very shallow depths below the vents and/or in breccias breaching the surface as a result of the sudden release of abnormally high fluid temperatures and pressures.

2) Stacked-Cell Convection Model: The deposition of precious metals was at and along the interface of a hot hydrothermal cell beneath a cooler, surficial cell resulting in lateral zoning.

3) Closed-Cell Convection Model: The deposition of precious metals was in sequential vertical zones due to vertical convection.

Similarities in appearance of all deposits exist, and there are gradations between all of these end-member models. The purpose in creating models in this study is to provide a framework for understanding the processes that form these deposits and to aid in the establishment of exploration guidelines for the discovery of new precious-metal resources.

The Hot-Springs Deposition Model

Most epithermal precious-metal mineral deposits were probably formed by hydrothermal systems that vented thermal waters to the surface in some manner. However, deposition from very hot solutions at or very near the surface results in a set of geological characteristics that set "hot springs" deposits apart from those epithermal deposits formed at greater depths. Examples of the hot springs type of deposit are Round Mountain, Nevada (Berger and Tingley, 1980); Hasbrouck Mountain, Divide district, Nevada (M.L. Silberman, 1981); DeLamar mine, Idaho; Sulphur, Nevada (Wallace, 1980); and McLaughlin mine, Knoxville district, California (c.f., Manhattan mine, Becker, 1888; Averitt, 1945). The hot-springs model presented here is largely based on the work of Berger and Tingley (1980), Berger et al. (1981), and M.L. Silberman (1981).

There are several important structural and mineralization features that occur in hot-springs precious-metal systems that need to be considered in any model constructed. These features are: (1) multiple episodes of silicification, brecciation, and resilicification; (2) quartz + adularia + hydromica (illite-sericite) veins; (3) two major classes of hydrothermal brecciation:cemented or matrix-supported and uncemented or nonmatrix-supported breccias; and (4) localized or extensive zones of hypogene argillization with kaolinite, alunite, jarosite, and iron oxides.

Fig. 3 shows a hypothetical cross-section through a hot-springs deposit. Depicted are a generalized sequence of alteration and geochemical zonation in a structural environment that becomes increasingly more complex upward in the system. The alteration generally follows the vertical sequence of silicification—quartz + adularia + hydromica − quartz + chlorite from the surface downward.

These types of deposits are associated with the deposition of silica sinter on the surface. Sinter is deposited intermittently during the course of the mineralization at depth, and values in gold, silver, antimony, arsenic, thallium, and mercury occur sporadically within the sinter. The highest values are in closest proximity to the vents, and may occur in bedded zones in the sinter which reflect multiple periods of precious metal transport in the hydrothermal fluids. Examples of sinter are observable at Gold Hill, Nevada; Hasbrouck Mountain, Nevada; and McLaughlin, California.

The sinter is underlain by a zone of intense silicification of the country rock. This silicified zone forms a seal or capping over the hydrothermal system and is characterized in ore-bearing metalliferous deposits (e.g., Round Mountain and Hasbrouck Mountain, Nevada) by multiple

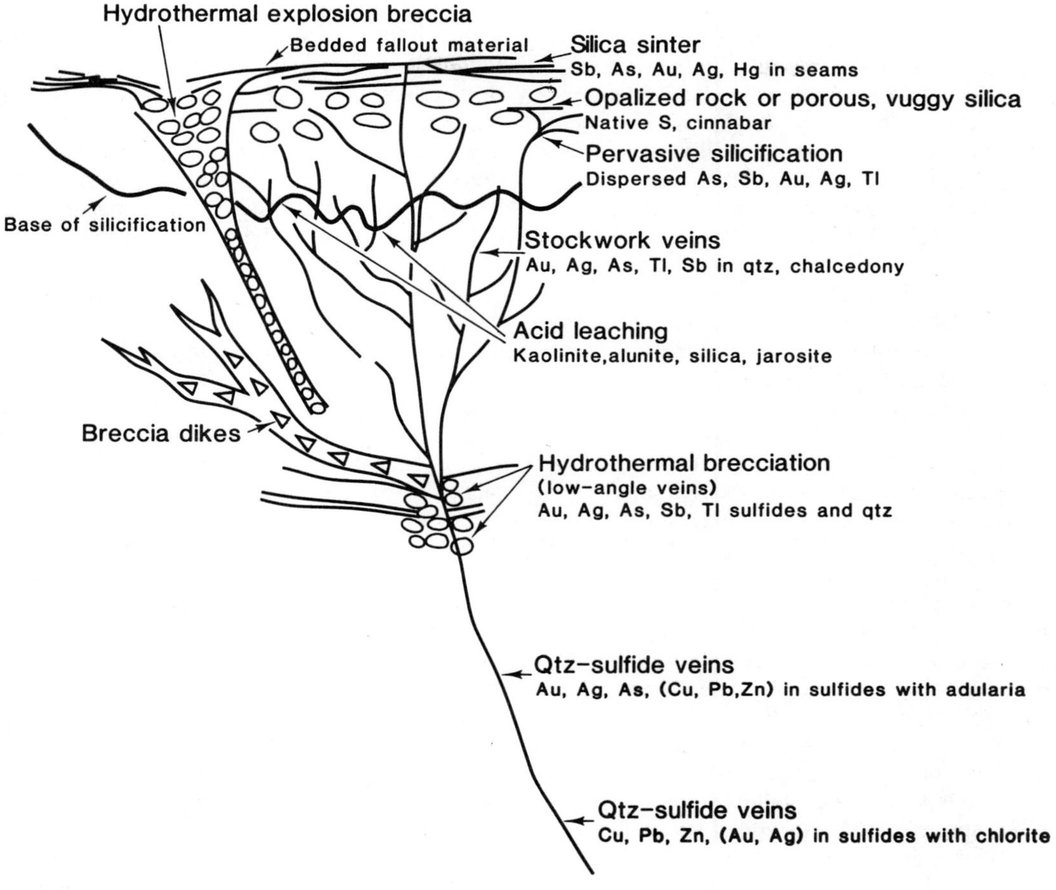

FIG. 3. Schematic cross section of the hot-springs deposition model showing the spatial relationships of alteration and trace-element geochemistry, and some of the more important structural features of this deposit type.

episodes of brecciation. Each episode of fracturing of the capping is followed by self-sealing with silica. The brecciated zones represent the main hydrothermal flues, and clasts in the breccia may be remnants of earlier periods of brecciation and self-sealing, indicating that repeated activity was focused in the same hydrothermal conduits. Sinter is deposited intermittently above the capping during periods that follow rupturing of the cap and before complete self-sealing. The number of episodes of repeated brecciation is difficult to document, but episodic rupturing appears to be important in the development of high concentrations of precious metals. The thickness of the silicified zone may be up to several hundred feet, depending upon the host-rock composition and permeability, amount of silica being transported in the system, and the duration of the thermal activity.

The breccia zones in the silica cap may generally be followed downward beneath the silica cap into stockwork veining (e.g., DeLamar, Idaho). The silica cap itself in some instances may consist essentially of numerous cross-cutting veins that act as a seal if the host rock is impermeable and brittle. The veins generally consist of either quartz or quartz + adularia with narrow selvages of hydromica alteration of mafic minerals adjacent to the veins. The veins contain sporadic precious metals and sulfides, and the ore zones may extend up into the silica cap such as is seen at Round Mountain and

Hasbrouck Mountain, Nevada. The veins only occasionally display chambered structure and normally have sharp contacts with the wallrock. The veins are generally narrow and seldom exceed a few inches in total thickness. The veins at Round Mountain are generally very thin, whereas many at DeLamar are several inches thick. The abundance of veining depends on the structural controls of the hydrothermal systems and the brittleness of the host rocks, as do the vertical and horizontal extent of individual veins. At Round Mountain the quartz-adularia veins are controlled largely by swarms of close-spaced joints while faulting and brittle fracture are more important at DeLamar. We feel that the veins represent a deeper part of the conduits feeding the cap and sinter, and the brecciated portions of the cap relate spatially to the veining.

Explosive brecciation is commonly in evidence beneath the silicified cap in the form of pipes, veins, and dikes. Some of these breccias consist wholly of uncemented rubbly host rock with clast sizes varying from large boulders to sand, the finer particles generally wrapping around the larger fragments. These nonmatrix-supported breccias often break through to the surface as seen at Round Mountain, Nevada, and they may disrupt the sinter, form slight depressions, and create "fallout" aprons. These fallout aprons appear as bedded sediments and consist of fragments of the silica cap, and any condensate carried in the explosive event may precipitate metals when trapped in the bedded material. In some of these breccias, there is little evidence of large-scale vertical transport of rock fragments in the breccias as there is commonly stratigraphic continuity through the breccia from wall to wall with little exotic material present. This is evident at Hasbrouck Mountain, Nevada, where fine, lamellar sediments are traceable across the breccias. Most of the nonmatrix-supported breccia fragments are very angular. The uncemented breccias may be from a few inches to many feet across and may extend for several hundred feet in the vertical dimension. The lack of cementing in these breccias means that the hydrothermal fluids were not able to recharge through the brecciated zone. We envision the breccias to have formed by predominantly gaseous discharges, and the aqueous fluids were possibly prohibited from flooding the breccias because of sealing by rapid mineral deposition at the focus of the brecciation or the breccias may have formed during a period of steam dominance in the geothermal system. At DeLamar, Idaho, some nonmatrix-supported breccias disrupt earlier siliceous ores. At Round Mountain, Nevada, one of the earliest recognizable breccia events is an uncemented breccia. Therefore, these breccias may form at different times during the life of a geothermal system. These open breccias only constitute ore where the rock was mineralized prior to brecciation.

Another type of hydrothermal breccia beneath the silica cap is a cemented or matrix-supported breccia. The cement is predominantly silica with or without accompanying sulfides and precious metals. These breccias are an important source of ore in all productive hot-springs deposits. At all of the example deposits, many of the cemented breccias show evidence of repeated brecciation and cementation as was evident in the silicified cappings. They may be pipes, podlike, or veins with sand-sized rock fragments in the quartz. Vuggy areas often occur between breccia fragments in the larger breccias, and these vugs can be filled with quartz and sulfides (with inclusions of free gold) or contain euhedral quartz and free gold. Quartz-adularia veins may cross cut the matrix-supported breccia or occur as clasts in the breccias. There are probably a considerable variety of matrix-supported breccias that are formed from a variety of different processes; however, further research is needed to fully understand these features and relate them to other aspects of hot-springs systems.

Where there are preexisting low-angle weaknesses in the host rock such as joints or bedding planes, the hydrothermal brecciation can extend laterally resulting in flat veins with cemented breccia and multiple injections of quartz veins with sulfides and precious metals. These low-angle veins tend to dip into centers of brecciation (Fig. 4) and the multiple episodes of brecciation at one center can result in a substantial dilation of the low-angle vein zone such that it is several feet thick adjacent to the breccia, decreasing in thickness away from the breccias. Stockwork quartz veins can occur in the hanging wall of the low-angle veins. Fig. 4 shows the general alteration patterns around the breccias and flat veins.

Explosive brecciation of all types that is near or breached the surface indicates that hot-springs precious-metal systems become overpressured, and consequently there is a violent and sudden

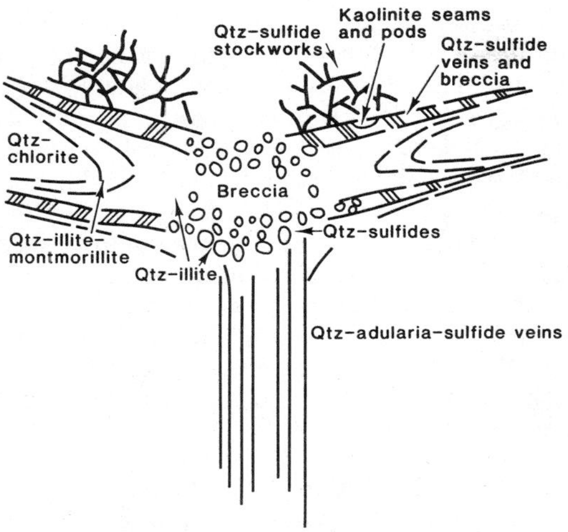

FIG. 4. Schematic cross section of the centers of violent boiling in the hot-springs depositional model showing the spatial relationship of alteration and structure.

release of energy. Thus, the development of a seal or cap on the system is a necessary requisite to the development of the rest of the system. The geological evidence is that this process of sealing followed by brecciation happens in more than one cycle in productive systems (e.g., Round Mountain, Nevada). Corroboration of the geological evidence for self-sealing and sudden pressure releases is found in the scant temperature data available from fluid inclusions. Nash (1972) measured fluid inclusion filling temperatures of about 260°C from a vein in the vicinity of Round Mountain, and temperatures of about 268°C (B.R. Berger, unpublished data, 1982) were measured from a vein at 500–600-ft depth in the Round Mountain deposit. Although only preliminary, these temperatures indicate that exceedingly high temperatures were maintained at near-surface locations and require that the system be overpressured.

Rupturing of the silica seal causes a sudden pressure decrease and the fluid boils. Chemical and temperature changes accompanying the rapid boiling cause the deposition of quartz, sulfides, and precious metals (Drummond, 1981). The volume expansion taking place at the same time causes brecciation. The gases evolved during boiling may recondense in cooler aquifers near or at the surface, and when the gases include H_2S and CO_2, they result in an acid-sulfate solution that may pervasively acid leach the host rock through downward percolation. The intensity of the acid leaching depends on the amount of condensate, the acidity of the condensate, and the level of the water table. A lowered water table may also result in acid leaching through the generation of a steam column beneath the silica cap and above the boiling water table. Recondensation along the under side of the silica cap results in localized acid leaching.

Minerals commonly found in the acid leaching parts of hot-springs precious-metal systems formed by water-rock interactions include kaolinite, alunite, jarosite, and iron oxides. Immediately adjacent to the silicified breccias the host rock may be pervasively argillically altered, and there may be seams and pods of massive kaolinite or illite that string out along fractures (Fig. 4) such as are found at Round Mountain. Following boiling, most of the fluid may have been removed and conductive heat loss from the super-heated host rocks may flash initial recharge waters to steam (R. Fournier, USGS, personal communication, 1981) which recondense locally along fractures and openings resulting in the kaolinite and pervasive argillic alteration.

Gold in the quartz-adularia veins and in the quartz-cemented breccias is commonly of a lower fineness than the gold in acid-leached areas. It is suggested that the acid leaching process also leaches silver and other metals from the gold, thereby upgrading the fineness of the gold. Although this is only one mechanism to explain two apparent generations of gold deposition, the association of high fineness gold and "acidic" alteration is not uncommon. Examples are Round Mountain, Nevada, and Pueblo Viejo, Dominican Republic (Kesler et al., 1981).

A schematic depiction of the spatial relationship of selected metals to alteration zones in hot-springs systems is shown in Fig. 5. Larger amounts of precious metals occur mainly beneath the silica cap in the quartz-adularia-hydromica zone either as veins, stockworks, or replacements, and in the quartz-sulfide cemented breccias; all of these occur above the quartz + chlorite zone. At Round Mountain, Nevada, arsenic, antimony, mercury, and thallium are highly enriched, and the abundances of these elements tend to decrease with depth.

Productive hot-springs systems were probably geothermal cells with an initially high water dis-

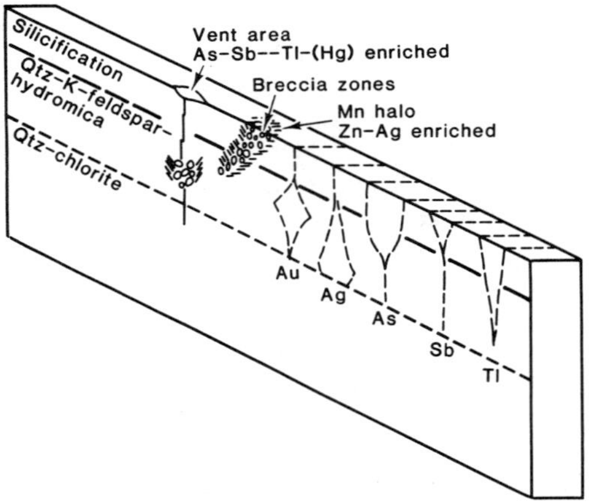

FIG. 5. Schematic depiction of the spatial relationship of gold and silver and selected trace elements to the generalized alteration zones in the hot-springs deposition model.

charge rate or rapid flow. The sinters do not constitute a significant resource base in terms of total tons of metal, but the silicified zones beneath the sinter may be a significant resource of very low-grade to low-grade precious metals. Detectable gold at 0.02 ppm was found in all samples taken at Round Mountain (Berger et al., 1981). The stockwork or vein zones beneath the silica cap make up the greatest resource potential, and moderate to high-grade ores in the tens of millions of tons may be present. Examples of this include McLaughlin, California, and Round Mountain, Nevada. The Homestake Mining Co. has published reserves of 20 million tons averaging 0.16 oz per ton for McLaughlin (Anderson, 1982), and the early underground production from Round Mountain had ores with grades approaching 1 oz per ton (Ferguson, 1921). Permeable rocks at or near the surface enhance the resource potential, and their presence beneath the silica cap may result in deposits with hundreds of millions of tons of ore. Round Mountain, Nevada, has published reserves of about 195 million tons containing more than 8 million oz of gold (c.f., *Wall Street Journal,* Jan. 5, 1982).

Open Vein Deposition Models

High-grade complex-sulfide vein systems differ markedly from hot-springs systems in that, in the vein systems (1) deposition took place at greater depths than in hot springs; (2) vein widths are substantial, up to several feet or more, resulting in large tonnages of high-grade ore; (3) there are higher total-sulfide concentrations, more complex mineralogies, and greater quantities of base metals; and (4) the systems lack many of the violent boiling breccia features found in hot springs deposits.

Differences between individual complex-sulfide vein systems are due to the hydrodynamics of the system, and the host rock fracture and permeability characteristics. Two types of open vein deposition systems, a stacked-cell convection system and a closed-cell convection system, are described in the following.

The Stacked-Cell Convection Model

Stacked-cell convection is exemplified by the Creede, Colorado, deposits, and this model is largely drawn from the work of P. Barton, P. Bethke, and R. Rye of the US Geological Survey and P. Giudici of Freeport Exploration Co. Fig. 6 shows a hypothetical longitudinal section of a stacked-cell deposit. Factors that appear to be essential to forming this type of deposit are:

1) Large moderately to steeply dipping structures open to lateral solution movement (generally faults).

2) Host rocks of low permeability, such that hot fluids are mostly restricted to the structures.

3) Relatively unrestricted flow of meteoric water into an area of continuous high heat flow.

4) The presence of a stratigraphic barrier of limited permeability, impermeable enough to restrict the upward movement of hot fluids, but open enough to permit a limited mixing with higher cooler waters and to allow hydrostatic boiling at depth, at temperatures of 200°–300°C.

These factors result in these features: (1) precious-metal deposition takes place through chemical changes accompanying boiling; (2) broad lateral and vertical precious- and base-metal zoning patterns (Fig. 6) due to various combinations of fluid mixing, hydrostatic boiling, and flash boiling in systems that are over pressured; (3) coarse-grained vein minerals with crustiform, comb, and other open-space filling textures; (4) hydrothermal brecciation throughout most of the deeper veins consisting of cross-cutting vein relationships and spalled blocks of host rocks; (5) stockwork veining and disseminated mineralization in the hanging wall in the upper parts of veins above bulges in the vein surface; and (6) disseminated and replacement

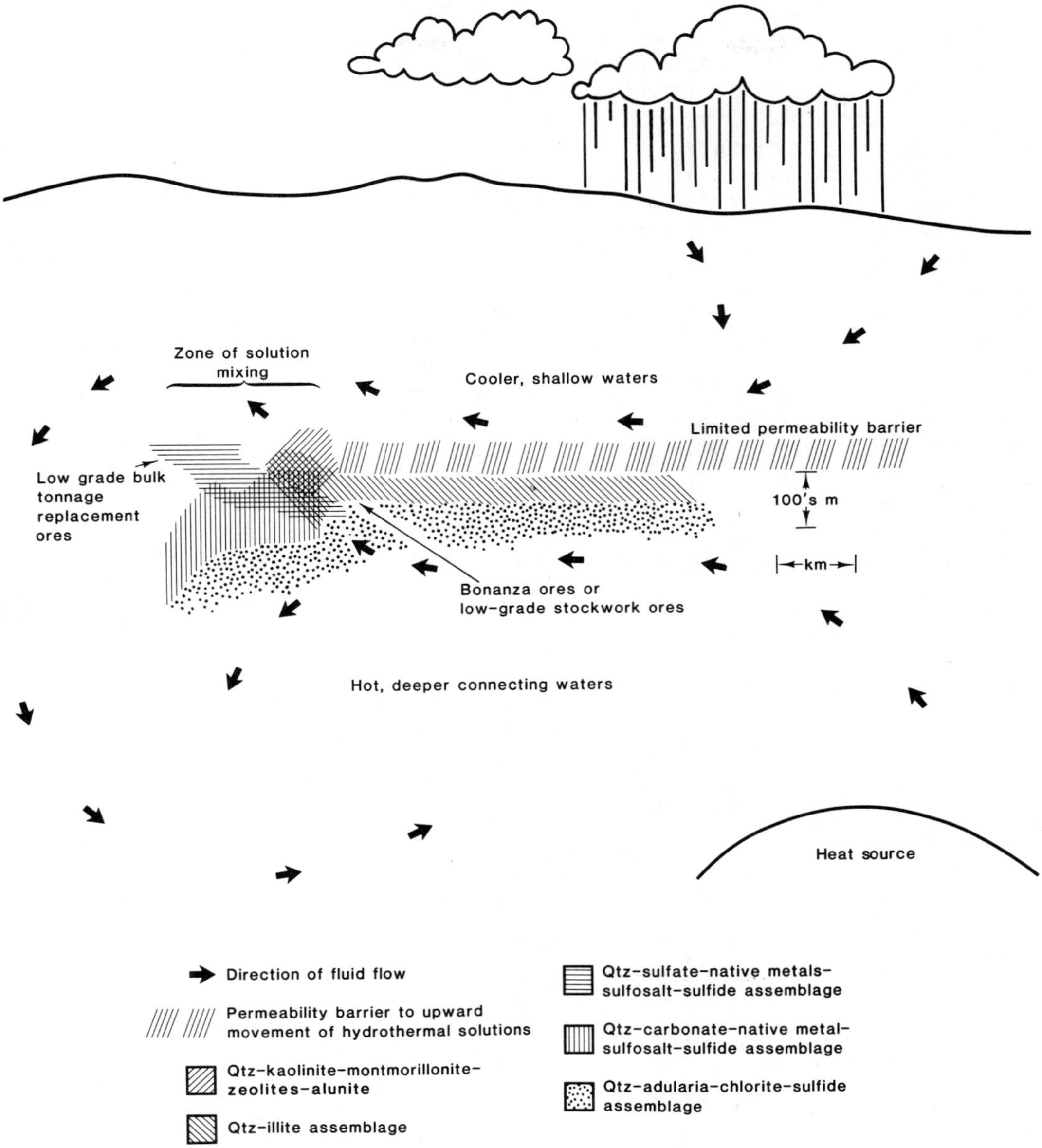

FIG. 6. A hypothetical cross section of fluid-flow patterns and lateral alteration zoning in the stacked-cell convection model of open-vein deposition.

mineralization in permeable horizons within or adjacent to the zones of mixing and boiling.

We envision the following scenario for stacked-cell deposits such as Creede, Colorado:

Ground waters decend along deep structures, become heated, and increase in salinity. Buoyancy moves the light higher chloride-bearing fluids upward where they encounter the limited permeability barrier. Boiling may take place intermittently along the barrier due to minor structural breaks in the barrier or an incomplete sealing of the system. Major mixing or boiling activity take place where the limited permeability interface between the hot fluids and cooler overlying waters no longer separates the two fluids. The hot fluids will rise into the overlying reservoir (Romagnoli et al., 1975). If violent boiling occurs in the upper more broken parts of the vein system and much of the water is removed by the escaping gases, then there may be a period of slow recharge of the open fractures during which residual heat in the rocks flashes initial recharge water to steam, thereby increasing the amount of acid-leaching alteration occurring in this part of the system. This mechanism of acid-leaching would occur most likely in those areas where the system was sealed and became over-pressured.

Stockwork and disseminated mineralization occur over bulges or major changes in strike and dip of the vein surface in the zone of mixing of the two reservoirs (Fig. 7). At Creede, this is exemplified by the Amethyst vein in the intersection area with the P and OH veins. The fracturing is created by faulting and hydraulic action. At Creede, extremely high-grade zones commonly exist in this zone of boiling. The process of boiling may be the cause for deposition of the metals in the "bonanza" ores for which epithermal districts are famous. Widespread lower-grade mineralization extends in irregular pods, both horizontally and vertically, into the hanging wall away from the main vein. Good grades commonly persist along the main vein adjacent to the stockwork zone, but the grades decrease upward although alteration persists for up to hundreds of feet. Disseminated mineralization also occurs where permeable beds intersect the zone of mixing. These permeable beds may in fact be a key factor in removing the limited permeability barrier. An example of this is the Creede formation at Creede, Colorado. The mineralogy of the veinlets in the stockwork and disseminated zones is similar to the adjacent main vein mineralogy though the textures are much finer grained and the silica

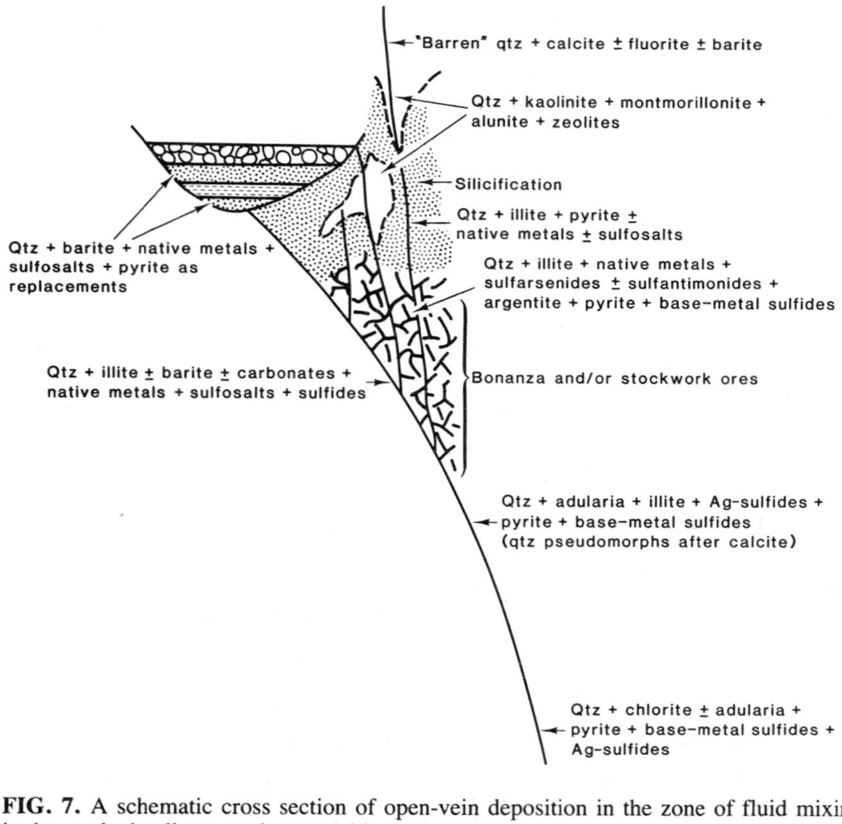

FIG. 7. A schematic cross section of open-vein deposition in the zone of fluid mixing in the stacked-cell convection model including areas of stockwork fracturing and replacement-type deposits.

concentration is locally higher. A schematic depiction of the alteration sequences in the upper part of the stacked-cell system is shown in Fig. 8.

Self-sealing is an important component of this type of vein system, but does not appear to be as significant overall as in the creation of vertically zoned epithermal veins.

The Closed-Cell Convection Model

Closed-cell convection differs from stacked-cell convection in that there is no limited permeability barrier that creates a strong lateral zoning pattern. This type of deposit is exemplified by Guanajuato and Pachuca, Mexico, and Chanarcillo, Chile. The model presented here is drawn largely from the work of Buchanan (1980) and is shown schematically in Fig. 9. Factors that appear to be essential to the formation of this deposit type are as follows:

1) A large, moderately to steeply dipping structure (generally a fault).

2) Host rocks of low permeability in the deep parts of the system, such that the hot fluids are mostly restricted to the structure.

3) The relatively unrestricted flow of meteoric water into an area of continuous high heat flow.

4) Periodic self-sealing of the upper parts of the system such that intermittent boiling takes place at intermediate depths in the temperature range of 200°–300°C.

These factors result in the following features:

a) Precious-metal deposition takes place through chemical changes accompanying boiling.

b) Vertical precious- and base-metal zonation (Fig. 10) due to hydrostatic and flash boiling (to a lesser extent fluid mixing).

c) Coarse-grained vein minerals with crustiform, comb, and other open-space filling textures.

d) Hydrothermal brecciation throughout most of the deeper veins consisting of cross-cutting vein relationships and spalled blocks of host-rocks.

e) Stockwork veining and disseminated mineralization in the hanging wall in the upper parts of veins above bulges in the vein surface.

f) Vertically separated or "stacked" ore bodies often separated by subeconomic vein filling.

g) Disseminated and replacement mineralization in permeable horizons within or adjacent to the zones of mixing and boiling.

h) Within a given district there is a tendency for boiling to have occurred at a regular altitude resulting in a general uniformity or regularity to the tops of ore bodies within a single district.

We envision the following scenario for closed-cell deposits such as Guanajuato and Pachuca:

. The deep convective circulation of meteoric waters along major channelways causes them to be heated by cooling intrusive bodies and rise as hot low-density saline solutions. In the absence of a horizontal impermeable barrier, the fluids will rise essentially vertically away from the heat source along open channelways. Decreases in the viscosity of water at high temperatures allows the migration of the fluids in the deeper more impermeable parts of the system (White, 1967), while increased silica solubility increases permeability.

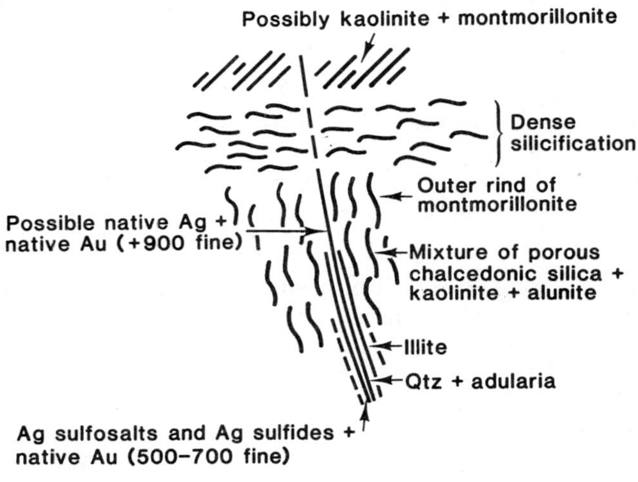

FIG. 8. A detailed cross section of alteration mineralogy in the zone of boiling in open-vein deposition showing the hypogene oxidative upgrading of the gold from 700 fine to 900 fine.

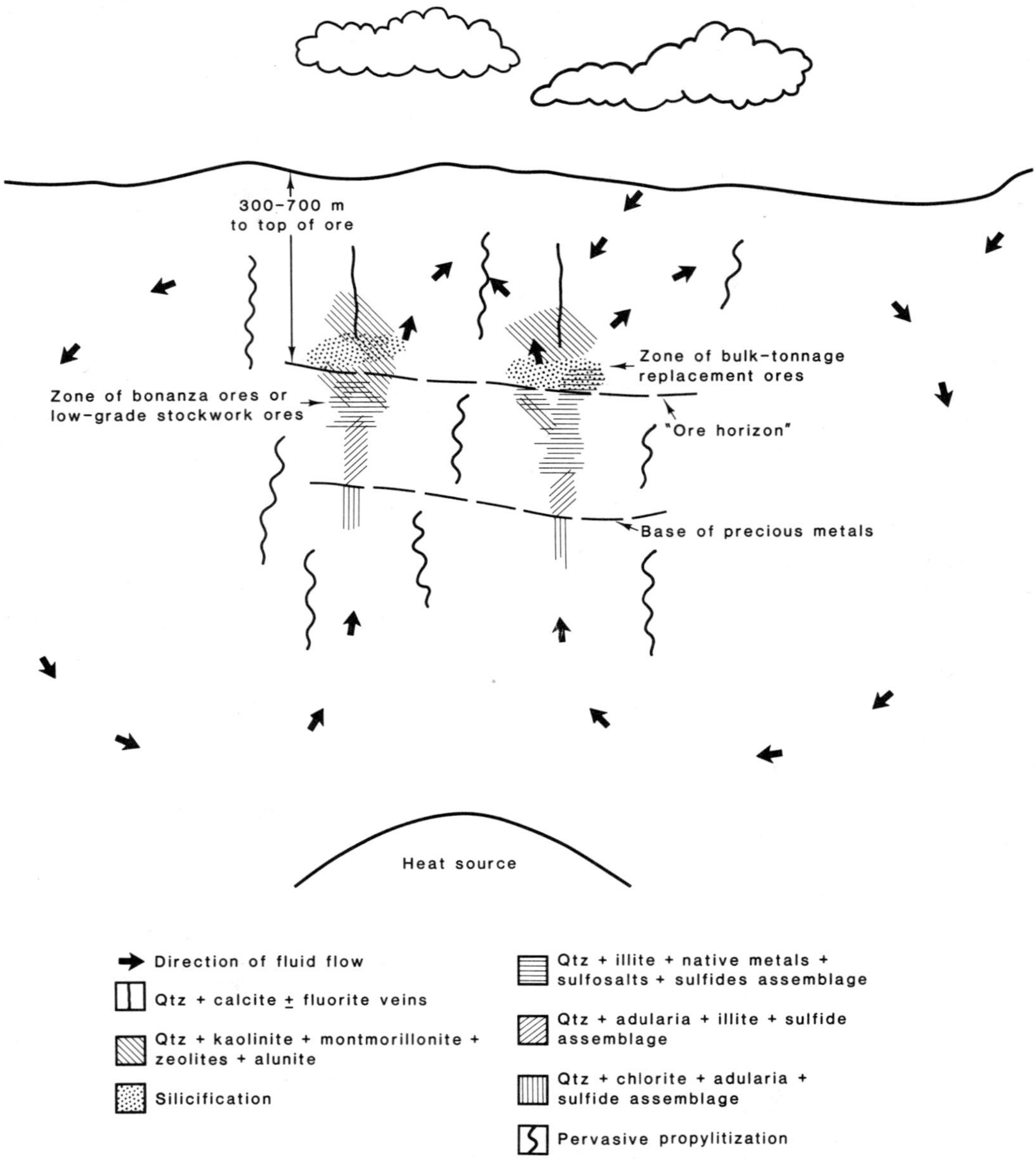

FIG. 9. A hypothetical cross section of fluid-flow patterns and vertical alteration zoning in the closed-cell convection model of open-vein deposition.

Boiling begins when the hydrostatic pressure becomes predominant over the lithostatic pressure. The removal of dissolved gases such as H_2S and CO_2 from the solution during boiling causes precipitation through a change in pH and a lowering of temperature depending on the original composition of the fluid and the temperature at which boiling occurs (Drummond, 1981). The precipitate, predominantly quartz, will block channelways or impede fluid flow. In addition, the residual water and evolved gases will alter preexisting mineral phases, further impeding the upward movement of solutions.

Cooler overlying water or preceding imperme-

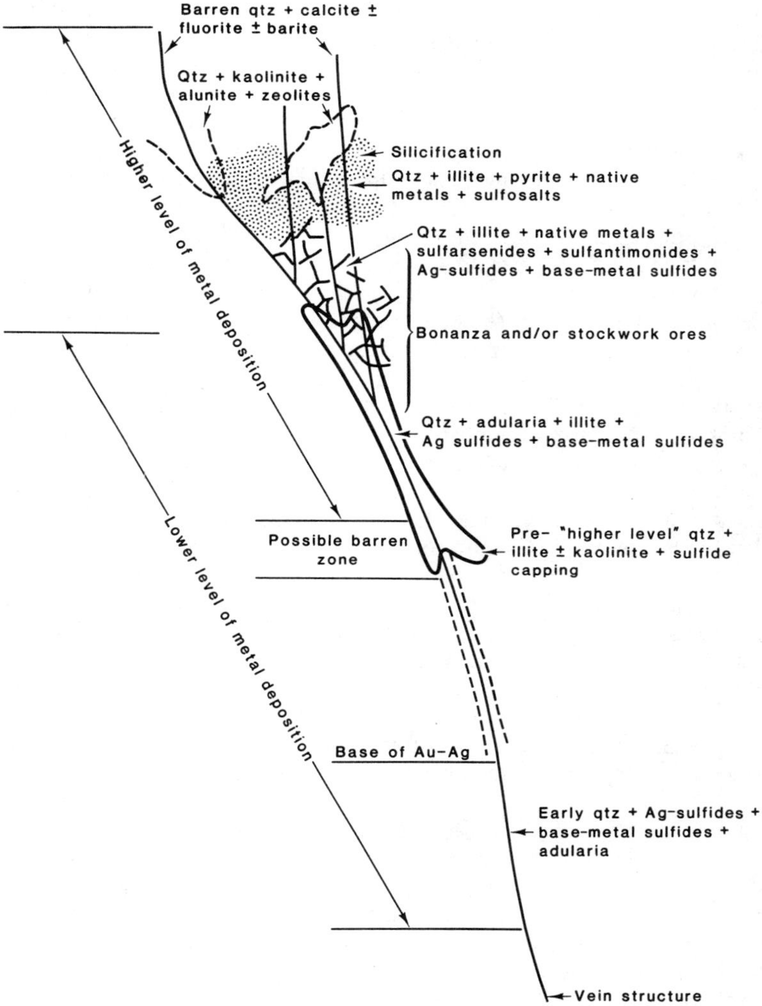

FIG. 10. A schematic cross section of open-vein deposition in the closed-cell convection model including two levels of mineralization resulting in stacked ore bodies, in this case separated by a barren zone.

able alteration types (e.g., dense stockworks of quartz veins such as at the Los Torres mine at Guanajuato, Mexico) may cause a recondensation of the escaped water vapor and other gases, resulting in an acid-leach solution that alters rock higher in the system and then, when possible, percolates downward reacting with the wallrocks and precipitating silica in the process. These precipitation mechanisms serve to seal off the hydrothermal system from further upward movement of the hot fluids. This is the very important process of self-sealing. When higher temperatures are maintained at levels successively closer to the surface before boiling, then stacked orebodies can be created. Stacked ores are exemplified by the Guanajuato, Mexico, and Hauraki, New Zealand, districts. A significant change in the level of boiling may actually lead to the separation of ore bodies by low-grade zones.

Individual ore bodies within a district tend to have a long vertical dimension relative to the width or thickness, except when replacement is an important mode of occurrence such as in carbonate terrain such as is found at Chanarcillo, Chile. The presence of multiple ore bodies aligned along a structural trend gives epithermal districts of the closed-cell type the appearance of having a greater horizontal than vertical dimension.

Disseminated and stockwork mineralization form in closed-cell convective systems in the same manner as they do in stacked-cell systems.

In the Calico district, California, the Barstow formation makes a host for disseminated ores similar to the Creede formation in Creede, Colorado. Additionally, carbonate-replacement deposits are important types of disseminated occurrences related to complex-sulfide vein systems.

SUMMARY AND CONCLUSIONS

Epithermal precious-metal deposits occur as veins, stockworks, and disseminations immediately below the paleosurface and at shallow depths. The nature of the occurrences is determined by the (1) geometry of the fracture system, (2) permeability and composition of the host rocks, and (3) hydro-dynamics of the fluid systems. Variations in these factors serve to make each district unique. General attributes considered to be significant in epithermal deposits are as follows:

1) All important districts occur within strong, persistent fractures (e.g., basin-range-type normal faults, caldera ring fractures, caldera-related graben structures, complexly faulted or domed areas).

2) Vein textures always include drusy cavities, crustification, comb structure, colloform banding, and vein breccias.

3) Alteration and metallization patterns tend to be systematic within and around the veins or hydrothermal vents (pervasive propylitization depends upon host rock composition and permeability). (a) In hot-spring systems evidence of mineralization always occurs at the surface, most commonly in the form of hyrothermal breccias (often silicified), silicification, and with or without overlapping argillic alteration—all containing anomalous concentrations of arsenic, thallium, antimony, and mercury; and (b) in complex-sulfide vein systems, vein tops often consist of multiple episodes of quartz stringers (commonly chalcedonic), "barren" fractures with only vague alteration (fracture systems often parallel to igneous dikes), quartz-barite or quartz-calcite-(fluorite) veins—all carrying anomalous concentrations of arsenic, antimony, and/or mercury.

4) The tops of ore bodies in complex-sulfide vein systems tend to apex at a common altitude, may be overlain by hundreds of feet of noneconomic quartz veining, and have discrete bottoms to the precious-metal ore bodies.

5) Intramineralization faulting is apparent in most deposits, and all high-grade districts show evidence of episodic precious-metal mineralization.

Boiling of the hydrothermal fluid is an important process in the deposition of precious metals. Fluid inclusion studies by Barton et al., (1977) at Creede, Colorado; Buchanan (1980) at Guanajuato, Mexico; and Clifton et al., (1980) at Oatman, Arizona, indicate that boiling accompanied each of these depositional events. Understanding the effects of boiling on mineralization processes aids in the understanding of base- and precious-metal zoning (including hypogene oxidation, a style of mineralization previously thought to be post-mineral supergene enrichment), alteration sequences and patterns, and the meaning of many breccias and "gouge" zones. Episodic boiling requires intermittent self-sealing of the hydrothermal system. This blocking of the fluid channelways and the formation of a subhorizontal capping (generally silica) allow fluid pressures and temperatures to build up beneath the seal.

Very high-grade bonanza ores occur in the upper, hydrothermally brecciated parts of complex-sulfide vein and hot springs systems. Bulk tonnage, lower-grade reserves occur in these same horizons, but also occur as disseminations or replacements in clastic, tuffaceous, or carbonate sediments or permeable tuffs. In the vein systems, the bulk tonnage ores are normally adjacent to or above the bonanaza ores such as at Creede, Colorado, and Calico, California. In hot-springs systems, they may also occur below the high-grade zones such as in the case at Round Mountain, Nevada.

Further advances in the understanding of epithermal ore deposits requires considerably more data derived from on-site examination than is generally available in the literature. More descriptive information on the inter-relationships of different alteration types, overlapping episodes of metallization, and hydrothermal brecciation accompanying alteration and metallization is needed. The geological and geochemical attributes of vein tops are inadequately documented. And finally, detailed trace-element zoning pattern studies of individual epithermal systems are needed. Exploration models become increasingly viable as these types of data are added to the concepts and developing body of knowledge about epithermal systems.

REFERENCES

Anderson, J.A., 1982, "Gold," *Mining Congress Journal,* pp. 51–58.

Averitt, P., 1945, "Quicksilver Deposits of the Knoxville District, Napa, Yolo, and Lake Counties, California," *California Journal of Mines & Geology,* Vol. 41, No. 2, pp. 65–89.

Barton, P.B., Jr., Bethke, P.M., and Roedder, E., 1977, "Environment of Ore Deposition in the Creede Mining District, San Juan Mountains, Colorado: Part III. Progress Toward the Interpretation of the Chemistry of the Ore-Forming Fluid for the OH Vein," *Economic Geology,* Vol. 71, pp. 1–24.

Becker, G., 1888, "Quicksilver Deposits of the Pacific Slope," Monograph 13, US Geological Survey.

Berger, B.R., and Tingley, J.V., 1980, "Geology and Geochemistry of the Round Mountain Gold Deposit, Nye County, Nevada (abstract)," *Precious Metals Symposium,* Nevada Bureau of Mines and Geology, Reno, NV, p. 18c.

Berger, B.R., et al., 1981, "Origin of Pathfinder Trace-Element Patterns Associated with Gold-Silver Mineralization in Late Oligocene Volcanic Rocks, Round Mountain, Nye County, Nevada (abstract)," *Precious Metals Symposium,* Association of Exploration Geochemists, Vancouver, B.C., Canada.

Bethke, P.M., and Rye, R.O., 1979, "Environment of Ore Deposition in the Creede Mining District, San Juan Mountains, Colorado: IV. Source of Fluids from Oxygen, Hydrogen, and Carbon Isotope Studies," *Economic Geology,* Vol. 74, pp. 1832–1851.

Buchanan, L.J., 1980, "Ore Controls of Vertically Stacked Deposits, Guanajuato, Mexico," SME-AIME, Preprint 80–82, AIME Annual Meeting, Las Vegas, NV, February, 26 pp.

Clifton, C.G., Buchanan, L.J., and Durning, W.P., 1980, "Exploration Procedure and Controls of Mineralization in the Oatman Mining District, Oatman, Arizona," SME-AIME Preprint 80–143, AIME Annual Meeting, Las Vegas, NV, February, 45 pp.

Drummond, S.E., Jr., 1981, "Boiling and Mixing of Hydrothermal Fluids: Chemical Effects on Mineral Precipitation," unpublished Ph.D. dissertation, The Pennsylvania State University, University Park, 380 pp.

Ferguson, H.G., 1921, "The Round Mountain District, Nevada," Bulletin 725, US Geological Survey, pp. 383–406.

Ferguson, H.G., 1929, "The Mining Districts of Nevada," *Economic Geology,* Vol. 24, pp. 131–141.

Grindley, G.W., 1970, "Subsurface Structures and Relation to Steam Production in the Broadlands Geothermal Field. New Zealand," *Geothermics*, Spec. Issue 2, Vol. 2, Pt. 1, pp. 248–261.

Kesler, S.E., et al., 1981, "Geology and Geochemistry of Sulfide Mineralization Underlying the Pueblo Viejo Gold-Silver Oxide Deposit, Dominican Republic," *Economic Geology,* Vol. 76, pp. 1096–1117.

Lindgren, W., 1933, *Mineral Deposits,* McGraw Hill Co., New York.

Nash, J.T., 1972, "Fluid Inclusion Studies in Some Gold Deposits in Nevada," Professional Paper 800-C, US Geological Survey, pp. C15–C19.

Nolan, T.B., 1933, "Epithermal Precious-Metal Deposits," *Ore Deposits of the Western States,* American Institute of Mining and Metallurgical Engineers, New York, pp. 623–640.

O'Neil, J.R., et al., 1973, "Stable Isotope and Chemical Relations during Mineralization in the Bodie Mining District, Mono County, California," *Economic Geology,* Vol. 68, pp. 765–784.

O'Neil, J.R., and Silberman, M.L., 1974, "Stable Isotope Relations in Epithermal Au-Ag Deposits," *Economic Geology,* Vol. 69, pp. 902–909.

Romagnoli, P., et al., 1975, "Aspectos Hidrogeologicos del Campo Geothermico de Ahuachapan, El Salvador," *Proceedings,* 2nd United Nations Symposium on the Development and Use of Geothermal Resources, San Francisco, Vol. 1, pp. 563–570.

Sigvaldason, G.E., and White, D.E., 1962, "Hydrothermal Alteration in Drill Holes GS-5 and GS-7, Steamboat Springs, Nevada," Professional Paper 450-D, US Geological Survey, pp. D113–D117.

Silberman, M.L., 1981, "Hot Spring Type, Large-Tonnage, Low-Grade Gold Deposits," *Characteristics of Mineral Deposit Occurrences,* R.L. Erickson, ed., Administrative Report, US Geological Survey, pp. 116–125.

Taylor, H.P., 1973, "O^{18}/O^{16} Evidence for Meteoric-Hydrothermal Alteration and Ore Deposition in the Tonopah, Comstock Lode, and Goldfield Mining Districts, Nevada," *Economic Geology,* Vol. 68, pp. 747–764.

Vikre, P.G., 1981, "Silver Mineralization in the Rochester District, Pershing County, Nevada," *Economic Geology,* Vol. 76, pp. 580–609.

Wallace, A., 1980, "Geology of the Sulphur District, Southwestern Humboldt County, Nevada," *Guidebook,* Field Conference on Epithermal Deposits, Society of Economic Geologists, pp. 80–91.

White, D.E., 1967, "Some Principles of Geyser Activity, Mainly from Steamboat Springs, Nevada," *American Journal of Science,* Vol. 265, pp. 641–684.

White, D.E., 1974, "Diverse Origins of Hydrothermal Ore Fluids," *Economic Geology,* Vol. 69, pp. 954–973.

White, D.E., 1980, "Steamboat Springs Geothermal Area," *Guidebook,* Field Conference on Epithermal Deposits, Society of Economic Geologists, pp. 44–51.

16

Unconventional Metals in Porphyry Deposits

Richard H. Sillitoe

Consulting Economic Geologists, London, England

ABSTRACT

Gold is an important byproduct and, locally, coproduct of porphyry copper deposits, and possesses potential for discovery as the major metal in porphyry deposits in either island arcs or Cordilleran belts. Tungsten and tin occur as major components of several potentially bulk-minable porphyry deposits, as well as being by-products from the Climax porphyry molybdenum deposit. Whereas porphyry tin systems are localized in Cordilleran back-arc settings, their tungsten counterparts appear to span a range of geotectonic settings, from Cordilleran principal arcs through back arcs to rifts. Although silver is a common byproduct of porphyry copper deposits, it tends to be preferentially concentrated with lead and zinc on the fringes of all types of porphyry deposits and/or as late-stage superpositions; the latter situation could be of some interest in the higher parts of porphyry systems. As byproducts, rhenium, uranium, platinoids, and magnetite are exploited occasionally, and rutile and the rare metals, such as tantalum and niobium, merit consideration.

INTRODUCTION

Porphyry deposits are traditionally considered as sources of copper and molybdenum, and account for approaching 60 and 95%, respectively, of world production and reserves of these two metals. Two broad categories of porphyry deposits may be recognized (Sillitoe, 1980; Bookstrom, 1981; Westra and Keith, 1981; White et al., 1981): standard porphyry copper-molybdenum deposits and Climax-type molybdenum deposits. The standard type ranges from widespread porphyry copper deposits, many carrying byproduct molybdenum, through to less common porphyry molybdenum deposits, and is emplaced in Cordilleran belts and island arcs above subduction zones. In contrast, the far less-abundant Climax type, hosted by alkali granite stocks, contains molybdenum and a number of other lithophile elements but lacks copper, and is emplaced in a variety of rift settings.

It has become recognized, especially during the last decade, that other metals occur as either principal and/or ancillary components of porphyry-type deposits. Although none of these metals, with the exception of rhenium, relies on porphyry deposits as its main, or even as an important source, some considerable potential for further discoveries of "unconventional metals" in porphyry deposits is believed to exist. This chapter aims to highlight this potential for gold, tin, tungsten, silver-lead-zinc, and several other less-important commodities. The treatment begins with a review of what is known, or may be reasonably inferred, of the unconventional metals in porphyry deposits, and is concluded with some suggestions on suitable petrogenetic and geotectonic settings in which to explore for the most likely unconventional-metal porphyry systems.

In preparing this review, a constant problem was the definition of a porphyry deposit. Even when the porphyry category is restricted to large volumes (>20 Mt) of stockwork and disseminated mineralization spatially, temporally and genetically connected to an intrusive mass, it proved difficult for a number of specific examples to decide whether or not they represent true porphyry deposits. Some of these borderline cases are described, particularly in the section on gold. Skarns, greisens, and vein swarms are excluded from consideration unless they accompany true porphyry-type mineralization.

It is assumed that the principal metals—e.g., copper, molybdenum, gold, tin or tungsten—present in a porphyry deposit primarily reflect the fundamental metal budget of the magmatic-hydrothermal system. It is also recognized, however, that metal zoning in a deposit is an additional important factor in determining the commodities that are commercially extractable. The zoning sequence in epigenetic deposits is fixed by several parameters, of which the free energies of the dissolved metal species concerned are particularly important (Susak and Crerar, 1982), and comprises a central (and earliest) to marginal (and latest) sequence of: molybdenum, tin, copper, zinc, lead, silver, and gold (Barnes, 1975). Although this entire idealized sequence is never encountered in a single porphyry deposit, compliance with its basic tenets is commonplace. Examples include: the widespread distribution of Ag-Pb-Zn around (and above) copper, copper-molybdenum and molybdenum centers; the presence of copper-rich halos around molybdenum cores in both porphyry copper-molybdenum (Bingham) and porphyry molybdenum (Hall and Buckingham, Nevada, and Opodepe, Mexico) deposits; and gold concentrations peripheral to copper-molybdenum centers (Andacollo, Chile). In the porphyry environment, as elsewhere, however, important exceptions to the idealized sequence occur, such as the presence of molybdenum halos around many gold-rich porphyry copper deposits (Fig. 2). Such exceptions may be attributed to major changes in the relative concentrations of metals, in the foregoing case gold vis-a-vis molybdenum, in the hydrothermal fluids. (See Barnes, 1975; Susak and Crerar, 1982.)

GOLD

Gold is an important byproduct of a number of porphyry copper deposits, and at least 16 deposits in the non-Communist world contain >0.4 ppm Au (Fig. 1). Byproduct gold from porphyry copper deposits is currently estimated to account for about 6% of non-Communist gold production. No copper-poor gold deposits with the same characteristics as porphyry copper-gold deposits have been recognized, although parts of at least two hypogene porphyry copper-gold systems might constitute exploitable gold deposits under current conditions. Moreover, a number of potentially bulk-minable gold deposits exist, which exhibit certain characteristics akin to, although not identical with, those of porphyry copper-gold deposits.

The gold-rich porphyry copper deposits, herein arbitrarily defined as those containing >0.4 ppm Au, possess a broad spectrum of geological characteristics, as well as showing several common features (Sillitoe, 1979). They are associated most commonly with either quartz-deficient stocks of high-K shoshonitic affinity, ranging from diorite to syenite in composition (e.g., Intermontane belt, British Columbia; Barr et al., 1976), or with low-K calc-alkaline quartz diorite or tonalite porphyries, as at most of the Philippine deposits, and at Saindak, Pakistan, and Panguna, Papua New Guinea. However, more felsic, probably calc-alkaline host stocks occur locally, with porphyries of adamellitic composition at Mamut, Sabah, and Ok Tedi, Papua New Guinea. Wallrocks are commonly andesitic volcanics, but a broad spectrum of other rock types is also present (Sillitoe, 1979). Emplacement ages and sizes of gold-rich deposits are comparable to their gold-poor counterparts.

Gold in porphyry copper-gold deposits is most commonly present as a component of centrally located zones of K-silicate alteration, carrying widespread hydrothermal biotite and, locally, alkali feldspar and other minerals [e.g., actinolite-tremolite, garnet (cf., Sinclair et al., 1982)]. These zones are enriched in chalcopyrite, with or without bornite, and are commonly relatively poor in pyrite. The gold-bearing potassium silicate cores may or may not be enveloped by sericitic alteration, which where present invariably possesses a lower gold content, except where it is present as an overprint to high-gold K-silicate alteration (e.g., Dizon, Philippines). The presence of a molybdenite-enriched halo to

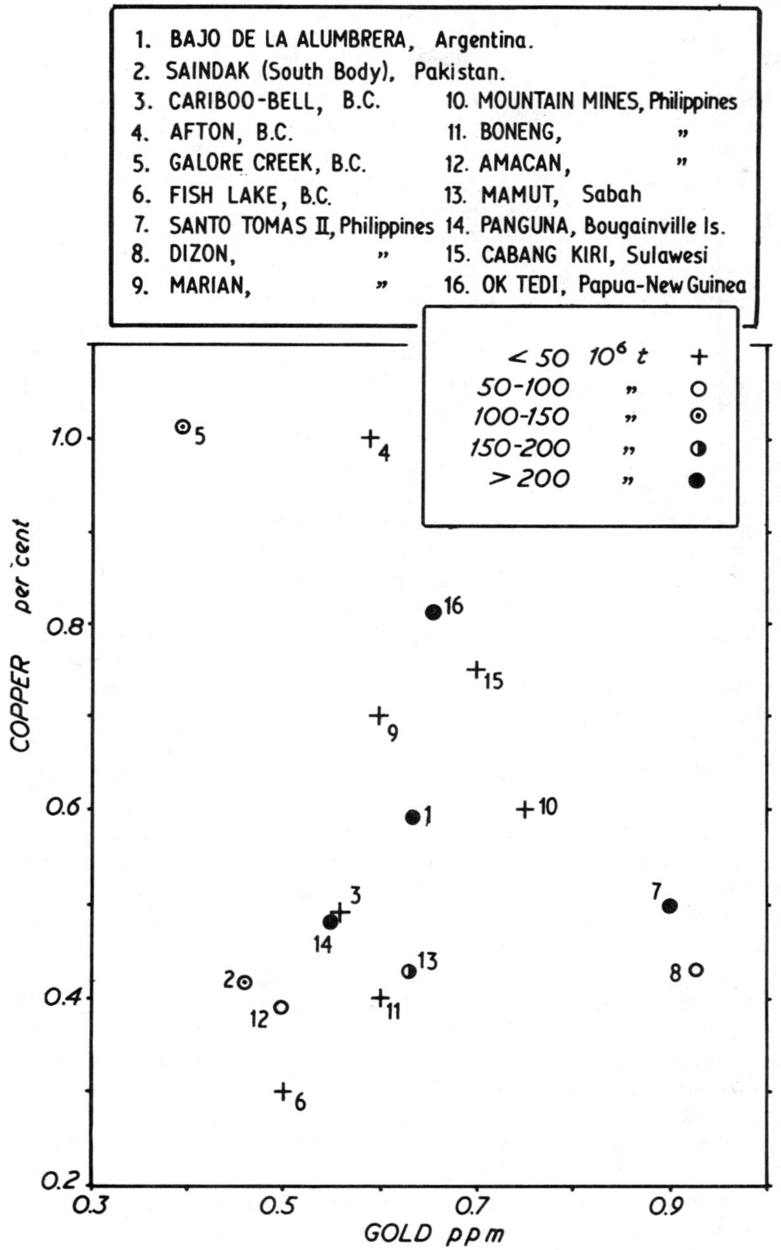

FIG. 1. Gold contents of selected gold-rich (>0.4 ppm) porphyry copper deposits. Further explanation in Sillitoe (1979). Reserves in metric tons.

the gold-rich core is common and well documented at Bajo de La Alumbrera, Argentina; Saindak, Pakistan (Fig. 2); and Dizon, Philippines. Sillitoe (1979) emphasized a frequent correlation of high gold values and above-average quantities of magnetite in potassium silicate alteration, with the hydrothermal magnetite occurring as veinlets, with or without transparent quartz, and as patches and disseminations. There is a positive correlation between gold and copper grades in many deposits (Panguna, Dizon, Santo Tomas II, Mamut, Bajo de La Alumbrera, Saindak), but generally no direct association of gold with pyrite. On the basis of limited available information, much gold is present as micron-sized particles of the native

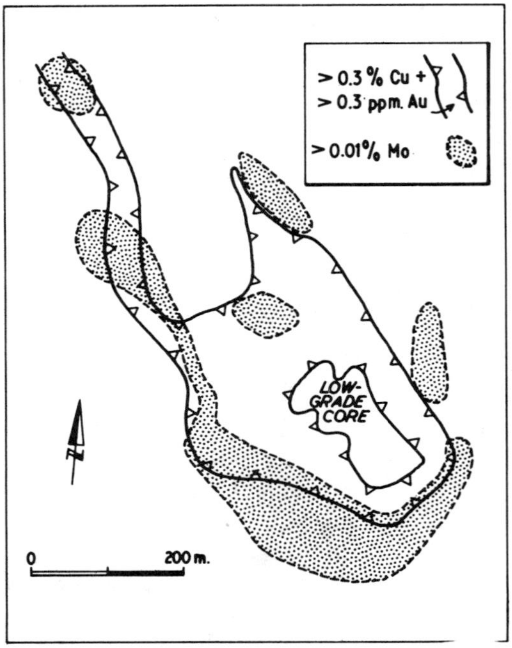

FIG. 2. Molybdenum halo (~30 Mt of 0.024% Mo) around copper-gold center, south body, Saindak, Pakistan. Weaker copper-gold mineralization continues into low-grade core. Extracted from Sillitoe and Khan (1977).

metal, with a substantial proportion of it closely associated with quartz, chalcopyrite, and, where present, bornite. Gold lost during flotation seems to be either coarse-grained; fine-grained and tied up in magnetite, quartz and hydrothermal silicate phases; or associated with pyrite introduced as a component of sericitic overprinting. It has also been observed that gold tellurides are present in porphyry copper deposits, as at Bucim, Yugoslavia, and Dos Pobres, Arizona, but they have yet to be described from the deposits depicted in Fig. 1.

On the basis of the writer's unpublished work in the Philippines and data supplied by Lowder and Dow (1978) for Cabang Kiri, Sulawesi, abnormally elevated gold values may also persist into the upper parts of high-gold porphyry systems, where gold accompanies high-pyrite advanced argillic assemblages. However, the suggestion that the Vunda gold prospect in Fiji represents such a high-level situation above a porphyry gold deposit (Lawrence, 1978) was not confirmed by the writer's field examination.

Portions of three porphyry copper-gold deposits are especially rich in gold: At current gold prices, central parts of the Dizon and Santo Thomas II deposits in the Philippines, carrying of the order of 2 ppm Au, could probably be economically exploited for their gold contents alone if they were present at or near surface. Moreover, 31 Mt of leached capping above the Ok Tedi ore body in Papua New Guinea assay 3.1 ppm Au, and are programmed to be mined during the early years of the operation (Howell et al., 1978). However, it is a moot point whether or not supergene gold enrichment took place at Ok Tedi. It is concluded that there is a high possibility of the existence of porphyry gold deposits, carrying only minor amounts of copper (and molybdenum), with similar geologic characteristics to the porphyry copper-gold deposits. Any such porphyry gold deposit is therefore likely to be present in the central part of a zone of K-silicate alteration, although advanced argillic alteration is also a potential host. The K-silicate alteration is likely to carry an anomalously high (up to 10 vol %) content of hydrothermal magnetite, accompanied in quartz-bearing host intrusives by abundant transparent grayish quartz as a dense stockwork and a replacement. Although a conventional leached-capping interpretation of such a hypothetical porphyry gold deposit would lead to a negative appraisal, magnetometry combined with gold geochemistry could act as an effective exploration tool.

The high fO_2/fS_2 suggested by abundant magnetite in many gold-rich porphyry copper deposits may combine with high initial gold contents of the systems to generate gold-rich core zones. Under low fO_2/fS_2 conditions, however, gold may tend to adopt its more usual external position in the zoning sequence (Barnes, 1975), as at Andacollo, Chile. There a porphyry copper-molybdenum deposit, albeit richer than other Chilean deposits in gold (0.25 ppm), possesses a substantial low-grade stratabound gold deposit hosted by four vesicular rhyolite flows in the outermost part of its propylitic halo (Llaumett, 1980; Sillitoe, 1983). Comparable is the case of the Battle Mountain district, Nevada, where the bulk-mined Tomboy gold deposit occurs in calcareous conglomerate distally with respect to the Copper Canyon stock, which itself carries copper-molybdenum-gold protore (Blake et al., 1978).

Several bulk low-grade gold deposits possess

numerous affinities with the porphyry deposit class (Sillitoe, 1983) but lack the characteristic zonal alteration pattern of porphyry copper-gold deposits, as well as containing unusual element associations. Porgera, Papua New Guinea; Salave, Spain; and Gilt Edge, South Dakota, are three such examples. At Porgera, some 125 Mt of 2.67 ppm Au, along with lead (~ 0.5%), zinc (~ 1%) and silver (12 ppm), are present in a multidirectional stockwork of veinlets in Upper Cretaceous shales and mudstones around irregular stocks and dikes of andesite and diorite porphyries (O'Dea, 1980) of probable shoshonitic affinity and late Cenozoic age. Alteration is dominantly propylitic although one veinlet generation possesses sericite-kaolinite selvages. Although associated with intramineral intrusions and ore-bearing hydrothermal breccias, the stockwork veinlets are typified by open-space filling and crustiform banding of epithermal character. At Salave, a gold-bearing stockwork, developed in an equigranular granodiorite pluton of Hercynian age, carries pyrite, arsenopyrite, stibnite, molybdenite, and minor sphalerite and chalcopyrite. Mineralization accompanies propylitic alteration, which in the high gold areas is rich in sericite, carbonate, and albite (Harris, 1980). At Gilt Edge, Tertiary latite and rhyolite porphyries, associated with linear bodies of hydrothermal breccia, transect Proterozoic metasedimentary rocks and host gold-molybdenum mineralization (Mukherjee, 1968). Breccias contain the highest-grade mineralization but appreciable tonnages of lower-grade gold mineralization are also present, with disseminated rather than stockwork pyrite in sericitized latite porphyry.

TIN

Porphyry tin deposits were first recognized from the central and southern parts of the Bolivian tin belt (Sillitoe et al., 1975) and similar deposits also occur elsewhere, as at Ardlethan, New South Wales, Australia (Sillitoe et al., 1975) and Khingan, USSR. Tin as cassiterite is also present in Climax-type molybdenum systems, where it tends to be concentrated with tungsten in the upper parts of and above the molybdenite-bearing ore shells, and is recovered as a by-product at Climax, Colorado (Wallace et al., 1968). At the enigmatic Majuba Hill prospect in Nevada, multiple intrusions of highly differentiated rhyolite porphyry carry tin in association with copper as well as molybdenum (MacKenzie and Bookstrom, 1976). Majuba Hill seems to possess similarities with the Bolivian porphyry tin deposits, as well as with Climax-type molybdenum deposits and copper-bearing tourmaline breccia pipes (Bookstrom and MacKenzie, 1978). Tin is also significant at the Mount Pleasant porphyry tungsten deposit in New Brunswick, Canada (see the following), where it is concentrated outward from central W-Mo-Bi zones (Dagger, 1972).

In Bolivia, porphyry tin deposits occur at Llallagua, Potosí, Oruro, Chorolque, San Pablo (Morococala district) and elsewhere. The porphyry-style mineralization is low-grade (0.1–0.3% Sn) and is only worked at Llallagua, where somewhat higher-grade brecciated parts of the ore body, which amounts to at least 80 Mt of 0.3% Sn, are extracted by block-caving methods. The Bolivian Mining Corp. (COMIBOL) has undertaken studies to determine the feasibility of open-pit mining the Llallagua porphyry tin resource.

Porphyry tin mineralization consists of fine-grained cassiterite and sulfides, mainly pyrite, and occurs as veinlets, disseminations, and breccia fillings within small (1–2 km^2) isolated upward-flaring stocks of pervasively altered felsic porphyry, probably of quartz latitic composition. Porphyries comprise phenocrysts of quartz, sanidine, sodic plagioclase, and biotite in a microcrystalline quartz-feldspar groundmass. Stocks were emplaced during the Lower and Middle Miocene (Grant et al., 1979) into Paleozoic sedimentary rocks, and at many localities, with the notable exception of the more deeply eroded Llallagua (Fig. 3) and San Pablo centers, are flanked by outward-dipping flows and pyroclastics of broadly similar composition.

The dispersed tin mineralization accompanies intense sericitic alteration, in which tourmaline is commonly widespread. Central areas at Chorolque were transformed to tin-poor quartz-tourmaline rock. The silver-rich upper parts of Potosí correspond to highly silicified advanced argillic alteration. Pyrite is ubiquitous, tends to be most abundant as an annulus around cassiterite-rich rocks, and is accompanied by minor pyrrhotite, arsenopyrite, chalcopyrite, stannite, and sphalerite. Weak propylitization surrounds the zones of feldspar-destructive alteration. Substantial portions of the tin-bearing stocks were transformed to a variety of syn- and late-mineral hydrothermal breccias, in which quartz and/or tourmaline replacement of rock-flour matrix is widespread. The highest-grade cassiterite mineralization at Llallagua is found in the more permeable breccias. Quartz-cassiterite or quartz-cassiterite-sulfide veins were superimposed on many of the porphyry tin-bearing stocks and their immediate wallrocks, and represent the waning stages of mineralization (Fig. 3). These veins support many of the major tin and tin-silver mining operations of central and southern Bolivia.

Fluid-inclusion and light-stable-isotope studies reported by Grant et al. (1980) have documented many similarities between the Bolivian porphyry tin deposits and porphyry copper-molybdenum deposits. Early tin-poor quartz-tourmaline alteration was developed as a product of high-temperature (>500°C) high-salinity boiling fluids of magmatic parentage, which became progressively diluted and cooled with meteoric

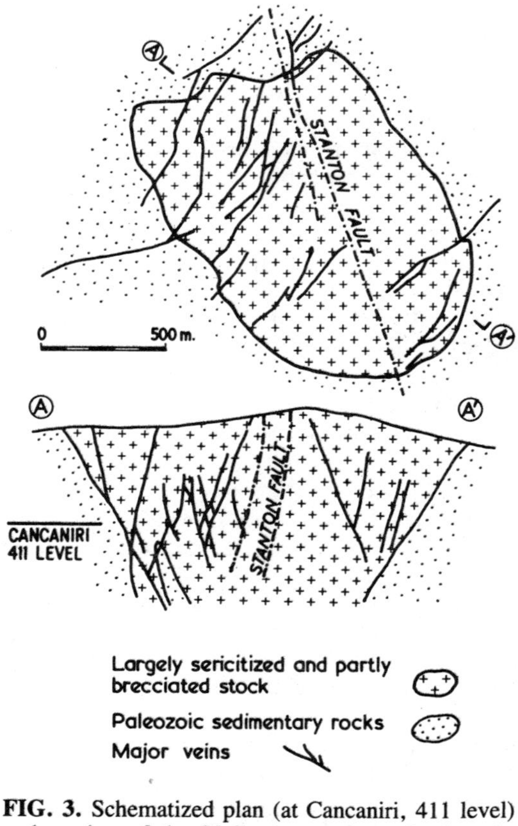

FIG. 3. Schematized plan (at Cancaniri, 411 level) and section of the Llallagua porphyry tin deposit, Bolivia. Taken from Turneaure (1935).

fluids during sericitization and porphyry-style cassiterite-sulfide deposition.

The features of the tin mineralization at Ardlethan are similar in many respects to the Bolivian tin deposits and, because of the abundance of hydrothermal breccias, resemble Llallagua in particular. Production to date plus reserves at Ardlethan amount to approximately 10 Mt of 0.45% Sn, although lowering the cutoff to about 0.1% Sn would probably quadruple this tonnage (R. Paterson, personal communication, 1980). Two-mica adamellite and quartz-feldspar porphyry dikes of late Silurian age are sericitized and chloritized, and carry abundant tourmaline and local topaz, siderite, and fluorite in a zone measuring some 1100 × 800 m at surface. Topaz is an early, centrally located alteration mineral and tends to be overprinted by chlorite and sericite (Scott, 1981). Within the alteration zone, hydrothermal breccias of various types and ages are widespread. Cassiterite is accompanied by pyrite, arsenopyrite, sphalerite, chalcopyrite, and galena as breccia fillings, veinlets, disseminations, and clots (Paterson, 1976). Since the Ardlethan deposit is associated with equigranular granite and adamellite as well as the later quartz porphyry dikes, it may represent a deeper level of erosion than that in the Bolivian deposits, although extrusive equivalents, including nearby garnetiferous quartz porphyry, do occur in the Ardlethan belt. A relatively deep level of emplacement might be supported by nearby stanniferous greisen.

The Khingan deposit in the USSR appears from Materikov's (1977) description to possess similarities to Ardlethan, with pervasively altered hydrothermal breccias hosting the best cassiterite ore.

TUNGSTEN

Tungsten as the principal metal occurs in at least three apparently different types of porphyry deposits, as exemplified by Yangchuling and other deposits, southeastern China; Mount Pleasant, New Brunswick, Canada; and Logjam Creek (Logtung), Yukon, Canada. Only Yangchuling currently produces, on a restricted scale, although Mount Pleasant is under joint development by Billiton Canada Ltd. and Brunswick Tin Mines Ltd. Tungsten, as either scheelite and/or wolframite, is also present as a minor constituent of several standard porphyry molybdenum and copper-molybdenum deposits in British Columbia and the Yukon (Endako, Kitsault, Adanac, Glacier Gulch, Casino), the Cascades (Glacier Peak, Van Epps), Peru (Toro Mocho, Santa Rosa) and elsewhere; one of the highest recorded grades is 0.024% W at Toro Mocho (Exploration Dept., Centromin-Perú, written communication, 1981). Tungsten is also present in zones within and immediately above hemispherical molybdenum shells in Climax type deposits, where it takes the form of huebnerite-wolframite; it is recovered as a byproduct at Climax, Colorado (Wallace et al., 1968). Scheelite tends to be preferentially concentrated in calc-silicate skarns abutting porphyry-copper-molybdenum deposits, and is now recovered as a byproduct at Twin Buttes, Arizona, where it commonly attains grades of 0.02–0.04% WO_3 in garnet skarn (Barter and Kelly, 1982).

Porphyry tungsten deposits should not be confused with potentially bulk-minable greisen-type tungsten deposits, such as that at Hemerdon, Devon, England (Horsnail, 1979), where wolframite and subsidiary cassiterite occur in a swarm of narrow greisen-bordered quartz veins. More equivocal cases do exist, however: at Cleveland, Tasmania, a substantial tin-copper deposit of Upper Devonian age occurs as massive replacements in carbonates and is underlain by a quartz-fluorite-wolframite-molybdenite stockwork (Collins, 1981), which at deeper levels is developed in a recently discovered quartz porphyry stock. The stockwork carries ~ 0.3% W, as well as minor molybdenum and tin values, but grades decrease downward in the stock (Anon., 1982). Although greisen-type tin deposits are present in granite a few kilometers to the south, the stockwork tungsten-molybdenum-tin mineralization at Cleveland has features more typical of a porphyry-type development.

At Yangchuling, Jiangxi province, the porphyry tungsten deposit occurs in Jurassic biotite granodiorite and granodiorite porphyry, of the petrological type commonly associated with porphyry copper deposits (Yan et al., 1980). The granodioritic complex was emplaced into Proterozoic clastic sediments and was associated with hydrothermal brecciation. Tungsten and molybdenum mineralization is concentrated in lenslike bodies on the margins and tops of granodiorite porphyry bodies, as well as in breccia pipes. Ore consists of veinlets and disseminations of scheelite, molybdenum-rich scheelite, and molybdenite, associated with chalcopyrite, bismuthinite, wolframite, sphalerite, pyrite, and pyrrhotite. Highest-grade tungsten-molybdenum ore occurs in quartz veinlets as an accompaniment to sericitic alteration and silicification, although scheelite and molybdenite deposition also occurred with K-feldspar during earlier, centrally located K-silicate alteration (Yan et al., 1980). There is a general tendency for molybdenum grades to increase downward in the deposit at the expense of tungsten. Similar to Yangchuling is the Xingluokeng deposit, Fujian province, which comprises wolframite, scheelite, and molybdenite mineralization in veinlets and disseminations in a small subvolcanic stock, which was subjected to sericitic and K-silicate alteration (Liu, 1981).

The Logjam Creek porphyry tungsten prospect, in southernmost Yukon, is superficially similar to Yangchuling but is genetically related to a petrochemically different type of intrusive rock, of quartz monzonitic composition, which elsewhere in the region gave rise to major tungsten-bearing skarn deposits. (See the following.) The quartz monzonite was emplaced into Mississippian sedimentary rocks, probably during the Cretaceous. The tungsten- and molybdenum-bearing quartz stockwork is centered on a felsic porphyry dike, which is an offshoot of the main quartz monzonite (Rowe, 1980). The quartz veinlets contain molybdenite, scheelite, and powellite, accompanied by accessory pyrite, pyrrhotite, sphalerite, garnet, fluorite, and beryl. Scheelite becomes more abundant outward in the stockwork at the expense of molybdenite, and K-silicate, sericitic, and propylitic alteration assemblages attend the mineralization (Rowe, 1980). In late 1980, a geological reserve of 163 Mt grading 0.12% WO_3 and 0.052% MoS_2 was announced, of which some 25% of the tungsten and 10% of the molybdenum occur in a swarm of late sheeted quartz veins and, to a lesser extent, in marginal skarns (Rowe, 1980).

In contrast, the Mount Pleasant, New Brunswick, porphyry tungsten deposits are associated with two partly flow-banded and brecciated rhyolite porphyry plugs of Mississippian age, which intruded their volcanic precursors, including ignimbrites. Each plug gave rise to W-Mo-Bi mineralization in a central alteration zone characterized by quartz (70%), topaz (10%), fluorite (5%), and sericite (Parrish and Tully, 1978) and, at depth, appreciable K-feldspar

(Dagger, 1972). This pervasive quartz-topaz alteration grades outward to a quartz-chlorite-sericite assemblage beyond the ore zone (Pouliot et al., 1978), in a manner reminiscent of that described previously at Ardlethan. Ore occurs as breccia fillings, disseminations, veinlets, and pods of wolframite, molybdenite, native bismuth, and arsenopyrite, surrounded but largely post-dated by tin minerals (cassiterite, stannite), arsenopyrite, sphalerite, chalcopyrite, and galena (Fig. 4); pyrite is deficient. The most important deposit, the Fire Tower zone, contains some 25 Mt, within which 9.15 Mt carrying 0.39% WO_3 and 0.12% MoS_2 constitute the proven ore reserve (Atkinson et al., 1981); bismuth grades average about 0.08%. The smaller North zone, 1 km distant, contains comparable tungsten, molybdenum, and bismuth grades, but is richer in tin (0.1%) (Parrish and Tully, 1978).

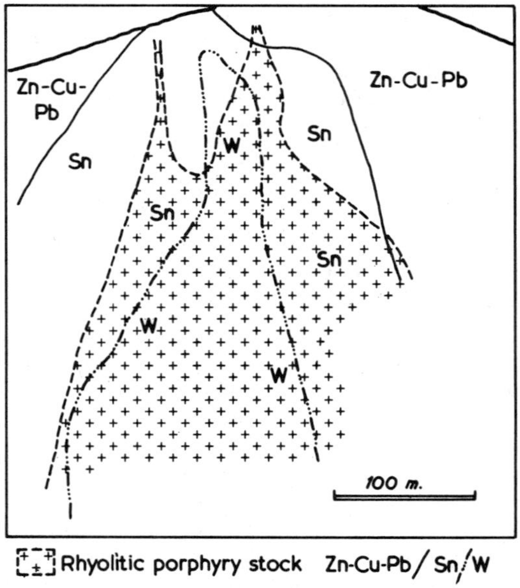

FIG. 4. Section (40 300 N) through the marginal part of the Fire Tower zone, Mount Pleasant porphyry tungsten deposit, New Brunswick, Canada. Taken from Dagger (1972).

SILVER-LEAD-ZINC

Data on the distribution of these three commonly associated metals in most porphyry systems are meager. In general, however, they are concentrated around or above porphyry copper, molybdenum, tin, or tungsten centers, as predicted theoretically (Barnes, 1975). They occur in well-defined veins, in contact-metasomatic replacements when wallrocks comprise carbonates, and as irregular veinlets, replacements, and disseminations, which locally, as at Sam Goosly, British Columbia, attain commercial proportions.

Silver is commonly extracted as a byproduct from porphyry copper-molybdenum and copper-gold deposits, in which values of 1 to 3 ppm are normal (Fig. 5). The highest silver contents known to the writer are 7.7 and 7.0 ppm at Galore Creek, British Columbia, and Toro Mocho, Peru, respectively, although the occurrence elsewhere of even higher values cannot be ruled out. However, even these high values are at least an order of magnitude too low to constitute an ore of silver. The precise mineralogical residence of the byproduct silver in porphyry deposits is poorly constrained, although small amounts are certainly present in solid solution in chalcopyrite and native gold (electrum), and some could occur as a solid solution in tetrahedrite-tennantite or galena (see the following), thereby accounting for the low silver recovery in some flotation circuits tailored to chalcopyrite ores (e.g., Dizon). Rare matildite, hessite, sylvanite, and native silver have also been reported.

When averaged over an entire porphyry copper ore body, lead and zinc values are generally trifling, although it is reported that zinc and lead as sphalerite and galena attain 1450 and 310 ppm, respectively, at Toro Mocho, Peru (Exploration Dept., Centromin-Perú, written communication, 1981), and zinc as sphalerite attains 900 ppm at Taysan, Philippines (Wolfe et al., 1978).

It is well documented that the occurrence of sphalerite, galena, and tetrahedrite-tennantite in porphyry copper-molybdenum/gold or Climax-type molybdenum ore zones is a result of late-stage superposition, commonly as an accompaniment to retrograde feldspar-destructive alteration. Such a situation accounts for the high zinc values at Taysan, where even late-mineral dikes are zinc bearing (Wolfe et al., 1978), and the argentiferous tetrahedrite-tennantite, sphalerite, and galena at Toro Mocho, which are concentrated in the peripheral sericitized parts of the copper-molybdenum-bearing stock as well as in superimposed vein systems (Exploration Dept., Centromin-Perú, written communication, 1981). Zinc, copper, and lead are also later

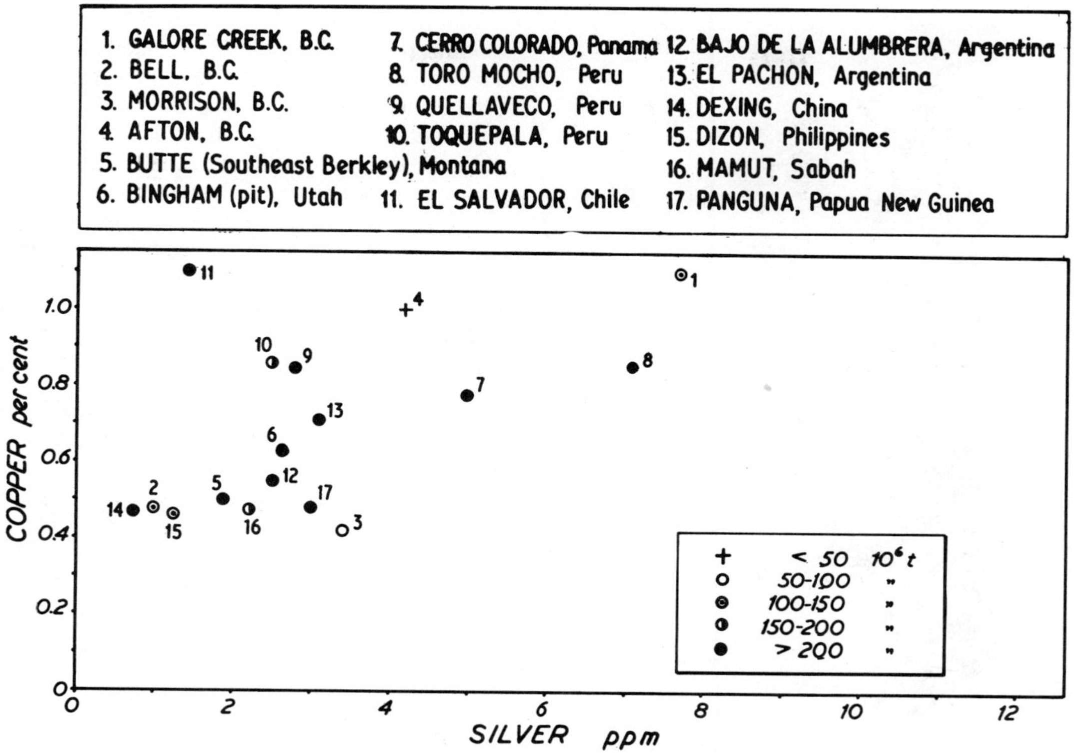

FIG. 5. Representative hypogene silver grades of porphyry copper deposits, including highest values known to the writer. Reserves in metric tons.

than (and peripheral to) W-Mo-Bi at Mount Pleasant, as noted previously.

It is proposed that most silver, lead, and/or zinc concentrations in porphyry systems (additional to peripheral veins or replacements) are likely to be products of late-stage superposition, which is progressively more probable as one goes higher in porphyry systems. A number of examples of appreciable lead-zinc accumulations, albeit subore grade, are indeed known from such high-level locales: At the Mi Vida porphyry copper-molybdenum prospect, Argentina, part of a late-mineral breccia pipe, some 2 km^2 at surface, underwent advanced argillic alteration in which sphalerite and galena accompany pyrite, marcasite, bornite, covellite, enargite, molybdenite, rhodochrosite, and native sulfur; combined lead-zinc values locally exceed 0.6% (Koukharsky and Mirré, 1976). Similarly, high-level portions of porphyry copper-molybdenum systems characterized by sericite-alunite alteration at Red Mountain, Arizona, and sericitic alteration at Morenci, Arizona, carry 0.15–0.2% and about 0.1% Zn, respectively (Corn, 1975; Moolick and Durek, 1966).

At the Redwell Basin Climax-type porphyry molybdenum prospect, Colorado, near-surface parts of a breccia pipe extending upward from two molybdenite-bearing shells contain ~ 0.8% combined lead-zinc as galena and sphalerite in veins, veinlets, and disseminations, (Fig. 6; Sharp, 1978). Sphalerite, galena, and rhodochrosite are also concentrated above the Henderson Climax-type molybdenum deposit (White et al., 1981). Furthermore, the low-grade galena-sphalerite-tetrahedrite mineralization in vugs and veinlets within the upper parts of a quartz latite porphyry stock at Hahns Peak, Colorado, which Young and Segerstrom (1973) considered as a disseminated silver-lead-zinc prospect, has been shown to be intimately tied to late-stage superposition of advanced argillic and sericitic alteration (Dowsett, 1980).

No strictly porphyry silver-lead-zinc deposit or prospect is known to the writer, although it should be emphasized that silver, lead, and zinc

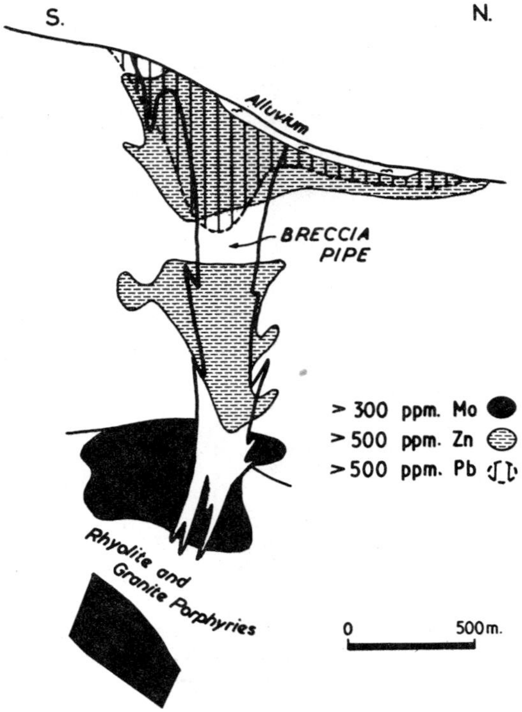

FIG. 6. Section through the Redwell Basin Climax-type porphyry molybdenum prospect, Colorado, to show zoning of lead and zinc with respect to molybdenum. Simplified from Sharp (1978).

accompany gold at the Porgera deposit mentioned previously, and bulk-tonnage stockwork silver deposits in volcanic host rocks are known at several places, such as Rochester, Nevada. Consequently the characteristics of any porphyry-type deposit containing exclusively silver, lead, and/or zinc cannot be surmized.

OTHER COMMODITIES

Rhenium

It is well known that appreciable (up to about 1500 ppm) rhenium substitutes for molybdenum in molybdenite from some porphyry copper-molybdenum deposits (Sutulov, 1976), with production of the element recorded from Island Copper, British Columbia; Kounrad, Kazakhstan; Ely, Nevada; Bingham, Utah; El Salvador, Chile; and elsewhere.

Uranium

Variable but small amounts of uranium are recovered from oxide-leach operations at several porphyry copper-molybdenum deposits, including Bingham, Utah; Twin Buttes, Arizona; Toro Mocho, Peru; and, from the contiguous exotic copper deposit, Chuquicamata, Chile. In a detailed examination of the Copper Cities porphyry copper deposit, Arizona, Still (1962) showed that uranium was present in the leached capping as metatorbernite and plumbogummite and that oxidation of hypogene uranothorite produced no discernable enrichment or impoverishment of uranium values, which averaged 0.005–0.006% U_3O_8. Some redistribution of uranium during oxidation does, however, seem likely.

Lithophile-element porphyries commonly contain substantially greater amounts of uranium, as documented for Climax and Henderson, Colorado (Desborough and Sharp, 1978; Desborough and Mihalik, 1980) and Llallagua, Bolivia, although some standard porphyry copper-molybdenum deposits, such as Larap, Philippines, are also unusually enriched in the element. Potential must also exist, therefore, for the commercial recovery of uranium from some unoxidized porphyry ores, perhaps by passing flotation tailings through a gravity circuit.

The possible existence of porphyry uranium deposits was proposed by Armstrong (1974) but, given that Rössing, Namibia, cannot be considered as a porphyry uranium deposit, their existence remains to be substantiated.

Magnetite

Magnetite was recovered for a time from flotation tailings at the gold-rich Santo Tomas II porphyry copper deposit in the Philippines (see previously), and is scheduled as a byproduct, along with important gold, when Bucim and Veliki Krivelj, Yugoslavia, and the south body at Saindak, Pakistan, come on stream. Much of the magnetite is a component of K-silicate alteration.

Platinoids

Although data are scanty, platinum and palladium recovery from anode slimes has resulted from refining of copper from a few porphyry copper deposits, including Ely, Nevada, which contained 0.012 ppm Pd and 0.003 ppm Pt. The recovery of platinoids seems to be commonplace from massive enargite-bearing pyrite ores situated high in porphyry copper systems, as at Butte, Montana, and Bor, Yugoslavia.

Rutile

Rutile is the principal form of titanium in both K-silicate and sericitic alteration zones of porphyry copper deposits (e.g., Lawrence and Savage, 1975). It is particularly abundant in K-silicate-altered stocks, where it is closely associated with low-titanium hydrothermal biotite (Czamanske et al., 1981). Its widespread occurrence in copper ore zones, in amounts ranging from 0.3–0.6% in several Arizona deposits, could render it amenable to recovery from flotation tailings as a source of titanium metal (Czamanske et al., 1981).

Rare Metals

Thorium, tantalum, columbium, scandium, and the light rare earths are some of the rare metals present in anomalously high trace amounts in Climax-type porphyry molybdenum deposits (e.g., White et al., 1981) and similar lithophile-element porphyries like Mount Pleasant. They occur as such accessory minerals as ilmenorutile, aeschynite, columbite, monazite, and thorite at Henderson and Climax, where they have been assigned a primary magmatic origin (Desborough and Sharp, 1978; Desborough and Mihalik, 1980). Some of these rare metals might be amenable to byproduct recovery at the Colorado porphyry molybdenum deposits or elsewhere in the future.

ENVIRONMENTS FOR EXPLORATION

Gold

Putative porphyry gold deposits, like their gold-rich copper counterparts, are likely to occur in association with shoshonitic or calc-alkaline stocks, which are I-types in the sense of Chappell and White (1974) and belong to the magnetite series of Ishihara (1977). Although the most promising search areas are undoubtedly in island arcs, particularly in the western Pacific region, they could also occur in Cordilleran (Andean-type) belts. The prospective volcano-plutonic arcs are those most directly related to subduction of ocean lithosphere (Fig. 7a) and characterized by ore deposits carrying copper, iron, gold, silver, and molybdenum, with or without tungsten (Sillitoe, 1981).

Tin

Not a great deal is known of the petrochemical affiliations of the Bolivian tin deposits, although a recent summary of available data (Sillitoe, 1981) strongly suggests that they belong to the S-type of Chappell and White (1974). The presence of accessory magmatic cordierite in the Salvadora stock at Llallagua (Grant et al., 1980) would support such a conclusion. The host rocks at Ardlethan, including garnet-bearing varieties, are part of the type suite of S-type granitoids (White et al., 1977). Intrusive rocks in the Bolivian deposits and at Ardlethan are highly evolved with low K/Rb ratios, high fluorine (at Ardlethan), and high lithium (in Bolivia). It may be concluded therefore that S-type terranes offer prospective search areas for porphyry tin deposits, where erosion has not been too deep. However, I-type igneous suites poor in, or devoid of, accessory magnetite [part of Ishihara's (1977) ilmenite series] are commonly stanniferous and could also possess porphyry tin potential.

If the Bolivian deposits and Ardlethan are typical of the class, then back-arc settings landward of principal volcano-plutonic arcs along Cordilleran margins are preferred locations for porphyry tin generation (Figs. 7b and 8b) (Sillitoe, 1981). However, the presence of accessory tin in Climax-type molybdenum deposits and at Mount Pleasant suggests that rift settings might also offer a suitable formational environment, as discussed further. Continent/continent collision belts (e.g., Higher Himalaya, Main Range of peninsular Malaysia) or megashears marking intracontinental transforms (e.g., Hercynian of Britanny), which are also characterized by S-type and/or ilmenite-series granitoids, are considered unfavorable for porphyry tin deposits because intrusion is generally deep-seated and lacks accompanying volcanism.

Tungsten

On the basis of available evidence, a wide variety of petrogenetic and geotectonic settings appears suitable for porphyry tungsten generation: The Yangchuling deposit, in common with neighboring porphyry copper deposits, and the Xingluokeng deposit, is associated with I-type, magnetite-series rocks (Li and Xu, 1981), and could have been emplaced as part of a subduction-related Cordilleran belt (Fig. 8a). Although superficially similar, the Logjam Creek prospect appears to be part of a tungsten-enriched suite of intrusives in the Yukon/Northwest Territories border region characterized by high initial strontium isotope ratios (Godwin et al., 1980; Kuran

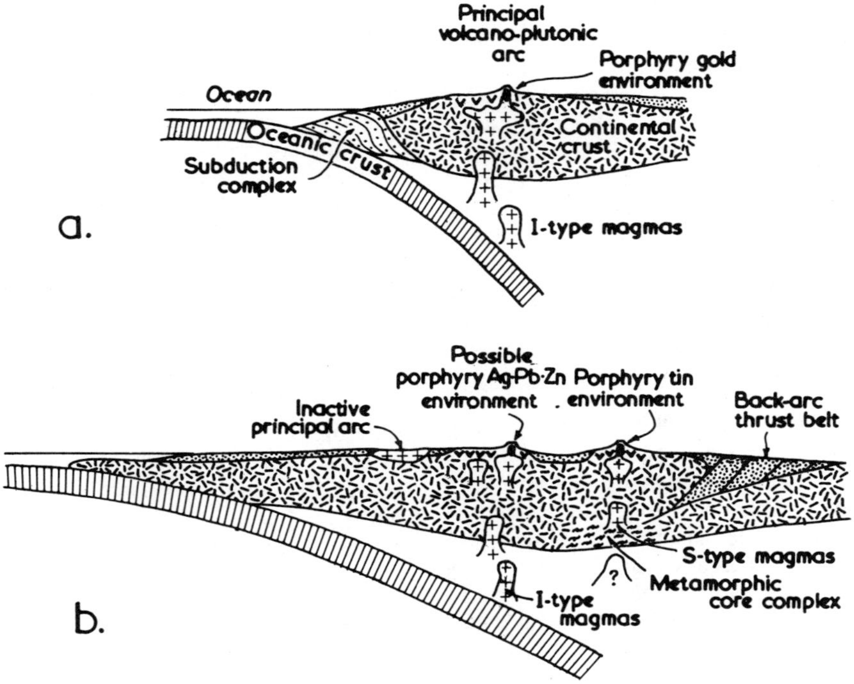

FIG. 7. Schematic representation of possible environments for generation of porphyry deposits at convergent plate margins: (a) Possible porphyry gold deposit in a principal arc above a moderate to steep subduction zone. Stress regime neutral. (b) Possible porphyry Ag-Pb-Zn deposit on inner side of principal arc and porphyry tin deposit in back-arc region, both above a shallow subduction zone. Stress regime compressional. See Sillitoe (1981) for further explanation.

et al., 1982), and presence of magmatic garnet and muscovite, both S-type characteristics. Sillitoe (1981) tentatively concluded that the tungsten deposits were emplaced in a back-arc setting bounded to the east by an east-directed fold-thrust belt similar to that immediately east of the Bolivian tin belt (Fig. 8b). Tempelman-Kluit (1979), however, linked tungsten-rich stock emplacement to crustal compression consequent upon collision of an arc terrane with the continental margin.

In contrast to these two scheelite-bearing porphyry tungsten deposits, the wolframite-bearing Mount Pleasant deposit is associated with high-silica rhyolite and exhibits enrichment in lithophile elements such as fluorine, columbium, rubidium, and thorium (Dagger, 1972). Such lithophile-element enrichment, especially of fluorine, is that typical of Climax-type molybdenum deposits in particular (Sillitoe, 1980) and of Loiselle and Wones' (1979) A-type granitoids in general. A variety of rift settings was proposed by Sillitoe (1980, 1981) for Climax-type molybdenum deposits (and A-type granitoids). They include rifting along a continental margin induced by slowing and eventual cessation of subduction (Fig. 8c), as in the western USA after 30 m.y.; intraplate rifting associated with hot-spot activity (Fig. 8d), which may abort or presage opening of an ocean basin; foreland rifting as a result of stress transmission from a zone of continent/continent collision (Fig. 8f); and pull-apart rifts along intracontinental transform fault zones (Fig. 8e). It is such a localized extensional situation associated with an early Carboniferous, north-northwest-trending dextral transform fault (see Ruitenberg, 1972), which triggered the Mount Pleasant magmatism and associated porphyry-type mineralization. It is tentatively suggested that any of these rift settings would be appropriate for generation of porphyry tungsten deposits or, for that mattter, of

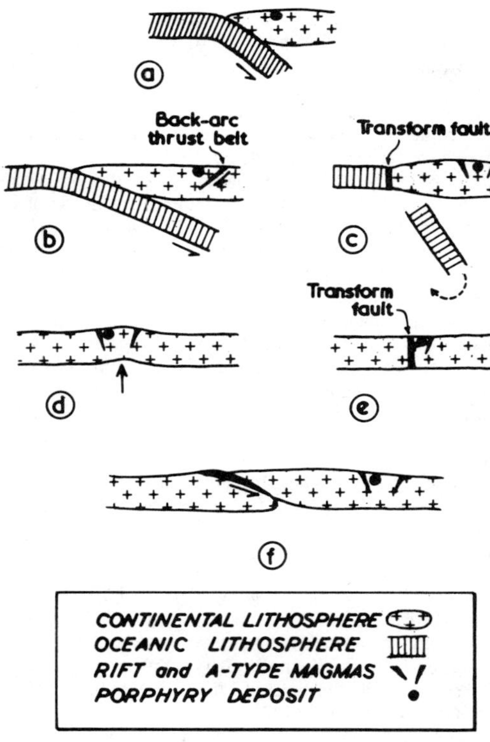

FIG. 8. Sketches illustrating possible geotectonic settings for generation of lithophile-element porphyry deposits: (a) in a principal arc above a moderate to steep subduction zone (e.g., possibly Yangchuling porphyry tungsten); (b) in a Cordilleran back-arc setting above a shallow subduction zone (e.g., Bolivan porphyry tin, possibly Logjam Creek porphyry tungsten); (c) in a back-arc setting during and after cessation of subduction [e.g., Climax-type porphyry molybdenum (+W-Sn) of Colorado]; (d) within a continental plate during rifting; (e) within a pull-apart rift along an intracontinental transform fault zone (e.g., Mount Pleasant porphyry W-Mo-Bi); and (f) within a rifted overthrust plate during continent/continent collision.

any lithophile-element porphyry systems.

Silver-Lead-Zinc

Since it appears that silver, lead, and zinc are capable of concentration in the upper parts of all the types of porphyry systems considered previously, no specific petrochemical signature or geotectonic setting is thought to be especially favorable for their generation. However, if true porphyry silver-lead-zinc deposits do exist, then it is proposed that the major provinces of these metals, which occur in association with I-type magnetite-related volcano-plutonic arcs (Fig. 7b), as in central Peru and northern Mexico (Sillitoe, 1981), offer the most prospective target areas. It is perhaps noteworthy that the Ag-Pb-Zn-rich Toro Mocho deposit occurs in central Peru.

CONCLUSIONS

Porphyry tin and tungsten deposits are known, and both are being exploited on a relatively small scale. In addition, exploration and development of porphyry tungsten deposits are currently active, and the likelihood of further discoveries is considered good. It might be further predicted that a porphyry-type source will become progressively more important for future world supplies of both tin and tungsten. Porphyry gold deposits, in the strictest sense, have not been identified although chances of their existence are thought to be high. Nevertheless, parts of some gold-rich porphyry copper deposits would be amenable to exploitation of gold alone, and bulk-tonnage gold deposits that have many affinities with porphyry deposits occur at several places. Silver-lead-zinc porphyries have not been defined, although porphyry-style accumulations of these metals may occur as late additions to most types of porphyry deposit, especially in their upper parts. However, these high-level accumulations or any true porphyry silver-lead-zinc prospect that might be encountered is likely to present a viable target only for silver at the present time; lead-zinc values are unlikely to attain the several percent required to permit their extraction as principal metals. Finally it may be noted that copper, molybdenum, gold, tin, and tungsten are all metals concentrated in skarn-type deposits adjoining unmineralized plutons as well as in porphyry deposits. Since silver, lead, and zinc are also concentrated in such skarn-type settings, a case could perhaps be made in favor of their occurrence somewhere as principal metals in a true porphyry deposit.

ACKNOWLEDGMENTS

Reviews by A.A. Bookstrom, D.P. Cox, and J.N. Grant are gratefully acknowledged.

REFERENCES

Anon., 1982, "Good Tungsten Grades in Tasmania," *Mining Journal,* Jan. 8, p. 29.

Armstrong, F.C., 1974, "Uranium Resources of the Future—'Porphyry' Uranium Deposits," *Formation of*

Uranium Ore Deposits, International Atomic Energy Agency, Vienna, pp. 625–634.

Atkinson, J.R., Kooiman, G., and Coates, H.J., 1981, "Geology of Mount Pleasant Tungsten, New Brunswick," *Canadian Mining Journal,* Vol. 102, No. 4, pp. 73–75.

Barnes, H.L., 1975, "Zoning of Ore Deposits: Types and Causes," *Transactions,* Royal Society of Edinburgh, Vol. 69, pp. 295–311.

Barr, D.A., Fox, P.E., and Northcote, K.E., 1976, "The Alkaline Suite Porphyry Deposits—A Summary," *Porphyry Deposits of the Canadian Cordillera,* A. Sutherland Brown, ed., Spec. Vol. No. 15, Canadian Institute of Mining & Metallurgy, pp. 359–367.

Barter, C.F., and Kelly, J.L., 1982, "Geology of the Twin Buttes Mineral Deposit: Pima Mining District, Pima County, Arizona," *Advances in Geology of the Porphyry Copper Deposits, Southwestern North America,* S.R. Titley, ed., University of Arizona Press, Tucson, pp. 407–432.

Blake, D.W., Theodore, T.G., and Kretschmer, E.L., 1978, "Alteration and Distribution of Sulfide Mineralization at Copper Canyon, Lander County, Nevada," *Proceedings,* Porphyry Copper Symposium, *Digest,* Arizona Geological Society, Vol. 11, pp. 67–78.

Bookstrom, A.A., 1981, "Tectonic Setting and Generation of Rocky Mountain Porphyry Molybdenum Deposits," *Relations of Tectonics to Ore Deposits in the Southern Cordillera,* W.R. Dickinson and W.D. Payne, eds., *Digest,* Arizona Geological Society, Vol. 14, pp. 215–226.

Bookstrom, A.A., and MacKenzie, W.B., 1978, "Majuba Hill, Pershing County, Nevada," *in* Conceptual Models of Exploration Geochemistry, T.S. Lovering and J.H. McCarthy, Jr., eds., *Journal of Geochemical Exploration,* Vol. 9, pp. 216–222.

Chappell, B.W., and White, A.J.R., 1974, "Two Contrasting Granite Types," *Pacific Geology,* Vol. 8, pp. 173–174.

Collins, P.L.F., 1981, "The Geology and Genesis of the Cleveland Tin Deposit, Western Tasmania: Fluid Inclusions and Stable Isotope Studies," *Economic Geology,* Vol. 76, pp. 365–392.

Corn, R.M., 1975, "Alteration-Mineralization Zoning, Red Mountain, Arizona," *Economic Geology,* Vol. 70, pp. 1437–1447.

Czamanske, G.K., Force, E.R., and Moore, W.J., 1981, "Some Geologic and Potential Resource Aspects of Rutile in Porphyry Copper Deposits," *Economic Geology,* Vol. 76, pp. 2240–2246.

Dagger, G.W., 1972, "Genesis of the Mount Pleasant Tungsten-Molybdenum-Bismuth Deposit, New Brunswick, Canada," *Transactions,* Institution of Mining & Metallurgy, Sec. B, Vol. 81, pp. B73–B102.

Desborough, G.A., and Mihalik, P., 1980, "Accessory Minerals in the Igneous Host of Molybdenum Ore, Henderson Mine, Colorado," Open-File Report 80-661, US Geological Survey.

Desborough, G.A., and Sharp, W.N., 1978, "Tantalum, Uranium, and Scandium in Heavy Accessory Oxides, Climax Molybdenum Mine, Climax, Colorado," *Economic Geology,* Vol. 73, pp. 1749–1751.

Dowsett, R.R., 1980, "Hydrothermal Alteration of the Hahns Peak Stock, Routt County, Colorado," *Economic Geology,* Vol. 75, pp. 30–44.

Godwin, C.I., Armstrong, R.L., and Tompson, K.M., 1980, "K-Ar and Rb-Sr Dating and the Genesis of Tungsten at the Clea Tungsten Skarn Property, Selwyn Mountains, Yukon Territory," *Bulletin,* Canadian Institute of Mining & Metallurgy, Vol. 73, No. 821, pp. 90–93.

Grant, J.N., et al., 1979, "K-Ar Ages of Igneous Rocks and Mineralization in Part of the Bolivian Tin Belt," *Economic Geology,* Vol. 74, pp. 838–851.

Grant, J.N., et al., 1980, "Evolution of the Porphyry Tin Deposits of Bolivia," *in* Granitic Magmatism and Related Mineralization, S. Ishihara and S. Takenouchi, eds., *Mining Geology,* Spec. Issue No. 8, pp. 151–173.

Harris, M., 1980, "Hydrothermal Alteration of Salave Gold Prospect, Northwest Spain," *Transactions,* Institution of Mining & Metallurgy, Sec. B, Vol. 89, pp. B5–B15.

Horsnail, R.F., 1979, "The Geology of Tungsten," *Tungsten,* Mining Journal Books Ltd., London, pp. 18–31.

Howell, W.J.S., et al., 1978, "History of the Ok Tedi Porphyry Copper Prospect, Papua New Guinea. II. The Years 1975 to 1978," *Economic Geology,* Vol. 73, pp. 802–809.

Ishihara, S., 1977, "The Magnetite-Series and Ilmenite-Series Granitic Rocks," *Mining Geology,* Vol. 27, pp. 293–305.

Koukharsky, M., and Mirré, J.C., 1976, "Mi Vida Prospect: A Porphyry Copper-Type Deposit in Northwestern Argentina," *Economic Geology,* Vol. 71, pp. 849–863.

Kuran, V.M., Godwin, C.I., and Armstrong, C.L., 1982, "Geology and Geochronometry of the Scheelite Dome Tungsten-Bearing Skarn Property, Yukon Territory," *Bulletin,* Canadian Institute of Mining & Metallurgy, Vol. 75, No. 838, pp. 137–142.

Lawrence, L.J., 1978, "Porphyry Type Gold Mineralization in Shoshonite at Vunda, Fiji," *Proceedings,* Australasian Institute of Mining & Metallurgy, No. 268, pp. 21–31.

Lawrence, L.J., and Savage, E.N., 1975, "Mineralogy of the Titaniferous Porphyry Copper Deposits of Melanasia," *Proceedings,* Australasian Institute of Mining & Metallurgy, No. 256, pp. 1–14.

Li, C.-Y., and Xu, J., 1981, "Genetic Classification of Tungsten Deposits of Jiangxi and Its Neighbouring Provinces," Symposium on Tungsten Geology, Oct., Nanchang, China, ESCAP-Regional Mineral Resources Development Centre, Preprint.

Liu, W., 1981, "The Geological Features of Mineralization of the Xingluokeng Tungsten-Molybdenum Deposit, Fujian," Symposium on Tungsten Geology, Oct., Nanchang, China, ESCAP-Regional Mineral Resources Development Centre, Preprint.

Llaumett, P., C., 1980, "Antecedentes Sobre Producción, Recursos y Expectativas de la Minería del Oro en Chile," *Minerales* (Chile), Vol. 35, No. 150, pp. 9–20.

Loiselle, M.C., and Wones, D.R., 1979, "Characteristics and Origin of Anorogenic Granites," *Abstracts with Programs,* Geological Society of America, Vol. 11, p. 468.

Lowder, G.G., and Dow, J.A.S., 1978, "Geology and Exploration of Porphyry Copper Deposits in North Sulawesi, Indonesia," *Economic Geology,* Vol. 73, pp. 628–644.

MacKenzie, W.B., and Bookstrom, A.A., 1976, "Geology of the Majuba Hill Area, Pershing County, Nevada," Bulletin 86, Nevada Bureau of Mines & Geology, 23 pp.

Materikov, M.P., 1977, "Deposits of Tin," *Ore Deposits of the USSR,* V.I. Smirnov, ed., Pitman Publishing, London, pp. 229–294.

Moolick, R.T., and Durek, J.J., 1966, "The Morenci District," *Geology of the Porphyry Copper Deposits, Southwestern North America,* S.R. Titley and C.L. Hicks, eds., University of Arizona Press, Tucson, pp. 221–231.

Mukherjee, N.E., 1968, "Alteration, Trace Element Dispersion, and Mineralization in the Intrusives of the Galena-Gilt Edge Area, Northern Black Hills, South Dakota," *Abstracts with Programs,* Geological Society of America, Rocky Mountain Section, p. 54.

O'Dea, T., 1980, "Gold Mineralization at Porgera, Papua New Guinea," *Proceedings,* New Zealand Conference, 1980, Australasian Institute of Mining & Metallurgy, Victoria, pp. 9–23.

Parrish, I.S., and Tully, J.V., 1978, "Porphyry Tungsten Zones at Mt. Pleasant, N.B.," *Bulletin,* Canadian Institute of Mining & Metallurgy, Vol. 71, No. 794, pp. 93–100.

Paterson, R., 1976, "Ardlethan Tin Mine," *Ore Deposits of the Lachlan Fold Belt, New South Wales,* L.B. Gilligan, N. Markham, and D.W. Suppel, eds., Excursion Guide No. 15C, 25th International Geological Congress, pp. 36–43.

Pouliot, G., et al., 1978, "Distribution of Alteration Minerals in the Fire Tower Zone at Brunswick Tin Mines Ltd., Mount Pleasant Area, New Brunswick," *Canadian Mineralogist,* Vol. 16, pp. 223–237.

Rowe, J.D., 1980, "Encouraging Results from the Logjam Creek Tungsten-Moly Studies," *Canadian Mining Journal,* Vol. 101, No. 9, pp. 71–73.

Ruitenberg, A.A., 1972, "Metallization Episodes Related to Tectonic Evolution, Rolling Dam and Mascarene-Nerepis Belts, New Brunswick," *Economic Geology,* Vol. 67, pp. 434–444.

Scott, K.M., 1981, "Wall-Rock Alteration in Disseminated Tin Deposits, Southeastern Australia," *Proceedings,* Australasian Institute of Mining & Metallurgy, No. 280, pp. 17–28.

Sharp, J.E., 1978, "A Molybdenum Mineralized Breccia Pipe Complex, Redwell Basin, Colorado," *Economic Geology,* Vol. 73, pp. 369–382.

Sillitoe, R.H., 1979, "Some Thoughts on Gold-Rich Porphyry Copper Deposits," *Mineralium Deposita,* Vol. 14, pp. 161–174.

Sillitoe, R.H., 1980, "Types of Porphyry Molybdenum Deposits," *Mining Magazine,* Vol. 142, pp. 550–553.

Sillitoe, R.H., 1981, "Ore Deposits in Cordilleran and Island-Arc Settings," *Relations of Tectonics to Ore Deposits in the Southern Cordillera,* W.R. Dickinson and W.D. Payne, eds., *Digest,* Arizona Geological Society, Vol. 14, pp. 49–69.

Sillitoe, R.H., 1983, "Styles of Low-Grade Gold Mineralization in Volcano-Plutonic Arcs," Report 36, Nevada Bureau of Mines & Geology, Reno pp. 52–68.

Sillitoe, R.H., Halls, C., and Grant, J.N., 1975, "Porphyry Tin Deposits in Bolivia," *Economic Geology,* Vol. 70, pp. 913–927.

Sillitoe, R.H., and Khan, S.N., 1977, "Geology of the Saindak Porphyry Copper Deposit, Pakistan," *Transactions,* Institute of Mining & Metallurgy, Sec. B, Vol. 86, pp. B27–B42.

Sinclair, A.J., et al., 1982, "A Preliminary Analysis of Gold and Silver Grades of Porphyry-Type Deposits in Western Canada," *Precious Metals in the Canadian Cordillera,* Association of Exploration Geochemists, pp. 157–172.

Still, A.R., 1962, "Uranium at Copper Cities and Other Porphyry Copper Deposits, Miami District, Arizona," Unpublished Ph.D. Thesis, Harvard University, Cambridge, MA.

Susak, N.J., and Crerar, D.A., 1982, "Factors Controlling Mineral Zoning in Hydrothermal Ore Deposits," *Economic Geology,* Vol. 77, pp. 476–482.

Sutulov, A., 1976, *Molybdenum and Rhenium, 1778–1977,* University of Concepción, Concepción, Chile, 257 pp.

Tempelman-Kluit, D.J., 1979, "Transported Cataclasite, Ophiolite and Granodiorite in Yukon: Evidence for Arc-Continent Collision," Paper 79-14, Geological Survey of Canada, 27 pp.

Turneaure, F.S., 1935, "The Tin Deposits of Llallagua, Bolivia," *Economic Geology,* Vol. 30, pp. 14–60, 170–190.

Wallace, S.R., et al., 1968, "Multiple Intrusion and Mineralization at Climax, Colorado," *Ore Deposits of the United States 1933/1967,* Vol. 1, J.D. Ridge, ed., AIME, New York, pp. 605–640.

Westra, G., and Keith, S.B., 1981, "Classification and Genesis of Stockwork Molybdenum Deposits," *Economic Geology,* Vol. 76, pp. 844–873.

White, A.J.R., Beams, S.D., and Cramer, J.J., 1977, "Granitoid Types and Mineralization with Special Reference to Tin," *Plutonism in Relation to Volcanism and Metamorphism,* N. Yamada, ed., Toyama, Japan, pp. 89–100.

White, W.H., et al., 1981, "Character and Origin of Climax-Type Molybdenum Deposits," *Economic Geology,* 75th Anniversary Vol., pp. 270–316.

Wolfe, J.A., Manuzon, M.S., and Divis, A., 1978, "The Taysan Porphyry Copper Deposit, Southern Luzon Island, Philippines," *Economic Geology,* Vol. 73, pp. 608–617.

Yan, M.Z., Wu, Y.L., and Li, C.Y., 1980, "Metallogenetic Systems of Tungsten in Southeast China and Their Mineralization Characteristics," *in* Granitic Magmatism and Related Mineralization, S. Ishihara and S. Takenouchi, eds., *Mining Geology,* Spec. Issue No. 8, pp. 215–221.

Young, E.J., and Segerstrom, K., 1973, "A Disseminated Silver-Lead-Zinc Sulfide Occurrence at Hahns Peak, Routt County, Colorado," Bulletin 1367, US Geological Survey, 33 pp.

Index

B

C

D

E

F

G

H

I

J

K

L

M

N

O

P

Q

R

S

T

U

V

W

X

Y

Z